Fluids and Electrolytes with Clinical Applications

A Programmed Approach

FOURTH EDITION

Joyce LeFever Kee

Associate Professor
College of Nursing, University of Delaware
Newark, Delaware

Delmar Publishers Inc.®

NOTICE TO THE READER

Publisher and author do not warrant or guarantee any of the products described herein or perform any independent analysis in connection with any of the product information contained herein. Publisher and author do not assume, and expressly disclaim, any obligation to obtain and include information other than that provided to them by the manufacturer.

The reader is expressly warned to consider and adopt all safety precautions that might be indicated by the activities described herein and to avoid all potential hazards. By following the instructions contained herein, the reader willingly assumes all risks in connection with such instructions.

The publisher and author make no representations or warranties of any kind, including but not limited to, the warranties of fitness for particular purpose or merchantability, nor are any such representations implied with respect to the material set forth herein, and the publisher and author take no repsonsibility with respect to such material. The publisher and author shall not be liable for any special, consequential or exemplary damages resulting, in whole or in part, from the readers' use of, or reliance upon, this material.

For information, address Delmar Publishers Inc.
2 Computer Drive West, Box 15-015
Albany, New York 12212

Printed in the United States of America
Published simultaneously in Canada
By Nelson Canada
A Division of the Thomson Corporation

10 9 8 7 6 5 4 3 2

ISBN: 0-8273-4263-2

Contributors

Sally M. Marshall, RN, MSN
Instructor, College of Nursing
University of Delaware
RENAL FAILURE: HEMODIALYSIS AND PERITONEAL DIALYSIS

Julie Waterhouse, RN, MSN
Instructor, College of Nursing
University of Delaware
CANCER

CONSULTANTS

Carolyn Freed, RN, PhD
Assistant Professor, College of Nursing
University of Delaware
SHOCK AND TRAUMA

Evelyn R. Hayes, RN, PhD
Chairperson, Department of Nursing Science
College of Nursing, University of Delaware
PHOSPHORUS

Elizabeth M. Jenkins, RN, MSN
Assistant Professor, College of Nursing
University of Delaware
CHAPTER 4: INTRAVENOUS THERAPY

Brent Thompson, RN, MSN
Instructor, College of Nursing
University of Delaware
CHILDREN WITH FLUID IMBALANCES

Julie Waterhouse, RN, MSN
Instructor, College of Nursing
University of Delaware
CHRONIC OBSTRUCTIVE PULMONARY DISEASE/COPD

Preface

Nurses are involved continually in the assessment of fluid and electrolyte imbalance. Medical advances and treatment have made these imbalances more prevalent; hence the expansion of the nurse's role. Every seriously ill person is likely to develop one or more imbalances, and even those who are only slightly or moderately ill may develop them. Nurses are responsible for maintaining homeostasis of fluid and electrolyte balance when caring for all patients. After completing this book the participant should understand more fully the effects of fluid and electrolyte balance and imbalance on the body in many conditions and clinical situations.

The fourth edition of this programmed text, *Fluids and Electrolytes with Clinical Applications*, develops the nurse's role in relation to nursing assessment of fluid and electrolyte imbalances. It includes a discussion of drugs that affect five electrolytes, hyperalimentation/TPN complications, shock symptoms and management, nursing diagnoses, and nursing actions. Chapter 2 contains a new section, Phosphorus, and Chapter 6 contains three new sections: Renal Failure: Hemodialysis and Peritoneal Dialysis; Cancer; and Chronic Obstructive Pulmonary Disease/COPD. Each chapter has been carefully reviewed, expanded, and updated.

Behavioral objectives introduce each chapter, clinical applications, case reviews, nursing diagnoses, and nursing actions conclude all chapters and sections and clinical management is discussed in most of them. Physiologic factors are presented in many sections of Chapters 5 and 6.

Twenty-five new tables have been added to make a total of 102 tables and diagrams, each followed by frames that will help participants to gain a deeper understanding of the material.

Twenty-eight sets of case reviews with answers contain case studies of patients with various fluid and electrolyte problems. Student and graduate nurses can apply these case studies in clinical practice. Throughout the text many practical nursing applications are given to help with nursing assessments.

The nursing diagnoses and nursing actions that follow each case review should help both student and graduate nurses to master the clinical applications of fluid, electrolyte, and acid-base concepts.

The content of this book has been geared to three levels within the nursing profession. First, it is intended for beginning students who have had some background in the biological sciences or who have completed an anatomy and physiology course. Second, it is for students who have a sufficient background in the biological sciences, chemistry, and physics but who need to learn about parenteral therapy and clinical conditions situations. Many of these students might wish to review the entire text to reinforce their previous knowledge and/or to increase their skills in handling practical nursing assessments and interventions. Finally, this book is intended to aid graduate nurses to review and increase their knowledge of fluid and electrolyte changes in order to assess their patients' needs and improve the quality of patient care.

The participant will work at his or her own pace while learning the principles, concepts, and applications of fluids and electrolytes as presented in this book. This self-instructional method of learning will also help the instructor to use class time to better advantage. It will also enable students to apply their knowledge to other clinical situations in their clinical practicum.

The first chapter discusses principles and concepts of fluid changes with clinical applications. Chapter 2 covers thoroughly six electrolytes, their functions, their causes and symptoms, and clinical applications with drug effects and electrolyte replacements. Regulatory mechanisms for pH control, causes and symptoms of acid-base imbalances, and their clinical applications are explained in Chapter 3. Chapter 4 deals with classifications of intravenous solutions with osmolality, hyperalimentation, and clinical applications, including rate and calculating intravenous administration and the nursing assessment and rationale in the management of IV therapy. Chapter 5 discusses four main clinical conditions: dehydration (extracellular fluid volume deficit), edema (extracellular fluid volume excess), water intoxication (intracellular fluid volume excess), and shock. Shift in extracellular fluid volume or fluid shift to the third space is explained. These four clinical conditions are programmed in detail to enable the student or graduate nurse to recognize these conditions as they occur. Included are physiologic factors, the causes and symptoms, clinical applications, a few clinical examples, and clinical management of the four conditions. Chapter 6 presents 11 clinical situations in which severe fluid and electrolyte imbalances can occur: the subjects are the aged, children, gastrointestinal surgery, trauma, renal failure: hemodialysis and peritoneal dialysis, burns and burn shock, cancer, COPD, CHF, cirrhosis of the liver with ascites, and diabetic acidosis. This chapter includes physiologic factors, clinical applications, clinical examples, and clinical management. Because these situations occur frequently, the nurse must understand the fluid and electrolyte changes that could mean the life or death of a patient.

Two appendices, Osmolality of Solutions and Fluid, Electrolyte, and Acid-Base Assessment Tool, act as guides for determining the osmolality of certain solutions and for assessing fluid, electrolyte, and acid-base imbalances.

A glossary covers words and terms used throughout the text. It should be useful to the student who has only some preparation in the biological sciences.

I would like to express my appreciation to the students who used this programmed text in its various forms before the first edition was published. By testing their acquired knowledge and ability to apply it after they had read the manuscript I was able to make necessary changes to improve the text. For the second edition, my deepest appreciation to Gale Buonanno, Susan Cross, Elizabeth Griffiths, Diane Hanna, Laurie

Jones, Kathy Kochan, Debra McCoy, Vickie Payne, Susan Skross, Tommie Lou Smith, June Taylor, Pamela Welch, and Barbara Witt. These nursing students tested the revisions for that edition and made valuable suggestions for additions to, omissions from, and clarification of the frames and material covered.

For the third edition I extend my deepest appreciation to the following professors and students in the College of Nursing, University of Delaware, for their helpful suggestions: Elizabeth Jenkins, Instructor, for Chapter 4; Carolyn Freed, Assistant Professor, for the section on trauma; Nancy S. Engel, Instructor, and Janis B. Smith, Instructor, and senior students—Sharon E. Baker, Bonita Barnett, Susan Bunting, Marianne Buzby, Martina M. Ciambella, Jeanne Marie Cost, Laureen A. Eick, Cynthia L. Flowers, Kim Frances, Kathy Glendenning, Alicia Halloran, Karen Hom, Mary Lou Jackson, April Nai, Susan Osworth, Sonya L. Peterson, Lois M. Showalter, Lorraine Shump, Judy Thornton, Robin Thornton, Sharon Turnbaugh, and Barbara Witmer—for the section on children with fluid imbalance.

For the fourth edition I extend my deepest appreciation to Sally Marshall, Julie Waterhouse, Elizabeth Jenkins, Evelyn Hayes, Carolyn Freed, and Brent Thompson, professors at the College of Nursing, University of Delaware, for their contributions and assistance.

I also offer my sincere thanks to Jane de Groot, nursing editor, and Anne Zielinski, editorial assistant, at John Wiley & Sons, for their helpful suggestions and assistance with this revision.

Joyce L. Kee

Newark, Delaware
September 1986

Instructions

TO THE INSTRUCTOR

Much class time is frequently spent on reviewing material or presenting new material that could easily be given through programmed instruction. This method of instruction enables the teacher to minimize the time spent in lecture on fluids and electrolytes and, thus, to devote more time to clinical conferences and seminar classes as they relate to clinical situations with fluid and electrolyte imbalance.

You may find it helpful to cover the material in this book by one of these three ways: (1) assigning the students a chapter at a time; (2) assigning the students the first three chapters to be completed by a certain date and the last three by a later date; or (3) assigning the students a given length of time to complete the entire text.

TO THE STUDENTS

Many students believe that the subject of fluids and electrolytes is very difficult to comprehend. This programmed book will provide you with important data on fluids and electrolytes from various aspects, and you will find that this material is not so difficult to understand and retain.

By taking the easy steps provided in this book, you will proceed through the chapters more quickly than you would expect. This book is written on a self-instruction basis so that you may proceed at your own speed. Each step is a learning process. Greater learning occurs if you either complete a chapter or spend a minimum of two hours at one sitting. Never end the study period without at least completing all the frames related to a single topic.

It would be helpful to begin your next study session with the final frames from the previous material, for it will enable you to check your retention of the material that was presented. The situational reviews throughout the chapters give immediate reinforcement of the data learned. The many sets of nursing actions should be useful for applying fluid, electrolyte, and acid-base concepts in various clinical settings. A glossary is included to assist you with words and terms used in the text.

Students, study the diagrams and tables before proceeding to the frames. If you make mistakes in the program, you need not be concerned so long as you rectify the mistakes. This book should increase your knowledge and understanding of fluids and electrolytes and be a great asset in applying this knowledge in your clinical practicum.

Contents

CHAPTER ONE

Body Fluid and Its Function

BEHAVIORAL OBJECTIVES

Upon completion of this chapter, the student will be able to:

Differentiate between the percentage of water found in the average adult body, newborn infant, and the embryo.
Name the three compartments of the body where body water is found.
Name the two classifications of body fluid and their percentages.
Define homeostasis and explain its state in maintaining body fluid equilibrium.
Explain how the body loses and maintains body fluid.
Define osmotic pressure, semipermeable and selectively permeable membranes, and osmol and osmolality, and relate the effects of each in the passage of body fluid.
Differentiate between isotonic (iso-osmolar), hypotonic (hypo-osmolar), and hypertonic (hyperosmolar) solutions, and explain the effects of these solutions on body cells.
Define milliequivalent and milligram and explain their symbols and significance in the body.
Define Starling's Law of Capillaries.
Explain the four measurable factors that determine the flow of fluid between the vessels and tissues and their effects on the exchange of fluid.
Explain the pressure gradient and the significance in colloid osmotic and hydrostatic pressure gradients.
Relate fluid changes that occur to patients in your clinical area.

INTRODUCTION

The human body is a complex machine that contains hundreds of bones and the most sophisticated systems of any structure on earth. Yet, the substance that is basic to the very existence of the body is the simplest substance known—water. In fact, it makes up almost two-thirds of an adult's body weight.

The body is not static—it is alive and solid particles within its framework are able to move into and out of cells and systems, and even into and out of the body, only because there is water.

The basis of all fluids is water, and as long as the quantity and composition of body fluids are within the normal range, we just take it for granted and enjoy being healthy. But, if the water content of the body for some reason departs from this range, the whole delicate balance is disrupted, and disease can find an easy target.

In this chapter, distribution of body fluid, fluid compartments of the body, physics terminology, fluid pressures, and clinical considerations are discussed. There are two situational reviews, two sets of nursing actions. and two sets of nursing diagnoses.

An asterisk(*) on an answer line indicates a multiple-word answer. The meanings for the following symbols are: ↑ increased, ↓ decreased, > greater than, < less than.

---

1 The greatest single constituent of the body is water, which represents about 60% of the total body weight in the average adult. In the early human embryo, 97% of body weight is water, in a newborn infant 77%.

Label the following drawings with the proper percentage of water to body weight.

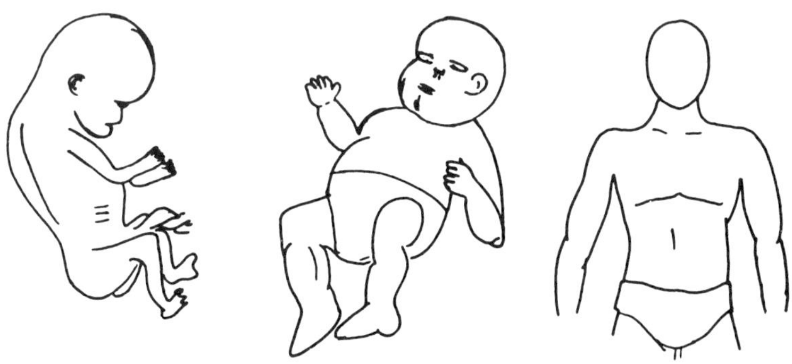

a. ____________ b. ____________ c. ____________

- - - - - - - - - - - - - - - - -

a. 97%; b. 77%; c. 60%

---

2 In the average adult, the proportion of water to *____________ is 60%.

- - - - - - - - - - - - - - - - -

body weight

---

3 Which has the highest percentage of water in relation to body weight?

____________. Which has the lowest? ____________.

- - - - - - - - - - - - - - - - -

embryo; adult

---

4 Can you think of any reason why the early human embryo and the infant have a higher proportion of water to body weight?

*________________________.

\- - - - - - - - - - - - - - - - - -

I asked what you thought. Many people think the extra water in infants acts as a protective mechanism. Since infants have larger body surface in relation to their weight, extra water acts as a cushion against injury.

---

5 Because fat is essentially free of water, the leaner the individual, the greater the proportion of water in total body weight.

Who has more water as body weight—a fat person or a lean one?

*________________________.

\- - - - - - - - - - - - - - - - - -

lean person

FLUID COMPARTMENTS

6 This body water is distributed among three types of "compartments:" cells, blood vessels, and tissue spaces between blood vessels and cells, which are separated by membranes.

Label the three compartments where body water is found.

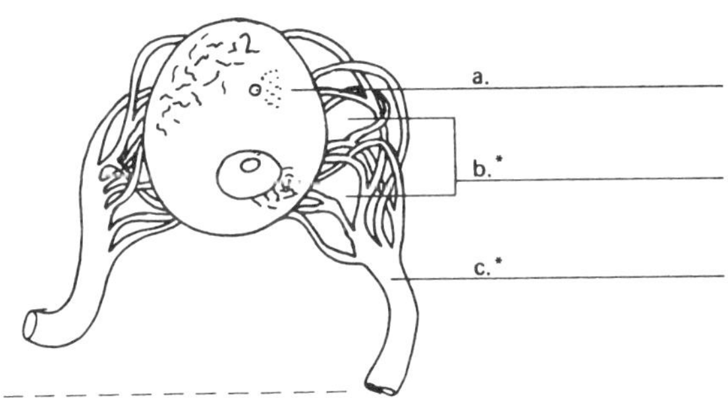

\- - - - - - - - - - - - - - - - - -

a. cell
b. tissue space
c. blood vessel

7 The term for the water in each type of "compartment" is as follows:

1. In the cell—*intracellular* water or *cellular* water.
2. In the blood vessels—*intravascular* water.
3. In tissue spaces between blood vessels and cells—*interstitial* water.

Label the complete diagram with the proper terms for body water in the three compartments.

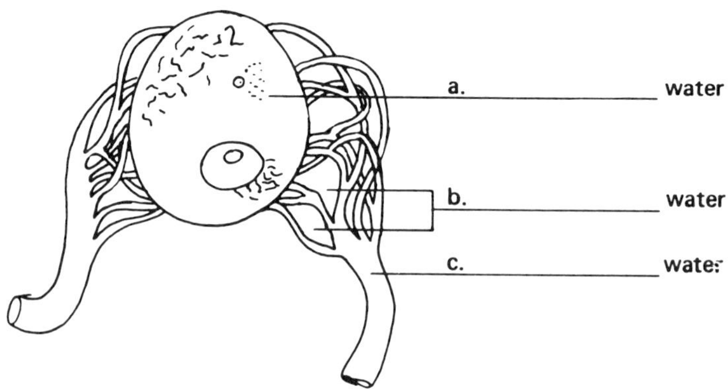

— — — — — — — — — — — — — — — — —

a. intracellular water
b. interstitial water
c. intravascular water

---

8 There are three prefixes that will be used frequently in this text:

inter-	between
intra-	within
extra-	outside of

The two classes of body water or body fluid are intracellular fluid and extracellular fluid. The corresponding areas or "compartments" are the intracellular space and the extracellular space.

What does the prefix *intra* mean?__________. What does the prefix *extra* mean? *__________.

— — — — — — — — — — — — — — — — —

within; outside of

9 Fluid within the cell is classified as intracellular fluid, whereas intravascular fluid and interstitial fluid are classified as extracellular.

The area within the cell is called the ______________ space, whereas the tissue spaces between blood vessels and cells, and the area within blood vessels are known as the ______________ space.

\- - - - - - - - - - - - - - - - -

intracellular; extracellular

10 Label the three "compartments" of body water and the two classes of body water.

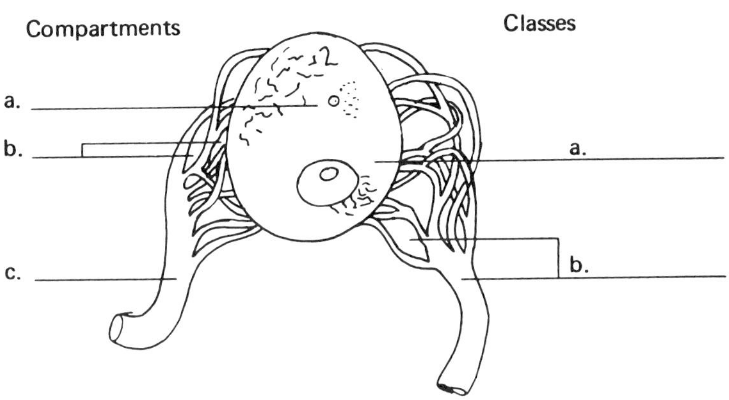

\- - - - - - - - - - - - - - - - -

Compartments

a. cell
b. tissue space
c. blood vessel

Classes

a. cellular or Intracellular
b. extracellular

11 Approximately two-thirds of the body fluid is contained in the intracellular space.

We have already said that the total water in the adult body is ________%; therefore, intracellular water must represent ________% of the total body weight, and extracellular water ________%.

\- - - - - - - - - - - - - - - - -

60; 40; 20

---

12 If one-fourth of the extracellular fluid is intravascular water, then three-fourths is *__________________.

— — — — — — — — — — — — — — — — — —

interstitial fluid

---

13 Therefore, extracellular fluid represents ________% of body weight, of which _____________________ is interstitial fluid.

Interstitial fluid represents ________ % of total body weight. Intravascular fluid represents ________ % of total body weight.

— — — — — — — — — — — — — — — — — —

20; three-fourths; 15; 5

INTAKE AND OUTPUT FOR HOMEOSTASIS

14 Since we already have learned that the percentage of body fluid varies with age and "fatness," then the proportion of intracellular and extracellular fluid in a fat person would be (greater/lesser) to his body weight.__________________.

— — — — — — — — — — — — — — — — — —

lesser

---

15 *Homeostasis* is a term you will be using. It means the state of equilibrium of the internal environment. Concerning body fluids, homeostasis is the maintaining of equilibrium or stability in relation to the physical and chemical properties of body fluid.

In a few words define the term *homeostasis*. *__________________. Explain the relationship of homeostasis to body fluid. *____.

— — — — — — — — — — — — — — — — — —

state of equilibrium. It maintains equilibrium of the physical and chemical properties of body fluid

---

16 The body normally maintains a state of equilibrium between the intake and loss of water.

If the body loses or gains water, it acts rapidly to compensate for this deficit or excess so that *____________________ will be maintained.

\- - - - - - - - - - - - - - - - - -

homeostasis, or equilibrium

---

17 When body water is insufficient, the urine volume diminishes, and the individual becomes thirsty. Therefore, he might *____________________________________ to make up the deficiency.

\- - - - - - - - - - - - - - - - - -

drink more water

---

18 Do you think man can go longer without food or without water? *______________

____

____.

Why do you think man cannot go without water? *_______________________

____

____.

\- - - - - - - - - - - - - - - - - -

If you said he can go longer *without food,* this is correct. Again water is the greatest single constituent of the body

Because water is contained in the various compartments of our body and is needed to carry nutrients, elements, and waste to and from our body tissues

19 If we drink an excessive amount of water, the urinary output would increase.

If you did not drink any fluids or if the body lost water, then the urinary volume would:

( ) a. increase
( ) b. decrease

If there were an excess of water in the body, then the urinary volume would:

( ) a. increase
( ) b. decrease

- - - - - - - - - - - - - - - - -

b. decrease; a. increase

---

20 Refer to Diagram 1. The normal sources of body water are __________, __________, and *__________.

- - - - - - - - - - - - - - - - -

liquid; food; oxidation of food

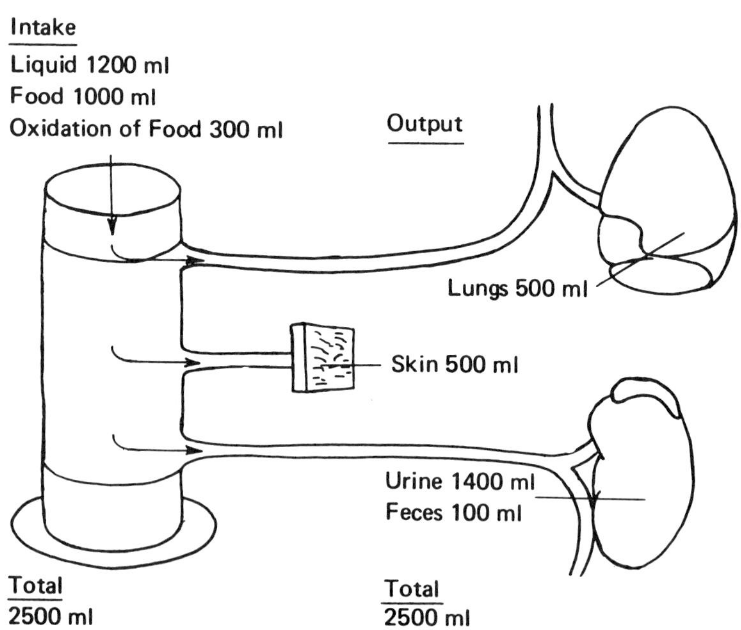

Diagram 1. Normal pattern of water intake and loss.

21 Refer to Diagram 1. The avenues for daily water loss are ________,

________, ________, and ________.

- - - - - - - - - - - - - - - - - -

lungs; skin; urine; feces

---

22 If your water intake amounted to 2500 ml for the day and your water output was 2500 ml, then your body has maintained a state of

*________________ of body fluid

- - - - - - - - - - - - - - - - - -

equilibrium or homeostasis

---

23 Would you think that the rate of water loss and gain is different in summer and winter, or the same all year around? *________________________.

- - - - - - - - - - - - - - - - - -

In the summer when the atmospheric temperature is high, water loss, via skin and lungs, increases

---

24 Evaporation of water from the skin, as we perspire, is a protective mechanism against overheating the body.

Can you explain how evaporation of water acts as a protective mechanism? *________________________________.

- - - - - - - - - - - - - - - - - -

It acts as a cooling system, keeping the body at a normal temperature

DEFINITIONS RELATED TO BODY FLUIDS

---

25 Do you know the meaning of the term *diffusion*?

Diffusion is the movement of molecules/solutes across a selectively permeable membrane along its own pathway, irrespective of all other molecules. Large molecules move *less* rapidly than small molecules. Molecules move faster from a higher concentration to a lower concentration.

Diffusion is the * ______________________________

across a selectively permeable membrane. Will small molecules move faster than large molecules? ______________________________.

Molecules/solutes tend to move faster from a * ______________________

______________________ to a * ______________________.

— — — — — — — — — — — — — — — — —

movement of molecules/solutes
Yes
higher concentration; lower concentration

---

26 Body water loss by diffusion through the skin which is independent of sweat gland activity is called *insensible perspiration.*

When sweat gland activity occurs and water appears on the skin, this is called *sensible perspiration.*

In a relatively comfortable temperature would insensible perspiration or sensible perspiration occur? ____________. Why? * ______________________

____.

— — — — — — — — — — — — — — — — —

Insensible.
There is not enough heat to cause sweat gland activity, so only the normal loss occurs, through insensible perspiration, with water diffusing through the skin and evaporating quickly

---

27 Give the meanings of the following words:

Insensible perspiration. * ______________________________.

Sensible perspiration. * ______________________________.

— — — — — — — — — — — — — — — — —

Insensible. Water loss by diffusion through the skin
Sensible. Water on the skin due to sweat gland activity

---

28 The volume of body water is primarily regulated by the kidneys. When water loss increases, e.g., through perspiration or diarrhea, the kidneys will conserve water by (increasing/decreasing) the urinary output.

______________________.

- - - - - - - - - - - - - - - - - -

decreasing

---

29 Some more definitions to know:

Membrane. A layer of tissue covering a surface or organ or separating spaces.

Osmosis. The passage of a solvent through a partition from a solution of lesser solute concentration to one of greater solute concentration.

Note: Osmosis may be expressed in terms of water concentration instead of solute concentration. Then water molecules will pass from an area of higher water concentration (fewer solutes) to an area of lower water concentration (more solutes).

Solvent. A liquid with a substance in solution.

Solute. A substance dissolved in a solution.

Permeability. A capability of a substance, molecule, or ion to diffuse through a membrane.

Early literature described membranes of body cells as semipermeable. Today *semipermeable* refers to artificial membranes, i.e., cellophane membrane, frequently described in the process of osmosis. *Selectively permeable membrane* refers to the human membrane.

Differentiate between:

Selectively permeable membrane. *______________________;

Semipermeable membrane. *______________________

- - - - - - - - - - - - - - - - - -

human membrane; artificial membrane

---

30 Do you recall the difference between a solvent and a solute? Explain.

Solvent. *______________________.

Solute. *______________________.

- - - - - - - - - - - - - - - - - -

Solvent. A liquid with a substance in solution
Solute. A substance dissolved in solution

---

31 In an effort to establish equilibrium, water in the body moves from a less concentrated solution (fewer solute particles per unit of solvent) to a more concentrated solution (more solute particles per unit of solvent) through a * ______________________ membrane.

— — — — — — — — — — — — — — — — —

selectively permeable or human

---

32 Osmotic pressure is the pressure or force that develops when two solutions of different strengths or concentrations are separated by a selectively permeable membrane.

In osmosis, to establish equilibrium, water would move from the (less/more) concentrated solution into the (less/more) concentrated solution.

The force that draws water across a selectively permeable membrane is called * ______________________.

— — — — — — — — — — — — — — — — —

less; more; osmotic pressure

---

33 In what direction does water flow? * ______________________ ______________________. Why? * ______________________.

— — — — — — — — — — — — — — — — —

From the lesser to the greater concentration. Because more solute particles have a "pulling" effect

---

34 Do you recall the meaning of *permeable*? If not, return to Frame 29.

A membrane is considered impermeable if an ion, substance, or molecule cannot diffuse freely across it.

Certain substances do not diffuse freely across the human membrane, so this membrane is considered ______________ to that substance.

— — — — — — — — — — — — — — — — —

impermeable

---

35 The kidney is influenced to excrete or conserve water by ADH, which is the antidiuretic hormone. This hormone is excreted from the posterior pituitary gland, also called the posterior hypophysis.

Do you know where the pituitary gland, also called the hypophysis, is? ______________.

- - - - - - - - - - - - - - - - - -

Yes? Fine.
No? Lies in a bony cavity beneath the base of the brain.

---

36 The antidiuretic hormone, or ADH, increases the permeability of the cells of the kidney tubules to water, thus allowing more water to be reabsorbed.

With a lack of this hormone, what would occur? *______________ ______________.

- - - - - - - - - - - - - - - - - -

an increased excretion of water from the kidney tubules

---

37 Two terms that need to be defined:

Serum. Consists of plasma minus fibrogen. It is obtained after coagulation of blood.

Plasma. Contains blood minus the blood cells. It is composed mainly of water.

Frequently these two terms are interchangeably used. Also the term blood *plasma* means plasma and blood *serum* means serum.

Serum and plasma are both found in what type of fluid? *______________ ______________.

- - - - - - - - - - - - - - - - - -

intravascular fluid or extracellular fluid

---

38 The posterior pituitary gland is influenced by the solute concentration of the plasma. If there is an increase of solute in the plasma, then the gland will release the hormone ADH, which will hold water in the body.

Explain how? *____.

For what reason should there be more water? *________________________________.

— — — — — — — — — — — — — — — — — —

It will absorb water from the kidney tubules
To dilute the solute

---

39 A small increase of solute concentration in the plasma, above the normal amount, would be sufficient to stimulate the posterior gland in releasing

__________________.

What two things would occur if there was less solute concentration in the plasma?

1. *____.

2. *____.

— — — — — — — — — — — — — — — — — —

ADH
1. ADH would not be released.
2. More water would be excreted from the body.

---

40 If you drank a lot of fluids, what would happen to the solute concentration of your plasma—would it become diluted or more concentrated? ________________________.

Then what would the posterior pituitary do? *________________________________.

— — — — — — — — — — — — — — — — — —

diluted. It would not release ADH

---

41 When the solute concentration increases, the thirst mechanism is stimulated, and the individual ingests water.

From the above statement how can homeostasis be maintained? *____.

— — — — — — — — — — — — — — — — — —

By drinking water or other liquids

---

42 An *osmol* is a unit of osmotic pressure. The osmotic effects are expressed in terms of osmolality. A *milliosmol* (mOsm) is 1/1000th of an osmol and will determine the osmotic activity.

Four terms to know:

Osmolality. Osmotic pull exerted by all particles per unit of water, expressed as osmols or milliosmols per kilogram of water.

Osmolarity. Osmotic pull exerted by all particles per unit of solution, expressed as osmols or milliosmols per liter of solution.

Ion. A particle carrying a positive or negative charge. (A further explanation of ion will be found in Chapter 2.)

Dissociation. Separation, i.e., a compound separating into many particles.

What is larger, the osmol or the milliosmol?____________________. Do you know the relationship between ion and dissociation? *____________________

______________________________.

\- - - - - - - - - - - - - - - - - -

osmol; Ion is a particle from dissociation of a compound

---

43 According to Frame 42, 1 milliosmol is 1/1000th of an osmol. Then 1 osmol would equal ________ milliosmols.

Milliosmols will determine the __________ activity of a solution.

\- - - - - - - - - - - - - - - - - -

1000; osmotic

---

44 The basic unit used to express the force exerted by the concentration of solute or dissolved particles is a(n) ____________.

The osmotic effect of a solute concentration in water is expressed as __________________, a property that depends on the number of osmols or milliosmols.

\- - - - - - - - - - - - - - - - - -

osmol; osmolality

45 Osmolality and osmolarity are frequently used interchangeably and both terms refer to concentration of solution. However, the term used to express the osmols or milliosmols per kilogram of water is ________________.

\- - - - - - - - - - - - - - - - - -

osmolality

---

46 Here are three more prefixes that will be frequently used in this text:

hypo- less than
iso- equal
hyper- excessive

The osmolality of a solution is *isotonic* or *iso-osmolar* if it has the same solute concentration within the solution as has plasma, *hypotonic* or *hypo-osmolar* if it has a lower concentration, and *hypertonic* or *hyperosmolar* if it has a higher concentration.

Early literature refers to the concentration of solutions as hypotonic, isotonic, and hypertonic. These terms are still in use; however, since the solute concentration is determined by the number of osmols or milliosmols in solution, hypo-osmolar, iso-osmolar, and hyperosmolar are the suggested terms.

A solution is classified by comparing its osmolality with that of

________________.

Plasma is considered to be a(n) (iso-osmolar/hypo-osmolar/hyper-osmolar) __________ fluid.

\- - - - - - - - - - - - - - - - - -

plasma; iso-osmolar

---

47 Match the following types of solutions with their solute concentrations.

___ 1. Iso-osmolar a. Higher solute concentration than plasma

___ 2. Hypo-osmolar b. Same solute concentration as plasma

___ 3. Hyperosmolar c. Lower solute concentration than plasma

\- - - - - - - - - - - - - - - - - -

1. b; 2. c; 3. a

48 The osmolality of plasma is 290 mOsm. Parenteral solutions or solutions for intravenous use having 290 mOsm with either +50 mOsm or −50 mOsm would be considered an isotonic or iso-osmolar solution.

A solution having less than 240 mOsm is considered ____________,

and a solution having more than 340 mOsm is considered __________.

\- - - - - - - - - - - - - - - - - -

hypotonic (hypo-osmolar); hypertonic (hyperosmolar)

---

49 The following chart contains a list of milliosmol values. Classify them as iso-osmolar, hypo-osmolar, or hyperosmolar.

Milliosmol Values (mOsm)	*Type of Osmolality*
220	__________
75	__________
350	__________
310	__________
560	__________

\- - - - - - - - - - - - - - - - - -

hypo-osmolar; hypo-osmolar; hyperosmolar; iso-osmolar; hyperosmolar

---

50 If two solutions have the same osmol values, their osmotic pressures will be the same.

Intracellular fluid in the cells is isotonic (iso-osmolar) fluid. Distilled water has a lower osmotic pressure than the cells; thus, it is considered to be __________ fluid

\- - - - - - - - - - - - - - - - - -

hypotonic (hypo-osmolar)

---

51 Extracellular hypertonic (hyperosmolar) fluid has a greater osmotic pressure than the cell, thus, intracellular water moves out of the cells and into the extra-cellular hypertonic (hyperosmolar) fluid by the process of ____________.

With the cells losing water, what would happen to their form and size?

* ____________________________.

\- - - - - - - - - - - - - - - - - -

osmosis. Cells would shrink and become small in size

52 If the extracellular fluid was hypotonic (hypo-osmolar), what would happen to the cells? *______________________.

— — — — — — — — — — — — — — — — — —

Cells would swell and enlarge in size.

53 A flask (liter) of 5% dextrose in water is 250 mOsm, and a flask (liter) of 0.9% sodium chloride or normal saline is 310 mOsm, having somewhat the same osmotic pressure as ______________________.

These solutions would be (iso-osmolar/hypo-osmolar/hyperosmolar) fluid.

______________________.

— — — — — — — — — — — — — — — — — —

plasma; iso-osmolar

54 The sum of 5% dextrose in normal saline would equal ____________ mOsm. This solution would be a(n) ______________ solution.

— — — — — — — — — — — — — — — — — —

560 mOsm; hyperosmolar

55 Name some iso-osmolar solutions:

1. *______________.

2. *______________.

3. *______________.

— — — — — — — — — — — — — — — — — —

1. 5% dextrose
2. 0.9% sodium chloride or normal saline
3. plasma

56 In studying serum chemistry alterations and concentrations, one is concerned with how much the ions or chemical particles weigh, which are measured in milligrams percent or mg % (mg % is the same as mg/100 ml or mg/dl), or with the number of electrical charged ions, which is measured in millequivalents per liter (1000 ml) or mEq/L.

The term *milliequivalent* involves the chemical activity of elements, whereas milliosmol involves the ______________________ activity of solution.

How do milligrams and milliequivalents differ? ______________________

- - - - - - - - - - - - - - - - -

osmotic
Milligrams. The weight of ions.
Milliequivalents. The chemical activity of ions.

---

57 Actually, milliequivalents are a better method of measuring the concentration of ions in the serum than milligrams.

Milligrams measure the ______________________ of ions and give no information concerning the number of ions or the number of electrical charges of ions.

- - - - - - - - - - - - - - - - -

weight

---

58 If you were having a party and wanted to invite equal numbers of boys and girls, which would be more accurate—inviting 1500 pounds of girls and 1500 pounds of boys or inviting 15 girls and 15 boys? *______________________.

Why? *______________________ ______________________

____.

- - - - - - - - - - - - - - - - -

15 girls and 15 boys. You would have an unequal number of boys and girls, for not every child weighs 100 pounds.

---

59 From the example in Frame 58, which would be more accurate in determining the serum chemistry of chemical particles or ions in the body—milli-

equivalents or milligrams? ______________________.

You will find both measurements used in this book and in your clinical settings for determining changes in our serum chemistry. However, when referring to ions, milliequivalents will be used in this book.

— — — — — — — — — — — — — — — — —

milliequivalents

---

60 Milligrams are reported in terms of mg % or mg/100 ml, and milliequivalents are reported in terms of mEq/L or mEq/1000 ml.

Complete the following chart, filling the uncompleted symbols and words.

Symbols		*Meanings*
5 m ________/L	=	____________ per liter
145 ________/1000 ______	=	____________ per 1000 ml
82 ________%	=	____________ per %
110 ______/100 ml (dl)	=	____________ per 100 ml (dl)

— — — — — — — — — — — — — — — — —

mEq/L = milliequivalent
mEq/1000 ml = milliequivalent
mg % = milligrams
mg/100 ml = milligrams; also, mg/dl = milligrams

CASE REVIEW

Mr. Kendall had been vomiting for several days. His urine output had been decreased. He was given 1 liter of 5% dextrose in water and then 1 liter of 5% dextrose in normal saline (0.9% NaCl).

1. In his adult stage, Mr. Kendall's body water represents __________ % of his total body weight. What percentage of his total body weight is in the intracellular compartment ____________ % and in the extracellular compartment ____________ %?
2. Explain why Mr. Kendall's urine output is decreased * ____.
3. The three sources for his water intake are ______________________, ______________________, and * ______________________.

   The four normal ways for his daily water loss are ______________________,

______________________ , ______________________ and

______________________ .

4. Because of vomiting, Mr. Kendall's body fluids were decreased and his urinary output was decreased. The solute concentration was increased due to less circulating body fluid. Would the posterior pituitary gland release (more/less) ADH?

   ______________________ .

5. Define osmolality and osmolarity.

   Osmolality. * ______________________ .

   Osmolarity. * ______________________ .

6. Mr. Kendall received 1 liter of 5% dextrose in water, which has a similar osmolality as plasma. A solution with osmolality similar to that of plasma is considered to be (iso-osmolar/hypo-osmolar).
7. The second liter he received was 5% dextrose in normal saline. This solution should be a(n) ______________________ solution.
8. The osmolality of plasma is ______________________ mOsm. A solution less than 240 mOsm is considered ______________________ .
9. One-half of normal saline (0.45% NaCl) solution has 155 mOsm. What is this type of solution? ______________________ .

— — — — — — — — — — — — — — — — —

1. 60; 40; 20
2. Mr. Kendall is losing body fluid from vomiting and a lack of fluid intake.
3. liquid; food; oxidation of food
   lungs; skin, urine; feces
4. more
5. *Osmolality.* Osmols or milliosmols per kilogram of water
   Osmolarity. Osmols or milliosmols per liter of solution
6. iso-osmolar
7. hyperosmolar
8. 290 mOsm; hypo-osmolar
9. hypo-osmolar

NURSING DIAGNOSIS

Potential for fluid volume deficit related to abnormal fluid loss secondary to episodes of vomiting.

NURSING ACTIONS

1. Recognize that infants and thin people have a higher proportion of body water than adults, older adults, and fat people.
2. Assess the intake and output status of the patient. Fluid intake and urine output are normally in proportion to each other.

3. Assess excess fluid loss from the skin and lungs. Diaphoresis (excess sweating) and tachypnea (rapid breathing) cause excess body water loss.
4. Assess the osmolality of intravenous solutions. Know that solutions with osmolality between 240–340 mOsm/L are iso-osmolar, a similar concentration to plasma. Hypo-osmolar solutions have < 240 mOsm/L (less than) and hyperosmolar solutions have > 340 mOsm/L (more than).
5. Recognize that 5% dextrose in water, normal saline (0.9% NaCl), and 5% dextrose in 0.2% NaCl are examples of iso-osmolar solutions.

STARLING'S LAW

61 E.H. Starling states that equilibrium exists at the capillary membrane when the fluid leaving circulation equals exactly the amount of fluid returning to circulation.

Factors regulating the flow of blood constituents between the interstitial and intravascular compartments are known as Starling's Law of Capillaries. There are four measurable factors that determine the flow of fluid between the two compartments.

Where does equilibrium exist when fluid leaves and returns to circulation?

* ______________________________.

Starling's Law of Capillaries is concerned with what two compartments?

______________ and ______________.

- - - - - - - - - - - - - - - - -

capillary membrane—emphasis on the equilibrium is on the fluid flow at the capillary membrane

intravascular; interstitial

62 Two terms to define:

Colloid. A nondiffusible substance, a solute suspended in solution.

Hydrostatic. A state of equilibrium of fluid pressures.

The measurable factors incorporated in Starling's Law of Capillaries are the *colloid osmotic pressure* and the *hydrostatic pressure* of both the blood in the intravascular compartment and the tissues surrounding the interstitial compartment.

The measurable factors that involve the intravascular compartment are the colloid osmotic pressure and the hydrostatic pressure of the (blood/tissue).

____________________.

The measurable factors involving the interstitial compartment are the colloid osmotic pressure and the hydrostatic pressure in the ____________.

\- - - - - - - - - - - - - - - - -

blood; tissues

---

63 What are the two measurable factors that determine the flow of fluid between the intravascular compartment or vessels and interstitial compartment or tissues?

1. * ____________.
2. * ____________.

\- - - - - - - - - - - - - - - - -

1. colloid osmotic pressure
2. hydrostatic pressure

---

64 Colloid means * ____________________; therefore, colloid osmotic pressure would be the amount of pressure exerted from

* ________________________.

Hydrostatic is the * ____________ of fluid; therefore, hydrostatic pressure would be the amount of pressure at ____________ of fluid.

\- - - - - - - - - - - - - - - - -

nondiffusible substances: nondiffusible substances;
state of equilibrium; equilibrium

---

65 The colloid osmotic pressure and the hydrostatic pressure of the blood and tissues move fluid through the ______________ membrane.

- - - - - - - - - - - - - - - - -

capillary

---

66 Do you know the meanings of the arterioles and venules? If not:

Arterioles. Minute arteries that lead into a capillary bed.

Venules. Minute veins that lead from the capillary.

Which would be larger, the arteriole or the artery? ______________.

The venule or the vein? ______________.

- - - - - - - - - - - - - - - - -

artery; vein

---

67 Fluid exchange occurs only across the walls of capillaries and not across the walls of arterioles or venules. Therefore, fluid moves into the interstitial space at the arteriolar end of the capillary and out of the interstitial space into the capillary at the *______________.

- - - - - - - - - - - - - - - - -

venular end

---

68 Gases move from an area of higher concentration to an area of lower concentration. This is the opposite of the movement of fluid in osmosis, in which fluid moves from the (less/more) ______________ concentrated solution into the (less/more) ______________ concentrated solution.

Since oxygen is in greater concentration in the capillaries, it moves into the interstitial fluid. Carbon dioxide passes in the opposite direction, from the higher concentration in the interstitial fluid to the ______________ in the capillaries.

- - - - - - - - - - - - - - - - -

less; more; lower concentration

---

69 In the capillaries, oxygen is in greater concentration and, therefore, moves

*____________________________________.

Carbon dioxide is in higher concentration in the interstitial fluid and, therefore, moves *____________________________________.

\- - - - - - - - - - - - - - - - -

into the interstitial fluid; into the capillaries

---

70 The capillary endothelium or capillary membrane acts as a selectively permeable membrane by permitting free passage of crystalloids. *Crystalloids* are diffusible substances and will dissolve in solution. They are noncolloid substances.

Albumin, protein, and gelatin are colloids of what type of substance?

____.

The amount of osmotic pressure that develops at a membrane depends mainly on the concentration of *____________________________________.

An example of a nondiffusible substance is __________________________.

\- - - - - - - - - - - - - - - - -

nondiffusible; nondiffusible substances or colloids; albumin, protein or gelatin

---

71 There is a current question as to whether proteins are colloids or crystalloids; however, for our purposes, we shall consider protein a colloid. According to Gibbs-Donnan theory of membrane equilibrium, the osmotic pressure of an ionized colloidal system, such as plasma, is also due in part to the unequal distribution of diffusible ions. Therefore, 30% of the osmotic pressure developed in normal plasma is due to the unequal distribution of sodium and chloride ions.

Osmotic pressure of an ionized colloidal system is due in part to the *____________________________________ of diffusible ions.

\- - - - - - - - - - - - - - - - -

unequal distribution

---

72 Blood contains blood cells and plasma. The red blood cell is normally bathed in plasma. Plasma and red blood cells have an equal quantity of nondiffusible solutes, so would there or would there not be an osmotic pressure at the cell membrane? * ______________________.

If the red blood cells were bathed in pure water, the presence of large quantities of nondiffusible substances inside the cell would cause * ______________________ ______________________.

- - - - - - - - - - - - - - - - -

there would not be; the cells to swell and maybe burst

---

73 Fluid flows only when there is a difference in pressure at the two ends of the system. This difference in pressure between two points is known as the *pressure gradient.*

If the pressure at one end was 32 and at the other end was 26, then the pressure gradient would be ______________________.

- - - - - - - - - - - - - - - - -

6

---

74 The plasma in the capillaries has hydrostatic pressure and colloid osmotic pressure. The tissue fluids have hydrostatic pressure and colloid osmotic pressure.

The difference of pressure between the plasma colloid osmotic pressure and the tissue colloid osmotic pressure is known as the * ______________________ ______________________.

The difference of pressure between the plasma hydrostatic pressure and the tissue hydrostatic pressure is known as the * ______________________ ______________________.

It is this difference in pressure that makes the fluid flow.

- - - - - - - - - - - - - - - - -

colloid osmotic pressure gradient: hydrostatic pressure gradient

---

75 Refer to Diagram 2. The plasma colloid osmotic pressure is 28 mm Hg (millimeters of mercury) and the tissue colloid osmotic pressure is 4 mm Hg.

The colloid osmotic pressure gradient would be *________________.

\- - - - - - - - - - - - - - - - - -

24 mm Hg

76 Refer to Diagram 2. The hydrostatic fluid pressure is 18 mm Hg in the capillary and the hydrostatic tissue pressure is -6 mm Hg; therefore, the hydrostatic pressure gradient is *________________.

\- - - - - - - - - - - - - - - - - -

24 mm Hg

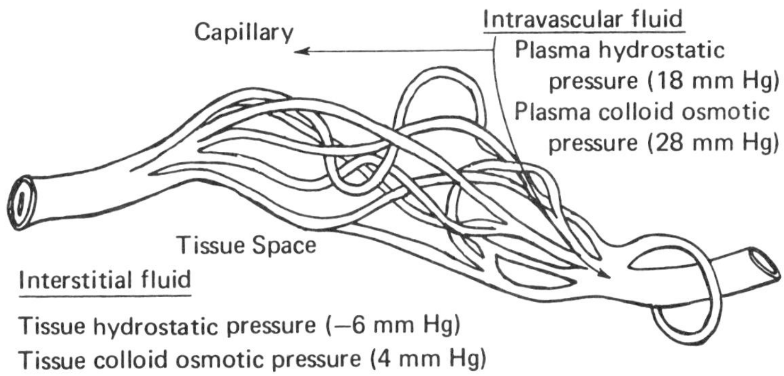

Diagram 2. The pressures in the intravascular and interstitial fluids.

77 The hydrostatic pressure gradient across the capillary membrane (24 mm Hg) is equal to the colloid osmotic pressure gradient across the membrane (24 mm Hg).

Thus, the two pressures are *________________.

\- - - - - - - - - - - - - - - - - -

equal or same pressure

---

78 The plasma hydrostatic pressure gradient tends to move fluid out of the capillary.

Why? * ____.

Refer to Diagram 2 if reply is unknown.

The colloid osmotic pressure gradient tends to move fluid into the capillary.

Why? * ______________________________________.

— — — — — — — — — — — — — — — — —

The plasma hydrostatic pressure is higher than the tissue

Plasma osmotic pressure is higher than tissue pressure

---

79 The balance between the two forces would keep blood volume constant for circulation. In this way fluid would not accumulate in the intravascular or the interstitial compartments.

Without the colloid osmotic forces, fluid (would/would not) ____________

be lost from circulation. Do you know the reason why? * ______________

____.

The blood volume would then be (sufficient/insufficient) ____________ to maintain circulation.

— — — — — — — — — — — — — — — — —

would

Fluid would stay in the tissues, causing accumulation and tissue swelling

insufficient

---

80 Name the man who formulated the Law of Capillaries and define this law in your own words.

Name. ______________________________.

Law. * ____

____.

— — — — — — — — — — — — — — — — —

Starling

Plasma and tissue colloid osmotic and hydrostatic pressures regulate the flow of blood constituents between the interstitial and intravascular compartments

---

CLINICAL APPLICATIONS

---

81 There are several diseases that affect the plasma colloid osmotic pressure due to the loss of serum protein.

Several terms need defining:

Protein. A nitrogenous compound, essential to all living organisms.

Plasma protein relates to albumin, globulin, and fibrinogen.

Serum protein relates to albumin and globulin.

Serum albumin. Simple protein. Contains the main protein in the blood—about 50%.

Serum globulin. A group of simple protein.

Patients with diagnoses of kidney and liver diseases or malnutrition will lose serum protein.

What are the two groups of simple proteins found in the serum?

____________________ and ____________________.

- - - - - - - - - - - - - - - - - -

albumin; globulin

---

82 The main function of serum albumin is to maintain the colloid osmotic pressure of blood.

Without colloid osmotic pressure, what would happen to the fluid in the tissues? *____.

- - - - - - - - - - - - - - - - - -

Fluid would accumulate and swelling would occur. This is known as edema.

---

83 Serum globulin is not fully understood, but one of its functions is to assist in maintaining colloid osmotic pressure of the blood.

The globulin molecule is larger than the albumin molecule, gram for gram, and it is less effective in maintaining osmotic pressure. Serum albumin will leak out of the capillaries in what types of diseases? * ______________________________

______________________________.

The serum globulin will be retained trying to compensate for the loss of albumin. As stated, globulin is not as effective as albumin in maintaining osmotic pressure, so what would happen to the colloid osmotic pressure in the capillaries? * ______________________________.

And to the fluid in the interstitial spaces or compartments?

* ______________________________.

— — — — — — — — — — — — — — — — — —

liver and kidney diseases and malnutrition

lower osmotic pressure

fluid accumulation and swelling would occur, edema

---

84 Frequently the physician orders an A/G ratio when the probable diagnosis is a kidney or liver disease.

From the previous frames, what do you think is the meaning of A/G?

* ______________________________.

A shift in the A/G ratio will aid the physician in diagnosing the patient's illness.

— — — — — — — — — — — — — — — — — —

albumin and globulin—very good

---

85 What do you think are some of the nursing responsibilities when caring for patients with abnormal serum albumin and serum globulin?

1. * ____________________.

2. * ____________________.

3. * ____________________.

— — — — — — — — — — — — — — — — — —

I asked what you thought—possible answers:

1. report abnormal serum findings immediately
2. observe and report physical findings of swelling or edema
3. keep an accurate record of fluid intake and output

---

86 Edema or swelling occurs when there is fluid retention, and dehydration occurs with excess fluid removal.

If the osmolality of intravascular fluid is greater than the osmolality of intracellular fluid, would the cells lose or gain water? ______________________.

Would edema or dehydration occur to the cells? ______________________.

\- - - - - - - - - - - - - - - - - -

lose; dehydration

---

87 With any venous obstruction or vein obstruction, there is an increased venous hydrostatic pressure. This in turn inhibits the fluid moving out of the tissues, causing the tissues to * ____

____.

\- - - - - - - - - - - - - - - - - -

accumulate fluid and swell

---

88 Normal circulation of blood is dependent on differences in hydrostatic pressure in the arteries, capillaries, and veins.

Increased hydrostatic pressure in the veins would * ______________________

____.

\- - - - - - - - - - - - - - - - - -

prevent circulation and cause swelling of the tissues

CASE REVIEW

Mr. Kendall developed edema of the lower extremities. His serum protein was low.

1. Factors regulating the flow of body constituents between the interstitial and intravascular compartments are stated by * ______________________________

   ______________________________.
2. Define:
   a. *Pressure gradient.* * ______________________________.

   b. *Crystalloids.* * ______________________________.

   c. *Colloids.* * ______________________________.

   d. *Albumin.* * ______________________________
3. Pressure gradients are responsible for the exchange of fluid between the capillaries and the ______________________________.

4. The amount of colloid osmotic pressure that develops depends on the concentration of nondiffusible substances such as * ____________________ .
5. The direction of the movement of fluid depends on the resultants of the opposing forces. The hydrostatic pressure is greater than the colloid osmotic pressure at the arterial end of the capillary; thus the fluid moves out of the ________________ and into the ____________________ .
   The osmotic pressure is greater than the hydrostatic pressure at the venous end of the capillary; thus the fluid moves out of the ____________________ and reenters the ____________________ .
6. Mr. Kendall's decrease in serum protein could account for his (edema/dehydration).
7. Mr. Kendall has venous obstruction due to varicosities. There is an increase in venous hydrostatic pressure, preventing fluid from moving out of tissues into circulation. Explain what will happen to the fluid. * ____________________ ______________________________ .

---

1. Starling's Law of Capillaries
2. a. *Pressure gradient.* Difference in pressure between two points in a fluid
   b. *Crystalloids.* Diffusible substances
   c. *Colloids.* Nondiffusible substances
   d. *Albumin.* Simple protein
3. tissues
4. protein or albumin
5. capillary, surrounding tissues; tissues, capillary
6. edema
7. Fluid accumulates in the tissue, causing swelling or edema

NURSING DIAGNOSIS

Potential alteration in fluid volume related to abnormal fluid retention or loss.

NURSING ACTIONS

1. Check the patient with cirrhosis of the liver, kidney disease, or malnutrition for pitting edema in the lower extremities. Press your finger in the edematous area; if the indented finger print exists for a minute or longer, pitting edema is present.
2. Check the serum albumin and serum protein levels of patients with malnutrition or cirrhosis. With low serum albumin and protein levels, the plasma/serum colloid pressure would be low, thus, fluid would remain in the tissue spaces (edema). Diuretics are helpful in decreasing edema, but they also can markedly decrease the circulatory fluid volume.
3. Assess fluid intake and output. Increased fluid intake with a decreased urine output can indicate extracellular fluid volume excess (ECFV excess).

REFERENCES

Abbott Laboratories: *Fluid and Electrolytes.* North Chicago, 1970, pp 4-13.

Beeson P, McDermott W: *Textbook of Medicine,* ed. 14. Philadelphia, WB Saunders Co, 1975, pp 1579-1581.

Best CH, Taylor NB: *The Physiological Basis of Medical Practice,* ed. 9. Baltimore, Williams & Wilkins Co, 1973, pp 4-113-116, 4-121-123.

Bland JW: *Clinical Metabolism of Body Water and Electrolytes.* Philadelphia, WB Saunders Co, 1963, pp 18-52, 65-79.

Burgess R: Fluids and electrolytes. *Am J Nurs* 65:90–93, 1965.

Garb S: *Laboratory Tests in Common Use,* ed. 6. New York, Springer Publishing Co, 1976, pp 91–92, 197–198.

Grant M, Kubo W: Assessing a patient's hydration status. *Am. J. Nurs* 75:1306–1311, 1975.

Guyton AC: *Textbook of Medical Physiology,* ed. 6. Philadelphia, WB Saunders Co, 1981.

Jacob SW, Francone CA: *Structure and Function in Man,* ed. 3. Philadelphia, WB Saunders Co, 1974.

Kleiner IS, Orten JM: *Biochemistry,* ed. 7. St. Louis, CV Mosby Co, 1966, pp 30-32.

Lancour, J: Two hormones: Regulators of fluid balance. *Monitoring Fluid and Electrolytes Precisely—Nursing Skillbook,* ed. 2. Springhouse PA, Springhouse Corp, 1983, pp 29–38.

Miller BF, Keane CB: *Encyclopedia and Dictionary of Medicine, Nursing, and Allied Health,* ed. 3. Philadelphia, WB Saunders Co, 1983, pp 325, 810.

Taber CW: *Taber's Cyclopedia Medical Dictionary,* ed. 13. Philadelphia, FA Davis Co, 1977.

Widmann, F: *Clinical Interpretation of Laboratory Tests,* ed. 9. Philadelphia, FA Davis Co, 1983.

CHAPTER TWO

Electrolytes and Their Influence on the Body

BEHAVIORAL OBJECTIVES

Upon completion of this chapter, the student will be able to:

Explain the relationship of nonelectrolytes, electrolytes, and ions.
Name the principal cation and anion of the extracellular and intracellular fluids.
Explain the physiological functions of potassium, sodium, calcium, magnesium, chloride and phosphorus.
Give the normal ranges of serum and urine potassium, sodium, calcium, magnesium, chloride, and phosphorus.
Explain the various clinical conditions causing potassium, sodium, calcium, magnesium, chloride, and phosphorus deficits or excesses.
Give the signs and symptoms of hypo-hyperkalemia, hypo-hypernatremia, hypo-hypercalcemia, hypo-hypermagnesemia, and hypo-hyperchloremia.
Relate the electrolyte imbalances to clinical conditions.
Relate the electrolyte imbalances to drug action and interaction.
Explain methods utilized in electrolyte replacement.
List the classes of food that are rich in potassium, sodium, calcium, magnesium, chloride, and phosphorus.

INTRODUCTION

Chemical compounds may behave in one of two ways when placed in solution. In one way, their molecules may remain intact as in urea, dextrose, and creatinine in the body fluid. These molecules do not produce an electrical charge and are considered non-electrolytes.

In the other type, the compound develops a tiny electrical charge when dissolved in water. The compound breaks up into separate particles known as ions, and this process is referred to as ionization. These compounds are known as electrolytes. Some electrolytes develop a positive charge when placed in water, whereas others develop a negative charge.

The chemical composition of seawater and human body fluid are very similar. The principal cations of seawater are sodium, potassium, magnesium, and calcium, and so it is with the body fluid. The seawater contains as principal anions chloride, phosphate, and sulfate; so does body fluid.

In this chapter six electrolytes (potassium, sodium, calcium, magnesium, chloride, and phosphorus are discussed in relation to human body needs (functions), normal serum and urine electrolyte levels, signs and symptoms of excesses and deficits, prevalent diseases causing or resulting from electrolyte imbalance, drugs and electrolyte interaction, replacement therapy, clinical applications, and foods that are rich in these electrolytes. A case review, nursing diagnoses, and nursing actions follow each electrolyte discussed.

An asterisk (*) on an answer line indicates a multiple-word answer.

The meanings for the following symbols are: ↑ increased, ↓ decreased, $>$ greater than, $<$ less than.

ELECTROLYTES: CATION AND ANION

1 *Electrolytes* are compounds that, when placed in solution, will conduct an electric current.

Pure water does not conduct electricity, but if a pinch of salt, which contains sodium and chloride is dropped into it, what do you think would happen to the water? *____________________________________

________________.

Refer to the introduction if needed to answer.

\- - - - - - - - - - - - - - - - -

Salt would produce an electrical charge. Sodium and chloride are ions found in seawater and in our body

---

2 *Ions* are dissociated particles of electrolyte that carry either a positive charge called a *cation* or negative charge called an *anion.*

Dissociated particles of electrolyte are called ____________.

These particles carry either a positive charge called ____________________

or a negative charge called ________________________.

\- - - - - - - - - - - - - - - - -

ion; a cation; an anion

---

3 What is the difference between a cation and an anion? *______________________

______________________________________.

– – – – – – – – – – – – – – – – –

cation: positive charge; *anion*: negative charge

Table 1 gives the principal cations and anions in human body fluid. Since we will be referring to these elements and their symbols throughout the program, take a few minutes now to memorize them. Be sure to note the + and – signs. There will not be a separate section on bicarbonate and phosphate in this chapter; however, they will be covered in the chapter on acid-base, so be sure to remember them. When you think you are ready, go ahead to the frames that follow, referring back to the table only when necessary.

TABLE 1. CATIONS AND ANIONS

Cations		Anions	
Na^+	(Sodium)	Cl^-	(Chloride)
K^+	(Potassium)	HCO_3^-	(Bicarbonate)
Ca^{++}	(Calcium)	HPO_4^{--}	(Phosphate)
Mg^{++}	(Magnesium)		

4 Place a C in front of the cations and an A in front of anions.

____ a. K ____ e. Na

____ b. Mg ____ f. Ca

____ c. Cl ____ g. HPO_4

____ d. HCO_3

– – – – – – – – – – – – – – – – –

a. C; b. C; c. A; d. A; e. C; f. C; g. A

5 If you have had chemistry, these ions and their symbols should be familiar to you. Complete the following chart using proper names and/or symbols.

Names of Ions	*Symbols*
Sodium	___
___	K
Calcium	___
___	Cl
Bicarbonate	___
___	HPO_4
___	Mg

- - - - - - - - - - - - - - - - - -

Names	*Symbols*
Potassium	Na
Chloride	Ca
Phosphate	HCO_3
Magnesium	

---

6 For electrical balance, the quantities of cations and anions in a solution, expressed in milliequivalents (mEq) always equal each other.

Electrolytes differ in their chemical activity, for sodium has one positive charge and calcium has *________________________.

- - - - - - - - - - - - - - - - - -

two positive charges

---

7 The term *milliequivalents* is used to express the number of ionic charges of electrolytes on an equal basis. It measures the __________ activity of ions or elements. Refer to Chapter 1, Frames 56 and 57 if needed to explain milliequivalents.

The total cations in milliequivalents must equal the total ________________ in milliequivalents.

- - - - - - - - - - - - - - - - - -

chemical; anions

---

8 Milliequivalents consider electrolytes in terms of their *_______________

_______________ rather than their weight.

- - - - - - - - - - - - - - - - - -

chemical activity

---

9 Electrolytes have different weights, but are considered during therapy in terms of their activity, which would be expressed as (milliequivalents/milligrams)

_______________.

- - - - - - - - - - - - - - - - - -

milliequivalents

Table 2 gives the weights and equivalences of five ions. Note how the weights of the named ions differ, but the equivalences remain the same according to their ionic charge. You are not expected to memorize the weights of these ions.

TABLE 2. ELECTROLYTE EQUIVALENTS

Kind of Ion	Weight (mg)	Equivalence (mEq)
$Sodium^{+}$	23	1
$Potassium^{+}$	39	1
$Chloride^{-}$	35	1
$Calcium^{++}$	40	2
$Magnesium^{++}$	24	2

---

10 An ion with two charges would have the same equivalence as *_______________

_______________.

- - - - - - - - - - - - - - - - - -

another ion with two charges

---

11 Name a cation and an anion with the same equivalence but different weights. *_______________.

- - - - - - - - - - - - - - - - - -

Sodium and chloride or potassium and chloride

---

12 The electrolyte composition of fluid differs within the two main classes of body fluid.

The two main classes of body water are ________________ and ________________.

- - - - - - - - - - - - - - - - -

intracellular; extracellular

Table 3 gives the ion concentrations in the intravascular fluid (which is frequently referred to as plasma), interstitial fluid, and intracellular fluid. Take a few minutes to memorize the fluids and their greatest concentration of ions. Memorizing the numbers is not necessary. Refer back to the table when necessary.

TABLE 3. ELECTROLYTE COMPOSITION OF BODY FLUID (mEq/L)

	Extracellular		
Ions	Intravascular or Plasma	Interstitial	Intracellular
Na^+	142	145	10
K^+	5	4	141
Ca^{++}	5	3	2
Mg^{++}	2	1	27
Cl^-	104	116	1
HCO_3^-	27	30	10
HPO_4^{--}	2	2	100

---

13 What are the three principal ions in intravascular fluid? ________________, ________________, and ________________.

What are the three principal ions in interstitial fluid? ________________, ________________, and ________________.

What are the three principal ions in intracellular fluid? ________________, ________________, and ________________.

- - - - - - - - - - - - - - - - -

sodium or Na; chloride or Cl; bicarbonate or HCO_3
same as in intravascular fluid;
potassium or K; phosphate or HPO_4; magnesium or Mg

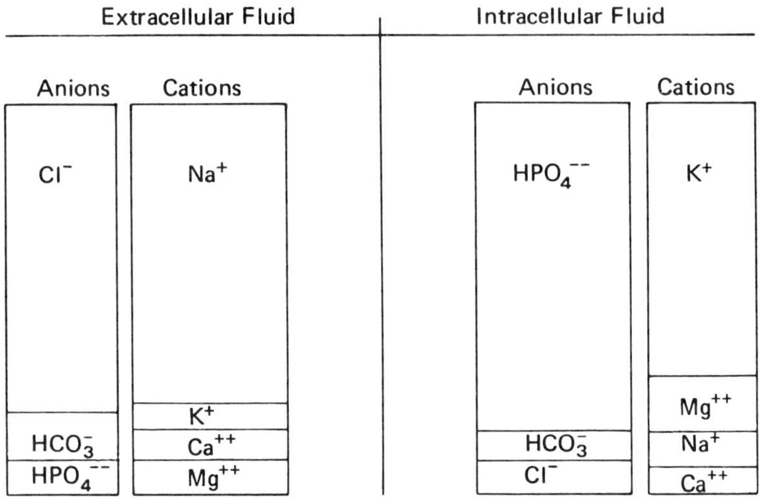

Diagram 3. Anions and cations in body fluid.

Diagram 3 shows the various cations and anions in extracellular and intracellular fluids. Pay special attention to the principal cations and anions in these fluids. Refer to the diagram only when necessary.

---

14 The principal cation in extracellular fluid is *____________________.

The principal cation in intracellular fluid is *____________________.

\- - - - - - - - - - - - - - - - - -

sodium or Na; potassium or K

---

15 The principal anion in extracellular fluid is *____________________.

The principal anion in intracellular fluid is *____________________.

\- - - - - - - - - - - - - - - - - -

chloride or Cl; phosphate or HPO_4

16 Check the three electrolytes having the greatest concentration in extracellular fluid.

___ Na	___ Cl
___ K	___ HCO_3
___ Ca	___ HPO_4
___ Mg	

Check the three electrolytes having the greatest concentration in intracellular fluid.

___ Na	___ Cl
___ K	___ HCO_3
___ Ca	___ HPO_4
___ Mg	

- - - - - - - - - - - - - - - - - -

Na	K
Cl	Mg
HCO_3	HPO_4

POTASSIUM

17 Although potassium is present in all body fluids, it is found mostly in

__________________ fluid.

What kind of ion is potassium? __________________.

- - - - - - - - - - - - - - - - - -

intracellular; cation

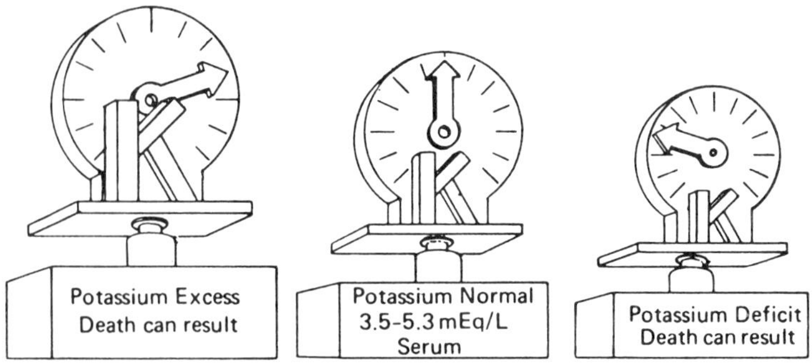

Diagram 4. Potassium—balance and imbalance.

Diagram 4 tells the effect of too much potassium or not enough in our body cells. Memorize the normal range of our serum potassium. You may wonder why the range of serum potassium and not cell potassium, since the cells have the highest concentration of potassium. It is easier to aspirate serum from the intravascular fluid than to aspirate body cells. When you are ready, go ahead to the frames following the diagram and refer to the diagram when necessary.

---

18 The normal mEq/L of serum potassium is *________________. Intracellular potassium is 150 mEq/L, but the concentration is not easily determined.

The kidneys excrete 80-90% of potassium loss. If the kidneys fail to function, what could result? *________________________________.

\- - - - - - - - - - - - - - - - - -

3.5-5.3; excess potassium leading to death

---

19 Either too much or too little potassium can cause cardiac arrest. The heart needs potassium for conducting nerve impulses and contracting the heart muscle.

How do you think too much potassium can cause cardiac arrest?

*____.

How do you think too little potassium can cause cardiac arrest? *____________

____.

\- - - - - - - - - - - - - - - - - -

Too much potassium causes irritability of the heart muscle, increasing and then decreasing the rate

Too little potassium weakens the heart muscle, causing the heart to beat irregularly

---

Table 4 gives the various functions of potassium according to body systems. Study the table and refer to it as needed.

TABLE 4. POTASSIUM AND ITS FUNCTIONS

Body System	Function
Neuromuscular	Transmission and conduction of nerve impulses Contraction of skeletal and smooth muscles
Cardiac	Nerve conduction and contraction of the myocardium
Cellular	Enzyme action for cellular energy production Deposits glycogen in liver cells Regulates osmolality of intracellular (cellular) fluids

20 Potassium is needed for transmission and conduction of *__________. Also potassium is needed for the contraction of *______________________________ and ______________________________.

— — — — — — — — — — — — — — — —

nerve impulses; skeletal and smooth muscles and myocardium

21 Name two cellular activities of potassium.

1. ______________.

2. ______________.

— — — — — — — — — — — — — — — —

1. enzyme action; 2. deposits glycogen in liver; also, regulates intracellular osmolality

22 Too much serum potassium is known as *hyperkalemia,* and too little serum potassium is known as *hypokalemia.*

A serum potassium of 3.0 mEq/L would be known as ______________________.

A serum potassium of 4.2 mEq/L would be known as ______________________.

A serum potassium of 5.8 mEq/L would be known as ______________________.

— — — — — — — — — — — — — — — —

hypokalemia; normal; hyperkalemia

---

23 The assimilative processes involved in the formation of new tissue (the synthesis of complex molecules from simple molecules) are referred to as *anabolism,* and the reactions concerned with tissue breakdown (the breakdown of complex molecules to simple molecules with a release of chemical energy) are referred to as *catabolism.*

When cellular activity is *anabolic* (state of building up), potassium will enter the cells. When cellular activity is *catabolic* (state of breaking down), potassium will leave the cells.

Potassium enters the cells in ________________ states and leaves the cells in ________________ states.

----

anabolic; catabolic

---

24 Potassium may leave the cells under various conditions. When tissues are destroyed as a result of trauma, starvation, or wasting diseases, large amounts of potassium * ______________________________.

Potassium leaves the cells in ________________ states.

----

will leave the cells; catabolic

---

25 During exercise, when muscles contract, the cells lose potassium and absorb a nearly equal quantity of sodium from the extracellular fluid. After exercise, when the muscles are recovering from fatigue, potassium reenters the cells and most of the sodium goes back to the extracellular fluid.

During exercise which ion may be more plentiful in the extracellular fluid—the potassium ion or the sodium ion? * ______________________________
______________________________.

The potassium ion. Of course it depends on how much exercise. The K ion has to go somewhere, so it goes into the extracellular fluid.

---

26 With exercise, potassium leaves the cells, causing muscular fatigue.

After exercise, potassium *____________________________________.

Potassium enters the cells in ____________________ states.

— — — — — — — — — — — — — — — — — —

reenters the cells; anabolic

---

27 The muscles, after releasing potassium from the cells, are like "half-filled water bottles," and are soft.

The soft muscles are a result of (hyperkalemia/hypokalemia).

____________________.

— — — — — — — — — — — — — — — — — —

hypokalemia

---

28 Name as many conditions as you can under which potassium might leave the cells.

1. ____________.
2. ____________.
3. ____________.
4. *____________.

— — — — — — — — — — — — — — — — — —

1. trauma; 2. exercise; 3. starvation; 4. wasting disease

---

29 In stress which can be caused by a harmful condition or emotional strain, an excessive amount of potassium is lost through the kidneys. The potassium leaves the cells, depleting the cells' needs. From the adrenal gland one of the adrenal cortical hormones, aldosterone, is produced in abundance during stress. This hormone influences the kidneys to excrete potassium and to retain sodium.

Frequently the cations, K and Na, have an opposing effect on each other in the extracellular fluid. When one is retained, the other is excreted.

Therefore, with an excessive production of aldosterone, what will happen to the cations—K and Na?

*____.

— — — — — — — — — — — — — — — — — —

K will be excreted and Na will be retained

---

30 When kidney function is normal, the excess potassium will be excreted by the kidneys. The range of potassium excreted daily by the kidneys is 20–120 mEq/L.

If potassium intake is either decreased, or if no potassium is taken orally or given intravenously, potassium will still be excreted by the kidneys. More potassium will be lost from the cells and the extracellular fluid (ECF) when potassium intake is diminished or absent. What type of potassium imbalance will occur? ________________.

— — — — — — — — — — — — — — — —

hypokalemia

---

31 With diminished or no potassium intake, the kidneys will:

( ) a. continue to excrete potassium
( ) b. inhibit potassium excretion

Why *____.

If the kidneys are injured or diseased and the urine output is markedly decreased, then

( ) a. the potassium piles up in the extracellular fluid
( ) b. the potassium piles up in the intracellular fluid
( ) c. the potassium is excreted through the skin

— — — — — — — — — — — — — — — —

a. X; b. –. Potassium excretion comes from the cells and ECF even when potassium intake is zero
a. X; b. –; c. –

Table 5 explains the various clinical conditions that cause too much potassium, or excess, and too little potassium, or deficit. Study this table carefully, noting the potassium changes with various conditions and reasons for these changes. You may have to refer to this table several times.

TABLE 5. CONDITIONS CAUSING POTASSIUM DEFICIT AND EXCESS

Condition	Potassium (K) Deficit (Hypokalemia)	Potassium (K) Excess (Hyperkalemia)
Vomiting, diarrhea, gastric suction, and laxative abuse	Gastric and intestinal secretions are rich in K. With the loss of these secretions, large amounts of K is lost.	
Dehydration	Loss of K from cells	
Anorexia and starvation	K is found in most food. However, lack of food, or diet of food lacking in K, will result in severe K deficit.	

TABLE 5. (Continued)

Condition	Potassium (K) Deficit (Hypokalemia)	Potassium (K) Excess (Hyperkalemia)
Steroids (cortisone) and excess licorice ingestion	Steroids promote K excretion and Na retention. Licorice contains glyceric acid which has an aldosteronelike effect (K excretion and Na retention).	
Diuretics (potassium-wasting)	Many diuretics, e.g., Lasix, HydroDIURIL, excrete K, and other electrolytes.	
Trauma, injury, surgery	K is lost due to tissue injury and cellular damage. Also, great amounts of K are needed to repair injured tissue.	
Burns	K is used in great quantities to repair burn tissues. Burns will cause reduction in K in the cells.	
Stress	Stress leads to increased production of adrenal cortical hormone, aldosterone. This results in loss of excessive amounts of K through kidneys.	
Metabolic alkalosis	K and Cl are lost.	
Adrenal gland disease	Kidney excretion of K is accelerated by ACTH and the adrenal cortical hormones, e.g., cortisone and aldosterone. Steroids excrete K.	Lack of adrenal cortical hormone will cause retention of K and excretion of Na.
Kidney dysfunction		Kidney dysfunction will cause retention of K in extracellular fluid.
Excess or rapid administration of K in solution		Excess or rapidly administered K (> 20 mEq/L per hour) can cause K to accumulate, especially if a severe K deficit is not present.
Transfusions of old blood (> 3 pints)		K is released from the cells in old blood. Serum level can be 5 × (times) its normal.
Diuretics (K-sparing) e.g. Aldactone, Dyrenium.		K is reabsorbed and Na excreted.
Metabolic acidosis		Acidosis causes K to be lost from the cells. If kidney function is decreased, K excess occurs in ECF.

---

32 Name two conditions that may cause potassium loss from the gastrointestinal tract.

1. ______________.

2. ______________.

— — — — — — — — — — — — — — — — — —

1. vomiting; 2. diarrhea. Others—gastric suction, laxative abuse.

---

33 Potassium excess or retention will occur with *________________________________

and *______________________________.

— — — — — — — — — — — — — — — — — —

kidney dysfunction; adrenal gland insufficiency

---

34 Trauma and injury to tissues as a result of burns and surgery can cause a potassium (deficit/excess). Why? *____

____.

— — — — — — — — — — — — — — — — — —

deficit; K is lost from the cells due to tissue injury. Also, great quantities of K are needed to repair these damaged tissues

---

35 Dehydration, diuretics, and starvation will result in a *____________________

______________________________.

— — — — — — — — — — — — — — — — — —

potassium deficit

36 List six clinical conditions causing a potassium deficit:

1. *__________.
2. *__________.
3. __________.
4. __________.
5. *__________.
6. __________.

\- - - - - - - - - - - - - - - - -

1. diarrhea, vomiting, gastric suction
2. dehydration, starvation
3. diuretics—potassium-wasting
4. burns
5. trauma or injury
6. surgery
7. stress
8. increase of adrenal cortical hormones
9. metabolic alkalosis

---

37 Potassium in intravenous (IV) fluids administered at a rate faster than __________ mEq/L per hour for 24 to 72 hours can cause __________.

Causes of potassium excess (hyperkalemia) are:

( ) a. Potassium-wasting diuretics
( ) b. Potassium-sparing diuretics
( ) c. Adrenal gland insufficiency
( ) d. Vomiting, diarrhea
( ) e. Multiple transfusions of old blood
( ) f. Metabolic acidosis with poor kidney function
( ) g. Renal shutdown

\- - - - - - - - - - - - - - - - -

20 mEq/L, hyperkalemia;
a. –; b. X; c. X; d. –; e. X; f. X; g. X

38 Although 98% of potassium is found in cells, focus is placed on the extracellular fluid, for it is more readily available for study.

The normal mEq/L or serum potassium (in extracellular fluid) is

* ______________________________.

— — — — — — — — — — — — — — — — —

3.5–5.3

Table 6 gives the signs and symptoms associated with hyper-hypokalemia. You should memorize the symptoms most commonly seen, which are marked with a dagger. You should also become familiar with symptoms without an asterisk, for they may also develop in patients. Your keen observation and assessment will aid the physician in making a more positive diagnosis. Please learn this table and refer to it as needed. Patients with hyperkalemia and hypokalemia can be found in many of our clinical situations. You may save a patient's life by recognizing and associating his symptoms as one of the potassium imbalances. If you are not familiar with these words, refer to the glossary.

TABLE 6. SIGNS AND SYMPTOMS RELATED TO HYPER-HYPOKALEMIA

Classes	Hyperkalemia	Hypokalemia
Gastrointestinal abnormalities	*Nausea *Diarrhea †Abdominal cramps	*Anorexia Nausea *Vomiting Diarrhea †Abdominal distention †Decreased peristalsis or silent ileus
Cardiac abnormalities	Tachycardia, later †bradycardia and cardiac arrest	†Arrhythmia †Dizziness Cardiac arrest
Urinary abnormalities	†Oliguria or anuria	Polyuria
Neurological abnormalities	Weakness, numbness, or tingling	†Malaise †Muscular weakness Mental depression Drowsiness Confusion Paralysis of respiration
Amount in extracellular fluid	Above 5.3 mEq/L	Below 3.5 mEq/L

†Most commonly seen symptoms of hyper-hypokalemia.

*Commonly seen symptoms of hyper-hypokalemia.

39 With hyperkalemia, the heart beats very fast, which is known as *tachycardia,* and then it slows down (*bradycardia*). The heart goes into a block, with little or no impulses being transmitted, and later cardiac arrest occurs.

You recall that the kidneys are responsible for excreting excessive amounts of potassium not needed by the body. If the kidneys excrete a small amount of urine, known as *oliguria,* or no urine, known as *anuria,*

what can occur to the potassium level? *_______________.

What would you expect the heart rate to be? *_______________

_______________.

- - - - - - - - - - - - - - - - - -

increase or rise in potassium (hyperkalemia);
tachycardia and later bradycardia

---

40 Name the three most commonly seen symptoms of hyperkalemia.

_______________, _______________, and _______________.

- - - - - - - - - - - - - - - - - -

Abdominal cramps; tachycardia and later bradycardia; oliguria or anuria

---

41 Can you name the other symptoms related to hyperkalemia?

1. _______________.

2. _______________.

3. _______________.

4. *_______________.

- - - - - - - - - - - - - - - - - -

1. nausea; 2. diarrhea; 3. weakness; 4. numbness or tingling

---

42 Hypokalemia causes the muscle to be soft, like half-filled water bottles, and weak. The abdomen becomes bloated due to smooth muscle weakness and not due to flatus. The blood pressure goes down (hypotension) and dizziness occurs. Malaise or uneasiness occurs.

The heart beat is irregular, known as ________. Eventually, if the irregularity of the heart beat is not corrected, bradycardia occurs and then cardiac arrest.

- - - - - - - - - - - - - - - - - -

arrhythmia

---

43 Polyuria or excess urine output can be a symptom of hypokalemia. State two *causes* of hypokalemia in which polyuria is involved. (Refer to Table 5 if needed.) *________ and ________.

- - - - - - - - - - - - - - - - - -

excess aldosterone; diuretics

---

44 Name the six most commonly seen symptoms of hypokalemia.

*________, *________, ________,

________, ________, and *________.

- - - - - - - - - - - - - - - - - -

abdominal distention; decreased peristalsis or silent ileus; dizziness; arrhythmia; malaise; muscular weakness

---

45 A weak grip, an irregular pulse, and dizziness upon standing may be signs of

________.

- - - - - - - - - - - - - - - - - -

hypokalemia

CLINICAL APPLICATIONS

46 Eighty to ninety percent of potassium excretion is lost in the urine, and only a very small percentage is lost in the feces.

Which of the following would have the greater loss of potassium?

( ) a. An individual taking a laxative

( ) b. An individual taking a diuretic.

— — — — — — — — — — — — — — — — —

a. –; b X

---

47 Hyperglycemia, an increased blood sugar, is a symptom of diabetes mellitus. Cells cannot utilize glucose; thus, catabolism (cellular breakdown) occurs, and potassium leaves the cells and is excreted by the kidney. If kidneys are not functioning adequately (> 600 ml per day), potassium can accumulate and serum potassium excess can occur.

When cells do not receive their proper nutrition, what happens to the cells?

* ____.

In hyperglycemia (hypokalemia/hyperkalemia) ____________ will occur due to cellular breakdown and polyuria. If there is kidney shutdown, (hypokalemia/hyperkalemia) ____________ occurs.

— — — — — — — — — — — — — — — — —

catabolism–cellular breakdown with loss of potassium;
hypokalemia; hyperkalemia

---

48 Rapid correction of abnormal cellular metabolism in a diabetic patient by administering glucose and insulin may lead to rapid transfer of potassium from the extracellular fluid to the cell. Therefore, the serum potassium would rapidly (increase/decrease) ____________.

— — — — — — — — — — — — — — — — —

decrease

---

49 When oliguria develops because of poor renal function potassium is no longer excreted, which results in a high level of serum potassium.

If there is poor renal function, do you think potassium should be administered? * ______________________________.

Why? * ____.

— — — — — — — — — — — — — — — — — —

No, NEVER with poor renal function. Hyperkalemia could be brought to a dangerous level.

---

50 Potassium therapy should not be administered to patients with untreated adrenal insufficiency and * ______________________________.

— — — — — — — — — — — — — — — — — —

renal failure or poor renal function

---

51 In the cirrhotic patient with degenerated liver cells, hypokalemia can precipitate hepatic coma or liver failure.

As a nurse caring for a patient with cirrhosis you would alert the physician of any low serum K levels, and you would watch for symptoms of * __________

____.

— — — — — — — — — — — — — — — — — —

hypokalemia and also hepatic coma

---

52 Potassium is most plentiful in the gastrointestinal tract. Vomiting and diarrhea can cause a potassium (deficit/excess) ______________________________.

— — — — — — — — — — — — — — — — — —

deficit

---

53 The serum levels of magnesium, chloride, and protein should be checked when correcting hypokalemia. Low serum levels of Mg, Cl, and protein inhibit potassium utilization by the body.

If hypokalemia and hypomagnesemia (Mg deficit) are present, would potassium deficit be corrected by giving potassium chloride orally or intravenously?

________________ Why? * ____________________________________.

\- - - - - - - - - - - - - - - - - -

No. Magnesium needs to be replaced with potassium

---

54 With prolonged hypokalemia, circulatory failure and heart failure can result. The electrocardiogram frequently shows a flat or inverted T wave. With potassium excess, the electrocardiogram shows a peaked T wave.

Serum potassium levels below 2.5 mEq/L and above 7.0 mEq/L are extremely dangerous and need immediate attention. Without correction, what can occur? * ____________________________________.

\- - - - - - - - - - - - - - - - - -

cardiac arrest

Diagrams 5 and 6 note electrocardiographic changes found with hypo-hyperkalemia. Students who have had a physiology course and/or a basic knowledge of electrocardiogram, also known as ECG or EKG, will find these diagrams most useful in their own clinical application when monitoring patients. Students who do not have this basic knowledge should refer to a physiology text and/or a text on electrocardiography. You may need this understanding of electrocardiographic changes in the future. Students who do not need this information to practice nursing may skip to Frame 63.

A brief resume of the electrocardiogram. The ECG measures the electrical activity from various areas of the heart and records this as P, QRS, and T waves.

P wave measures the electrical activity initiating contraction of the atrium or the atrial muscle.

QRS wave complex measures the electrical activity initiating contraction of the ventrical, which is the thickest part of the heart muscle responsible for forcing blood from the heart into the circulation. "Heart attack," also known as myocardial infarction, frequently affects this part of the heart muscle.

T wave is the electrical recovery of the ventricles.

Abnormal potassium levels affect the T wave of the electrocardiogram. Note the normal T wave structure in Diagram 5 and compare the normal with the abnormal, with Patterns 1 and 2. Study this diagram and then proceed to the frames.

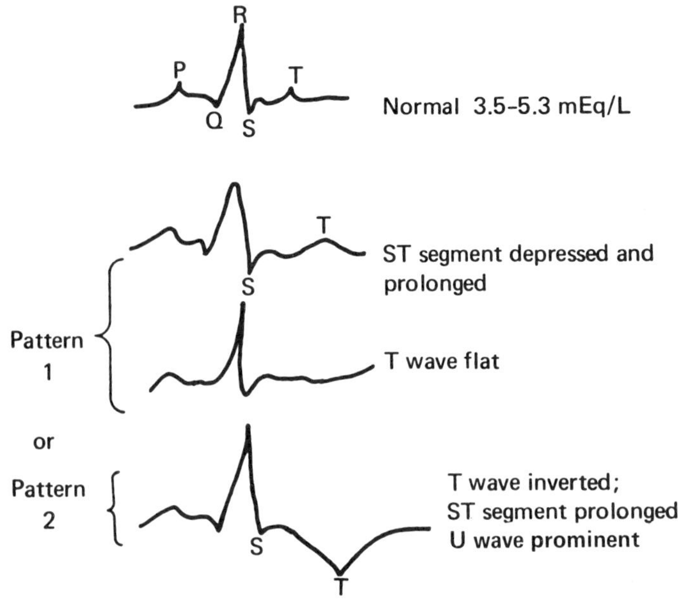

Diagram 5. Electrocardiographic changes in serum potassium deficit. (Adapted from Harry Statland: *Fluid and Electrolytes in Practice.* JB Lippincott Co., Philadelphia, PA, 1963, p 120.)

---

55 Name the abnormal changes of the T wave comparing with the normal.

* ______________________ and ______________________.

The abnormal T waves would indicate a potassium (excess/deficit). ______________________.

- - - - - - - - - - - - - - - - - -

flat T wave; inverted T wave; deficit

---

56 The ST segment is prolonged in both the patterns. This change also relates to

* ______________________.

- - - - - - - - - - - - - - - - - -

potassium deficit

---

57 With serum potassium *deficit* the following electrocardiographic changes may occur:

( ) a. Flat T wave
( ) b. Inverted T wave
( ) c. High-peaked T wave
( ) d. ST segment depressed and prolonged
( ) e. Absence of the P wave
( ) f. U wave prominent

— — — — — — — — — — — — — — — —

a. X; b. X; c. —; d. X; e. —; f. X

High-peaked T waves are an early electrocardiographic sign of hyperkalemia. Heart block can result from severe hyperkalemia, e.g., 8-10 mEq/L of serum potassium. Study Diagram 6 carefully, noting especially the T waves, QRS complex, and P wave. If any of the words are unfamiliar, please refer to a physiology text and/or a text on electrocardiography.

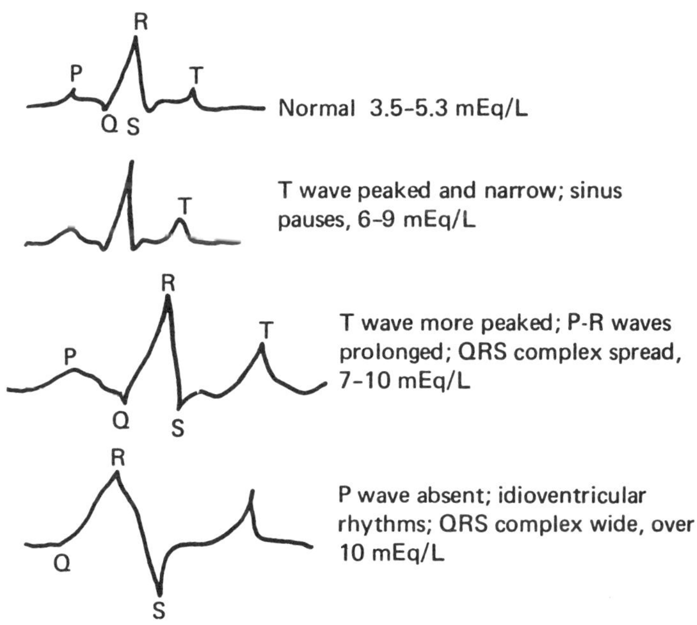

Diagram 6. Electrocardiographic changes in serum potassium concentration. Changes that do occur are most marked in the preordial leads over the right side (V_1-V_4 position) of the heart. (Adapted from Harry Statland: *Fluid and Electrolytes in Practice,* JB Lippincott Co, Philadelphia, PA, 1963, p 116.)

58 Name the abnormal change of the T wave occurring with hyperkalemia. *________

____.

- - - - - - - - - - - - - - - - - -

high-peaked T wave

---

59 A flat or inverted T wave on an electrocardiogram frequently indicates a ________

____________________ state, whereas a high-peaked T wave can indicate a

____________________ state.

- - - - - - - - - - - - - - - - - -

hypokalemic; hyperkalemic

---

60 As the serum potassium concentration increases, 7-10 mEq/L, the T wave becomes *____________________, the P-R waves are ____________________, and the QRS complex is ____________________.

- - - - - - - - - - - - - - - - - -

more peaked; prolonged; spread

---

61 The following electrocardiographic changes can occur with a high serum potassium:

( ) a. Flat T wave
( ) b. Inverted T wave
( ) c. High-peaked T wave
( ) d. ST segment depressed and prolonged
( ) e. QRS complex spread
( ) f. P-R waves prolonged

- - - - - - - - - - - - - - - - - -

a. –; b. –; c. X; d. –; e. X; f. X

62 Place KE for potassium excess or KD for potassium deficit. Which could cause the following electrocardiographic changes?

_________ a. Flat T wave

_________ b. Inverted T wave

_________ c. High-peaked T wave

_________ d. ST segment depressed and prolonged

_________ e. QRS complex spread

_________ f. P-R waves prolonged

- - - - - - - - - - - - - - - - -

a. KD; b. KD; c. KE; d. KD; e. KE; f. KE

DRUGS AND THEIR EFFECT ON POTASSIUM BALANCE

63 Diuretics are divided into two categories: potassium-wasting and potassium-sparing. Potassium-wasting diuretics excrete potassium and other electrolytes such as sodium and chloride. Potassium-sparing diuretics retain potassium but excrete sodium and chloride.

Indicate the electrolytes that will be lost when potassium-sparing diuretics are taken:

( ) a. Potassium
( ) b. Sodium
( ) c. Chloride

- - - - - - - - - - - - - - - - -

a. –; b. X; c. X

Table 7 lists the trade and generic names of potassium-wasting and potassium-sparing diuretics and a combination of potassium-wasting and potassium-sparing diuretics. Study the types of diuretic in each category and refer to the table as needed.

TABLE 7. POTASSIUM-WASTING AND POTASSIUM-SPARING DIURETICS

Potassium-Wasting Diuretics	Potassium-Sparing Diuretics
Thiazides:	Aldosterone-Antagonist:
Chlorothiazide/Diuril	Spironolactone/Aldactone
Hydrochlorothiazide/Hydrodiuril	Triamterene/Dyrenium
Loop Diuretics:	Amiloride/Midamor
Furosemide/Lasix	**Combination: K-Wasting and K-Sparing Diuretics**
Ethacrynic acid/Edecrin	Aldactazide
Mercurials:	Spironazide
Mercaptomerin sodium/Thiomerin	Dyazide
Carbonic Anhydrase Inhibitors:	Moduretic
Acetazolamide/Diamox	
Osmotic Diuretic:	
Mannitol	

---

64 Name the potassium imbalance that is most likely to occur in people taking a potassium-sparing diuretic and who have poor kidney function. ______________

______________________________________.

- - - - - - - - - - - - - - - - -

potassium excess or hyperkalemia

---

65 Potassium-wasting diuretics can cause (hypokalemia/hyperkalemia).

______________________.

Name the potassium imbalance that is most likely to occur in people who take potassium-sparing diuretics and who have poor kidney function.

____.

- - - - - - - - - - - - - - - - -

hypokalemia; potassium excess or hyperkalemia

66 Enter W for potassium-wasting, S for potassium-sparing, and C for a combination of potassium-wasting and -sparing diuretics for the following drugs. Refer to the table as needed:

( ) a. Chlorothiazide/Diuril
( ) b. Aldactazide
( ) c. Triamterene/Dyrenium
( ) d. Acetazolamide/Diamox
( ) e. Mercaptomerin sodium/Thiomerin
( ) f. Mannitol
( ) g. Dyazide
( ) h. Spironolactone/Aldactone
( ) i. Hydrochlorothiazide/Hydrodiuril
( ) j. Furosemide/Lasix
( ) k. Ethacrynic acid/Edecrin
( ) l. Amiloride/Midamor

---

a. W; b. C; c. S; d. W; e. W; f. W; g. C; h. S; i. W; j. W; k. W; l. S

Laxatives, corticosteroids, antibiotics, and potassium-wasting diuretics are the major drug groups that can cause potassium deficit or hypokalemia. The drug groups attributed to potassium excess or hyperkalemia are oral and intravenous potassium salts, CNS agents, and potassium-sparing diuretics. Table 8 lists the drugs that affect potassium balance.

TABLE 8. DRUGS AFFECTING POTASSIUM BALANCE

Potassium Imbalance	Drugs	Rationale
Hypokalemia (serum potassium deficit)	Laxatives Enemas (hyperosmolar)	Laxative abuse can cause potassium depletion.
	Corticosteroids: Cortisone Prednisone	Ion Exchange Agent: Steroids promote potassium loss and sodium retention.
	Licorice	Licorice action is similar to aldosterone—promotes K loss and Na retention.
	Levodopa/L-Dopa	Increases potassium loss via urine.
	Antibiotics I: Amphotericin B Polymyxin B Outdated tetracyclines Gentamicin Neomycin	Tosic effect on renal tubules, thus decreasing potassium reabsorption.

TABLE 8. Continued

Potassium Imbalance	Drugs	Rationale
	Antibiotics II: Penicillin Ampicillin Carbenicillin Ticarcillin Nafcillin	Potassium excretion is enhanced by the presence of nonreabsorbable anions.
	Barium Lithium Terbutaline/Brethine	These agents promote potassium into cells, thus lowering the serum potassium level.
	Potassium-wasting diuretics	See Table 7.
Hyperkalemia (serum potassium excess)	Potassium chloride (oral or IV) Potassium salt (No-Salt) K penicillin K PO_4 enema	Excess ingestion or infusion of these agents could cause potassium excess.
	Indomethacin/Indocin Captopril/Capoten heparin	These decrease renal excretion of potassium.
	CNS Agents: Barbiturates Sedatives Narcotics Heroin Amphetamines	These CNS agents are usually characterized by muscle necrosis and cellular shift of potassium from cells to serum.
	Succinylcholine/Anectine	Loss of potassium from cells.
	Potassium-sparing diuretics	See Table 7.

Adapted with permission from A. Nanji: Drug-induced electrolyte disorders. *Drug Intelligence and Clinical Pharmacy*, 1983, pp 177–179.

---

67 Enter KD for potassium deficit/hypokalemia and KE for potassium excess/hyperkalemia beside the drugs that can cause a potassium imbalance. Refer to Table 8 as needed:

_______ a. Laxatives

_______ b. Corticosteroids

_______ c. Barbiturates

_______ d. Narcotics

_______ e. Indomethacin/Indocin

_______ f. Licorice

_______ g. Antibiotics

_______ h. Levodopa

_______ i. Heparin

_______ j. Potassium chloride

_______ k. Succinylcholine/Anectine

_______ l. Terbutaline/Brethine

- - - - - - - - - - - - - - - - - -

a. KD; b. KD; c. KE; d. KE; e. KE; f. KD; g. KD; h. KD; i. KE; j. KE; k. KE; l. KD

---

68 Digitalis is a drug that strengthens the heart muscle and slows down the heart beat. A serum potassium deficit or hypokalemia, enhances the action of digitalis and causes the drug to become more potent. Digitalis toxicity or intoxication (slow and irregular pulse, nausea and vomiting, anorexia) can result from low serum potassium.

Thiazides and loop diuretics can cause (hypokalemia/hyperkalemia) _______________________.

The nurse needs to be alert for what type of drug toxicity when a patient is taking potassium-wasting diuretics and digitalis? *_______________________.

- - - - - - - - - - - - - - - - - -

hypokalemia; digitalis toxicity

---

69 Symptoms of digitalis toxicity are bradycardia, slow heart beat, and/or arrhythmia.

Can you name two (2) other symptoms of digitalis toxicity?

a. *____________________.

b. ____________________.

— — — — — — — — — — — — — — — — —

a. nausea and vomiting; b. anorexia

---

70 A serum potassium excess or hyperkalemia will decrease the action of digitalis. If a person has a serum potassium of 5.8 mEq/L, will (more/less) digitalis be needed to obtain the appropriate digitalis dosage? ____________________.

— — — — — — — — — — — — — — — — —

more

---

71 Quinidine is an antiarrhythmic drug used to correct irregular heart rate. Hypokalemia blocks the effect of quinidine; therefore more quinidine may be needed to cause the appropriate action. Hyperkalemia enhances the action of quinidine and can produce quinidine toxicity and myocardium depression.

Explain the effect of hypokalemia on digitalis. *____________________ ____________________; on quinidine *____________________ ____________________.

— — — — — — — — — — — — — — — — —

It enhances the action of digitalis; it blocks the action of quinidine

---

72 Cortisone causes excretion of potassium and retention of sodium. If a person takes digoxin (digitalis), hydrochlorothiazide/Hydrodiuril, and prednisone/cortisone daily, what type of severe electrolyte imbalance could result?

____________________.

Explain the effect this imbalance has on digitalis. *____________________ ____________________.

— — — — — — — — — — — — — — — — —

hypokalemia or potassium deficit; hypokalemia precipitates digitalis toxicity or it enhances the action of digitalis.

---

73 Calcium gluconate emphasizes the effect of potassium excess on the heart muscle. The effect is transient. When the serum potassium level is high (hyperkalemia), calcium gluconate will decrease the cardiac effect. Remember calcium gluconate helps the heart rhythm that results from hyperkalemia but it will not correct the potassium level.

How do you think it affects the ECG? * ______________________________

______________________________.

It improves the ECG and decreases the T wave

POTASSIUM CORRECTION

Oral potassium supplements help to replace potassium lost from taking large doses of potent diuretics or due to debilitating disease.

Table 9 contains examples of frequently ordered oral potassium supplements.

TABLE 9. ORAL POTASSIUM SUPPLEMENTS

Preparation	Drug
Liquid	Potassium chloride 10% = 20 mEq/15ml; 20% = 40 mEq/15ml
	Kay Ciel (potassium chloride)
	Kaochlor 10% (potassium chloride)
	Kaon Cl 20 (potassium chloride)
	Potassium Triplex (potassium acetate, bicarbonate, citrate)
Tablet/capsule	Potassium chloride (enteric-coated tablet)
	Kaon–plain (potassium gluconate)
	Kaon Cl (potassium chloride)
	Slow K (potassium chloride–8 mEq)
	Kaochlor (potassium chloride)
	K-Lyte–plain (potassium bicarbonate-effervescent tablet)
	K-Lyte/Cl (potassium chloride)

74 Name the drug that would replace serum potassium and serum choride deficits.

* ______________________________.

---

potassium chloride (liquid or tablet).

---

75 Oral potassium is extremely irritating to the gastric mucosa and should be diluted in at least 4 ounces of water or juice and, preferably, 8 ounces of fluid.

Name two oral potassium supplements that contain bicarbonate.

1. *______________________.
2. ______________________.

- - - - - - - - - - - - - - - - -

1. potassium triplex; 2. K-Lyte—plain. Also Kaon—plain. The gluconate in Kaon is converted to bicarbonate. Kaon comes with or without Cl.

---

76 There have been reports of deaths related to hyperkalemia caused by oral potassium supplements. These agents are recommended when taking digitalis and potassium-wasting diuretics.

Would oral potassium supplements be recommended for a person with poor kidney function? ______________________.

Why? ____

______________________________________.

- - - - - - - - - - - - - - - - -

No

Because 80-90% of potassium is excreted from the body by the kidneys; hyperkalemia might result

---

77 Severe serum hyperkalemia may occur from administering intravenous potassium in solution rapidly, not giving time for potassium to pass into the cells.

The normal rate of intravenous flow for potassium is 20–40 mEq in a liter in 8 hours or no more than 10 mEq/L per hour. What might result from administering 40 mEq of potassium per hour? ______________________.

Why? *____.

- - - - - - - - - - - - - - - - -

hyperkalemia; potassium accumulation in the ECF

---

78 Intravenous (IV) potassium is irritating to blood vessels (can cause phlebitis) and tissues (can cause sloughing and necrosis).

Potassium should NEVER be given as a bolus (injected directly into the vein).

Do you know what could happen? * ____________________________.

The nurse should assess the injection site when the patient is receiving IV KCl in solutions for __________________ and __________________.

— — — — — — — — — — — — — — — — — —

Cardiac arrest. GOOD. Potassium concentration is extremely irritating to the myocardium (heart muscle).
phlebitis and infiltration or tissue sloughing or necrosis

---

79 For severe hypokalemia, 40–60 mEq of KCl can be diluted in 1 liter of IV fluids, and no more than 20 mEq per hour should be given.

What is the recommended KCl dosage diluted in 1 liter of IV fluids?

* ____________________________.

— — — — — — — — — — — — — — — — — —

20-40 mEq/L

---

80 In hypokalemia, if the serum potassium level is 3.0–3.5 mEq/L, 100–200 mEq of KCl are needed to raise serum level 1 mEq/L. Remember, don't administer it all at once; high concentration is toxic to the heart muscle and irritating to the blood vessels.

If the serum potassium level is below 3.0 mEq/L, 200–400 mEq of KCl is needed to raise serum level 1 mEq/L.

An individual has a serum potassium level of 2.7 mEq/L. How much KCl, administered IV, will be needed to raise the serum level to 3.7 mEq/L? * ____________________________.

— — — — — — — — — — — — — — — — — —

200–400 mEq KCl

---

81 Daily potassium requirement is 40–60 mEq. An individual with a serum potassium of 3.4 mEq/L needs (more/less) daily potassium requirement. __________.

— — — — — — — — — — — — — — — — — —

more

Table 10 explains the methods used for correcting potassium excess (hyperkalemia). Study the table carefully and refer back to it as needed.

TABLE 10. CORRECTION OF POTASSIUM EXCESS (HYPERKALEMIA)

Treatment Methods	Rationale
Potassium restriction	Restriction of potassium intake will slowly lower the serum level. For mild hyperkalemia (slightly elevated K levels), i.e., 5.5–5.8 mEq/L, potassium restriction is normally effective.
IV sodium bicarbonate ($NaHCO_3$)	By elevating the pH level potassium moves back into the cells, thus lowering the serum level. This is a temporary treatment.
10% Calcium gluconate	Calcium decreases the irritability of the myocardium resulting from hyperkalemia. It is a temporary treatment and does not promote K loss. Caution: Administering calcium to a patient on digitalis can cause digitalis toxicity.
Insulin and glucose (10–50%)	The combination of insulin and glucose moves potassium back into the cells. It is a temporary treatment, effective for approximately 6 hours, and is not always as effective when repeated.
Kayexalate (sodium polystyrene) and Sorbitol 70%	Kayexalate is used as a cation exchange for severe hyperkalemia and can be administered orally or rectally. Approximate dosages are as follows: *Orally:* Kayexalate—10–20 g 3 to 4 times daily Sorbitol 70%—20 ml with each dose *Rectally:* Kayexalate—30–50 g Sorbitol 70%—50 ml; mix with 100–150 ml water (Retention enema—20–30 minutes)

---

82 To correct mild hyperkalemia, restriction of potassium intake is suggested. Could you correct a hyperkalemia of 7.0 mEq/L by restricting potassium intake? ________. Why? * ____.

- - - - - - - - - - - - - - - -

No; it is severe hyperkalemia, and the method is too slow

---

83 For temporary correction of moderate potassium excess, indicate which methods can be used.

( ) a. Potassium restriction diet
( ) b. IV sodium bicarbonate
( ) c. 10% Calcium gluconate
( ) d. Insulin and glucose
( ) e. Kayexalate and Sorbitol

- - - - - - - - - - - - - - - - - -

a. –; b. X; c. X; d. X; e. –

---

84 If a patient is taking digitalis and has a serum potassium of 7.4 mEq/L, would 10% calcium gluconate be indicated for temporary correction of hyperkalemia?

________________ Explain. *________________

________________________________.

- - - - - - - - - - - - - - - - - -

No. Calcium excess (hypercalcemia) enhances the action of digitalis, causing digitalis toxicity

---

85 Drugs, Kayexalate (sodium polystyrene sulfonate), a cation exchange resin, and Sorbitol 70%, are given for severe hyperkalemia. They cause a sodium-potassium ion exchange, and the potassium is excreted.

What is suggested for mild hyperkalemia? *________________.

What is suggested for severe hyperkalemia? *________________.

- - - - - - - - - - - - - - - - - -

Restrict potassium intake
Kayexalate and Sorbitol–ion exchange

CASE REVIEW

Mr. Johnson, 68 years old, has been vomiting and has had diarrhea for two days. He takes Digoxin, 0.25 mg, and HydroDIURIL, 50 mg, daily. His serum potassium level is 3.2 mEq/L. He complains of being dizzy. The nurse assesses his physiological status and notes that his muscles are weak and flabby, his abdomen is distended, and peristalsis diminished.

1. What was his potassium imbalance? ________________.
2. "Normal" range of potassium balance is *________________.

3. Should the nurse have checked his pulse rate, since he was receiving Digoxin?

_______________. Explain. *_______________

_______________.

4. Vomiting will cause a potassium (deficit/excess).
5. Name Mr. Johnson's sign and symptoms of potassium deficit.

 a. _______________.

 b. *_______________.

 c. *_______________.

 d. *_______________.

Mr. Johnson's heart activity was monitored on ECG. He received 1 liter of 5% dextrose in water with 40 mEq/L of KCl.

6. A flat T wave would be indicative of _______________.

7. The daily potassium requirement is *_______________.

8. A higher concentration of KCl in IV fluids can cause *_______________

_______________.

9. List at least three common symptoms found with hyperkalemia.

 a. *_______________.

 b. *_______________.

 c. *_______________.

A week after his acute illness, his serum potassium was 3.7 mEq/L and his serum chloride was in the "normal" range. The physician ordered Mr. Johnson to take an oral potassium supplement with his daily Digoxin and HydroDIURIL (hydrochlorothiazide). For Mr. Johnson's arthritis, prednisone 4 times a week was ordered.

10. Name an oral potassium supplement that the physician can prescribe. _______________

_______________. Explain your reason.

*_______________.

11. Explain the effect of cortisone on potassium in the body.

*_______________.

12. If a potassium supplement was not prescribed, name a potassium-sparing diuretic that can be taken in conjunction with hydrochlorothiazide _______________

_______________.

- - - - - - - - - - - - - - - - -

1. hypokalemia
2. 3.5-5.3 mEq/L
3. Yes. Hypokalemia will enhance the action of Digoxin, causing digitalis toxicity

4. deficit
5. dizziness; muscles weak and flabby; distended abdomen; diminished peristalsis
6. hypokalemia
7. 40–60 mEq/L
8. hyperkalemia, which is toxic to heart muscle and can cause phlebitis (irritated blood vessel)
9. Abdominal cramps; tachycardia and later bradycardia; oliguria
10. Potassium triplex, Kaon, or K-Lyte; since the chloride level is normal, potassium chloride
11. excretes potassium
12. Aldactone or Dyrenium

NURSING DIAGNOSES

Potential alteration in nutrition related to insufficient intake of foods rich in potassium.

Potential alteration in electrolyte balance (potassium) related to vomiting, diarrhea, injury, surgery, burns, drugs (potassium-wasting diuretics, steroids), adrenal gland disease, or renal failure.

Knowledge deficit related to lack of understanding of disease process causing potassium imbalance.

Potential for noncompliance to treatment regime for potassium imbalance related to inadequate explanation.

Potential alteration in cardiac output related to potassium imbalance, hypo-hyperkalemia.

Potential for injury to vessels, tissues, or gastric mucosa related to phlebitis from concentrated potassium solution, infiltration of potassium solution into subcutaneous tissues, or ingestion of concentrated oral potassium irritating and damaging to the gastric mucosa.

NURSING ACTIONS: POTASSIUM

1. Recognize clinical conditions and problems that can cause hypokalemia, e.g., vomiting, diarrhea, gastric suction, tissue injury and trauma, diuretics, steroids, etc., and can cause hyperkalemia, e.g., rapidly administered IV potassium, renal failure (oliguria and anuria), adrenal gland insufficiency, and potassium-sparing diuretics.
2. Check the serum potassium levels daily or as ordered. When potassium imbalance is suspected, electrolytes are ordered more frequently. Serum levels below 3.5 mEq/L indicate hypokalemia, and above 5.3 mEq/L, hyperkalemia. Immediate medical care is essential when serum potassium is below 2.5 mEq/L or above 7.0 mEq/L to avoid potential cardiac arrest.
3. Assess for signs and symptoms of hypokalemia, e.g., dizziness, arrhythmias, abdominal distention, soft muscles, and decreased peristalsis or ileus, and hyperkalemia, e.g., tachycardia and then bradycardia, abdominal cramps, and oliguria or anuria.
4. Teach patients to eat foods rich in potassium, when hypokalemia is present or

when they are taking potassium-wasting diuretics and steroids. Examples of such foods are fresh fruits, fruit juices, dry fruits, vegetables, meats, nuts, Coca Cola, and cocoa.

5. Dilute oral potassium supplements in at least 4 ounces of water or juice. Concentrated potassium is irritating to the gastric mucosa.
6. Check intravenous site for phlebitis or infiltration when KCl is given IV in a liter of solution. Potassium is irritating to blood vessels and subcutaneous tissue. NEVER administer potassium intravenously as a bolus or IV push.
7. Report ECG (EKG) changes (with the T wave). A flat or inverted T wave can indicate hypokalemia, and a high-peaked T wave can indicate hyperkalemia.
8. Monitor the serum potassium level on patients receiving potassium-wasting diuretics and steroids (cortisone preparations).
9. Assess for signs and symptoms of digitalis toxicity, i.e., nausea, vomiting, anorexia, bradycardia, arrhythmia, when patient is receiving a potassium-wasting diuretic and/or steroids with a digitalis preparation. Hypokalemia enhances the action of digitalis.
10. Recognize other drugs and substances, i.e., glucose, insulin, laxatives, lithium carbonate, salicylates, tetracycline, and licorice, that will decrease serum potassium levels.
11. Monitor serum magnesium, chloride, and protein when hypokalemia is present. Correcting potassium deficit is not effective when hypomagnesemia, hypochloremia, and hypoproteinemia are also present.
12. Irrigate gastrointestinal tube with normal saline solution to prevent electrolyte loss. Gastrointestinal fluid loss from GI suctioning, vomiting, and diarrhea should be measured.
13. Check urine output hourly or every 8 hours according to the patient's clinical status. Urine output should be at least 25 ml per hour, or 600 ml per 24 hours, when patient has been given oral or IV potassium. The kidneys excrete 80–90% of the potassium loss.
14. Regulate the flow rate of intravenous fluid with potassium so that no more than 10 mEq/L of KCl is administered per hour. Rapidly administered KCl can cause hyperkalemia.
15. Administer fresh blood (blood transfusion) to patients with hyperkalemia. The serum potassium level of fresh blood is 3.5–5.5 mEq/L. With blood that is 3 weeks old, the serum potassium level could be as high as 25 mEq/L.
16. Monitor the medical treatments for hyperkalemia. Know which corrective treatments are used for mild, moderate, and severe hyperkalemia.
17. Note if the patient is on digitalis when calcium gluconate is ordered for temporary correction of hyperkalemia. Hypercalcemia enhances the action of digitalis, causing digitalis toxicity.
18. Administer Kayexalate and Sorbitol orally or rectally, according to the amount prescribed by the physician. The serum potassium should be checked frequently during treatment to prevent hypokalemia resulting from overcorrection of hyperkalemia.

SODIUM

86 Sodium is the main cation found in the * ______________________ fluid.

— — — — — — — — — — — — — — — —

extracellular or intravascular

---

87 Sodium loss from the skin is negligible under normal conditions, but with increased environmental temperature, fever, and/or muscular exercise, the loss rises.

If an individual runs a race and the atmospheric temperature is 100, what do you think would happen to the sodium in his body?

* ______________________.

— — — — — — — — — — — — — — — —

sodium loss

---

88 The normal concentration of sodium in the extracellular fluid is 135-146 mEq/L.

The normal concentration of sodium in perspiration is 50–100 mEq, which is about half the concentration found in the * ______________________.

— — — — — — — — — — — — — — — —

extracellular fluid.

---

89 When the body's sodium level is elevated, perspiration is not a means of regulating sodium excretion, for it is regarded as a by-product of temperature regulation.

The concentration of sodium in the extracellular fluid is * ______________________.

— — — — — — — — — — — — — — — —

135-146 mEq/L

---

90 Bones contain as much as 800 to 1000 mEq of sodium, but only a portion of the sodium is available for exchange with sodium in other parts of the body.

The concentration of sodium in the extracellular fluid is * ______________________.

— — — — — — — — — — — — — — — —

135-146 mEq/L

---

91 Bones contain (more/less) sodium than extracellular fluid. ________.

\- - - - - - - - - - - - - - - - -

more

---

92 Thirst often leads to the replacement of water, but not of sodium;

One (can/cannot) replace sodium by drinking lots of water. ________.

\- - - - - - - - - - - - - - - - -

cannot

---

93 Ocean water is about three times as salty as our body fluid—far too salty for our body organs, i.e., stomach and intestines.

Ocean water would be considered a (hypo-osmolar/hyperosmolar) fluid.

________. Therefore, in cases of ocean water ingestion, the water is drawn from the body fluid into the stomach and intestines by the process of (osmosis/diffusion). ________.

\- - - - - - - - - - - - - - - - -

hyperosmolar; osmosis

---

94 As the stomach and intestines accumulate huge volumes of water, vomiting occurs. Explain the reasons for: (1) the accumulation of stomach fluid, and (2) vomiting.

1. * ________.
2. * ________.

\- - - - - - - - - - - - - - - - -

1. Body fluids accumulate in the stomach due to osmosis (lesser to the greater concentration)
2. Fluids overextend the stomach

---

95 An elevated serum sodium is known as sodium excess or *hypernatremia* and a decreased serum sodium is known as sodium deficit or *hyponatremia.*

Hypernatremia is known as * ________. Hyponatremia is known as * ________.

\- - - - - - - - - - - - - - - - -

sodium excess; sodium deficit

---

96 One of the main functions of sodium is to influence the distribution of water in the body. Water will accompany sodium.

A name for sodium excess is ______________________.

A name for sodium deficit is ______________________.

A function of sodium is to influence the distribution of * ______________________

______________________.

— — — — — — — — — — — — — — — — — —

hypernatremia; hyponatremia; body water (water accompanies sodium)

Table 11 explains how one organ and two glands influence serum sodium. The kidneys have an important role in maintaining homeostasis of body sodium. The hypothalamus produces ADH (antidiuretic hormone) and the posterior hypophysis (posterior pituitary gland) stores and secretes ADH. This hormone will absorb large quantities of water from the kidneys. The adrenal glands are composed of two sections, the cortex and the medulla, each secreting its own hormones. The hormones from the adrenal cortex are frequently referred to as steroids. Study this table carefully. Refer to a physiology text for any further clarification.

TABLE 11. INFLUENCES AFFECTING SERUM SODIUM

Organ	
Kidneys	The regulators. Kidneys maintain homeostasis through excretion or absorption of sodium from renal tubules according to excess or deficit of serum sodium.
Glands	
1. Posterior hypophysis or posterior pituitary gland	Pituitary antidiuretic hormone (ADH) favors water absorption from distal tubules of kidneys, and thus sodium excretion is promoted to lesser extent.
2. Adrenal cortex of the adrenal glands	Adrenal cortical hormones, e.g., cortisone and aldosterone, favor sodium absorption from renal tubules. These steroids influence kidneys to absorb sodium and excrete potassium.

---

97 The chief regulation of sodium occurs within the ______________________.

— — — — — — — — — — — — — — — — — —

kidneys

---

98 The amount of water absorbed from the kidneys depends on the amount of ADH being secreted. If less water is absorbed from the renal tubules, what happens frequently to the sodium? *____

__________________________.

\- - - - - - - - - - - - - - - - -

less sodium absorption

---

99 Explain the effect of cortisone and aldosterone on sodium and potassium.

*____.

\- - - - - - - - - - - - - - - - -

They influence the kidneys to absorb sodium and excrete potassium

---

Table 12 explains the functions of sodium. The two most important functions of sodium are for water balance and neuromuscular activity. Study the table carefully and refer to the table as needed.

TABLE 12. SODIUM AND ITS FUNCTIONS

Body System	Functions
Neuromuscular	Transmission and conduction of nerve impulses (sodium pump—see Cellular).
Body fluids	Largely responsible for the osmolality of vascular fluids. Doubling Na gives the approximate serum osmolality.
	Regulation of body fluid (sodium causes water retention).
Cellular	Sodium pump action. Sodium shifts into cells as potassium shifts out, repeatedly, to maintain water balance and neuromuscular activity. When Na shifts into the cell, depolarization occurs (cell activity); and when Na shifts out and K shifts back into the cell, repolarization occurs.
	Enzyme activity.
Acid-base	Assists with the regulation of acid-base balance. Sodium combines readily with chloride (Cl) or bicarbonate (HCO_3) to promote acid-base balance.

---

100 An important function of sodium is to aid neuromuscular activity. Name another electrolyte responsible for neuromuscular activity. ______________.

\- - - - - - - - - - - - - - - - -

potassium

---

101 The concentration or tonicity (osmolality) of vascular fluids is determined by which electrolyte? ______________.

To get a rough estimate of the serum osmolality, what can you do?

*______________________________.

\- - - - - - - - - - - - - - - - -

sodium; double the serum sodium

---

102 Explain the action of the sodium pump. *______________________________

______________________________.

Name two purposes for the sodium pump.

1. *______________________________.
2. *______________________________.

\- - - - - - - - - - - - - - - - -

Sodium shifts in as potassium shifts out of cells; depolarization and cell activity occur. Then K shifts in and Na shifts out for repolarization (cell rest).
1. water balance; 2. neuromuscular activity

---

103 What are the two anions that combine with sodium to help regulate acid-base balance. ______________ and ______________.

\- - - - - - - - - - - - - - - - -

chloride and bicarbonate

---

104 Large amounts of sodium are contained within the following body secretions—saliva, gastric secretions, bile, pancreative juice, and intestinal secretions.

Indicate which of the following body secretions contain large quantities of sodium:

( ) a. Saliva
( ) b. Thyroid secretions
( ) c. Gastric secretions
( ) d. Bile
( ) e. Parathyroid secretions
( ) f. Pancreative juice
( ) g. Intestinal secretions

---

a. X; b. –; c. X; d. X; e. –; f. X; g. X

Table 13 lists specific conditions causing either a sodium deficit or excess or both deficit and excess. If you are not familiar with any words in this table, refer to a physiology text or a nursing text or the glossary. Study this table carefully. Refer to it as needed.

TABLE 13. CONDITIONS CAUSING SODIUM DEFICIT AND SODIUM EXCESS

Conditions	Na Deficit (Hyponatremia)	Na Excess (Hypernatremia)
Vomiting	Sodium concentration is high in gastric mucosa. Vomiting will decrease sodium concentration in alimentary tract.	With severe vomiting, loss of water is greater than loss of Na. Saltiness of remaining body fluids becomes dangerously high.
Sweating, increased environmental temperature, fever, muscular exercise	Large amounts of sodium are lost from the skin, which frequently results from increased environmental temperature, fever, muscular exercise, and sweating.	
Diarrhea	There is loss of Na from intestinal secretions.	When babies have diarrhea, their loss of water can be greater than their loss of Na.
Tap-water enema	Tap-water enemas wash out salty intestinal contents from body.	
Burns	Great quantities of sodium are lost from burn wounds and from oozing of the burn surface.	
Surgery	There is loss of Na from postoperative wound drainage, bleeding, and vomiting.	

TABLE 13 (Continued)

Conditions	Na Deficit (Hyponatremia)	Na Excess (Hypernatremia)
Gastric suction	Tube inserted through nose or mouth into stomach and intestines for purpose of drainage will cause "salty" gastric and intestinal secretions to pass out gastrointestinal tract.	
Potent diuretics and low sodium diet	Diuretics and low Na diet are prescribed frequently by physicians for edema. Over a period of months, a severe Na deficit could result.	
SIADH (syndrome of inappropriate antidiuretic hormone)	ADH promotes water reabsorption from the distal renal tubules. After surgery, pain, narcotics, or IPPB, more ADH is produced, which causes more water to be reabsorbed and the ECF to be diluted.	
Water intake (excess or decrease) and electrolyte-free solutions	Drinking great quantities of plain water will dilute extracellular fluid. Administering continuous 5% dextrose in water will also dilute ECF and can cause water intoxication.	When there is a slight Na deficit and no water intake, Na in ECF can be greater than normal.
Adrenal cortex dysfunction	Adrenal cortical hormones, e.g., cortisone and aldosterone, having sodium-retaining effect on renal tubules. Adrenal insufficiency will cause loss of Na across renal tubules. Addison's disease is example of adrenal insufficiency.	Excessive adrenal cortical hormone will cause excess of sodium in body. Cushing's syndrome is example of overproduction of adrenal cortical hormone.
Kidney, heart, and circulatory dysfunctions	With advanced renal disorders, renal tubules do not respond to antidiuretic hormone; therefore, there is loss of sodium and water.	Reduced glomerular filtration will result in Na retention. Congestive heart failure, shock, or renal arterial insufficiency will result in Na retention.

105 Place a D for sodium deficit or E for sodium excess concerning the following:

( ) a. Moderate vomiting
( ) b. Tap-water enema
( ) c. Increased environmental temperature fever and muscular exercise
( ) d. Decrease of adrenal cortical hormone
( ) e. Increase of adrenal cortical hormone
( ) f. Burns
( ) g. Great volumes of water intake
( ) h. Little to no water intake
( ) i. Potent diuretics and low Na diet
( ) j. Surgery
( ) k. Gastric suction
( ) l. Diarrhea
( ) m. Severe diarrhea in babies
( ) n. Reduced glomerular filtration (kidney dysfunction)
( ) o. Congestive heart failure and shock
( ) p. SIADH

- - - - - - - - - - - - - - - - - -

a. D: b. D; c. D; d. D; e. E; f. D; g. D; h. E; i. D; j. D; k. D; l. D; m. E; n. E; o. E; p. D

---

106 Vomiting will (increase/decrease) the sodium concentration in the alimentary tract. __________________.

- - - - - - - - - - - - - - - - - -

decrease

---

107 Name the conditions causing a large sodium loss through the skin.

1. ________________..
2. ________________.
3. ________________.
4. ________________.

- - - - - - - - - - - - - - - - - -

1. sweating
2. increased environmental temperature
3. fever
4. muscular exercise

---

108 What effects do cortisone and aldosterone have on the renal tubules?

*______________________________________.

\- - - - - - - - - - - - - - - -

sodium retention and potassium excretion

---

109 Addison's disease occurs when there is an adrenal cortical hormone (insufficiency/overproduction), whereas Cushing's syndrome occurs when there is an adrenal cortical hormone (insufficiency/overproduction) ______________.

\- - - - - - - - - - - - - - - -

insufficiency; overproduction

---

110 With Addison's disease, there is a sodium ______________.

With Cushing's syndrome, there is a sodium ______________.

\- - - - - - - - - - - - - - - -

loss; retention

---

111 With burns, there is a great sodium ____________. Why? *______________

______________________________________.

\- - - - - - - - - - - - - - - -

loss; Due to oozing at the burn surface (or because sodium replaces potassium in the damaged cells)

---

112 Repeated tap-water enemas can result in a *______________. Why?

*______________________________________.

\- - - - - - - - - - - - - - - -

sodium loss. They wash away salty intestinal secretions

---

113 Postoperative wound drainage, bleeding, and vomiting can cause a sodium (deficit/retention) ____________.

SIADH may occur following surgery. Explain how it can cause a sodium deficit. * ____

____.

— — — — — — — — — — — — — — — — —

deficit. Excess or continuous ADH causes water to be reabsorbed from the kidney, thus diluting ECF

---

114 The use of gastric suction for the purpose of drainage can cause (hypernatremia/hyponatremia). Why? * ______________________________.

— — — — — — — — — — — — — — — — —

hyponatremia. Gastric and intestinal secretions pass out through the tube

---

115 Excessive intake of plain water or continuous administration of 5% dextrose in water can cause a (hyperkalemic/hypokalemic) state ____________

Why? * ____.

What type of water imbalance could occur? * ______________________.

— — — — — — — — — — — — — — — — —

hypokalemic. It will dilute extracellular fluid. Water intoxication. Very good.

---

116 If a patient is on a potent diuretic, i.e., Lasix, and low sodium diet (< 1.5 g sodium) for several months, what type of sodium imbalance would occur?

____________________.

— — — — — — — — — — — — — — — — —

hyponatremia. If the serum sodium is less than 125 mEq/L, then dietary sodium should not be restricted.

117 Congestive heart failure, shock, or obstruction of the arterial supply to the kidney will cause sodium ________________. Why? *________________.
________________________________.

— — — — — — — — — — — — — — — — —

retention. Reduced glomerular filtration

Table 14 gives the signs and symptoms associated with hypo-hypernatremia. You should memorize the symptoms commonly seen, which are marked with an asterisk. Specific gravity may be a new word to you. It is a weight of a substance, e.g., urine, in comparison with an equal volume of water. Water has a specific gravity of 1.000. Study this table carefully. Refer to the glossary for any unknown words and refer to this table as needed.

TABLE 14. SIGNS AND SYMPTOMS RELATED TO HYPO-HYPERNATREMIA

Class	Hyponatremia	Hypernatremia
Central nervous system or CNS, muscular and skin	*Abdominal cramps *Muscular weakness Apprehension *Headache Convulsion	*Flushed skin *Elevated body temperature Excitement
Gastrointestinal or GI	*Nausea *Vomiting	*Rough, dry tongue
Cardiac		Blood pressure ↓ or ↑ *Tachycardia
Serum osmolality	↓ 280 mOsm/L	↑ 295 mOsm/L
Specific gravity of urine	↓ 1.010	↑ 1.025
mEq/L	135 and lower	146 and higher

*Commonly seen symptoms of hypo-hypernatremia.

118 In hyponatremia, the serum sodium level is below ____________. What is the serum value in hypernatremia? *________________________.

— — — — — — — — — — — — — — — — —

135 mEq/L; above 146 mEq/L

119 Name the commonly seen symptoms of hyponatremia.

1. *_______________.
2. *_______________.
3. _______________.
4. _______________.
5. _______________.

– – – – – – – – – – – – – – – – – –

1. abdominal cramps
2. muscular weakness
3. headache
4. nausea
5. vomiting

---

120 A serum osmolality below 280 mOsm/L could indicate (hyponatremia/hypernatremia) _______________. What could a serum osmolality above 295 mOsm/L indicate? ____________________.

– – – – – – – – – – – – – – – – – –

hyponatremia (also indicates ECF dilution caused by a sodium deficit or excess water retention); hypernatremia

---

121 To estimate roughly the serum osmolality, the nurse could *____.

Grant and Kubo* used a formula for calculating serum osmolality.

$$\text{Serum osmolality} = 2 \times \text{serum Na} + \frac{\text{BUN}}{3} + \frac{\text{Blood glucose}}{20}$$

What would the serum osmolality be if the serum sodium was 145, BUN 24, blood glucose 110? _______________ mOsm/L. Would the serum osmolality be indicative of (hypo-osmolar/iso-osmolar/hyperosmolar)? _______________.

– – – – – – – – – – – – – – – – – –

double the serum sodium level; 303.5 mOsm/L; hyperosmolar. Good.

*MM Grant. WM Kubo: Assessing a patient's hydration status. *Am J. Nurs*, 75: 1306, August 1975.

---

122 Name the commonly seen symptoms of hypernatremia.

1. *____________.
2. *____________.
3. *____________.
4. ____________.

Later in this program when you are studying dehydration, you will notice that many symptoms of hypernatremia resemble the symptoms of dehydration.

\- - - - - - - - - - - - - - - - -

1. flushed skin
2. elevated body temperature
3. rough, dry tongue
4. tachycardia

---

123 Hypernatremia, when associated with dehydration, causes a decrease in blood pressure. However, if an individual has hypernatremia, "normal" fluid balance, and a history of hypertension, the blood pressure might be (decreased/increased) ____________________.

\- - - - - - - - - - - - - - - - -

increased

---

124 Place D for symptoms of sodium deficit and E for symptoms of sodium excess:

( ) a. Abdominal cramps
( ) b. Muscular weakness
( ) c. Flushed skin
( ) d. Headache
( ) e. Elevated body temperature
( ) f. Rough, dry tongue
( ) g. Tachycardia
( ) h. Nausea and vomiting
( ) i. Serum Na, 135 mEq/L and below
( ) j. Serum Na, 146 mEq/L and above

\- - - - - - - - - - - - - - - - -

a. D; b. D; c. E; d. D; e. E; f. E; g. E; h. D; i. D; j. E

CLINICAL APPLICATIONS

125 The majority of laboratory analyses of electrolyte content are carried out on the plasma, which represents less than one-twelfth of the total body fluid; therefore, results may occasionally be misleading.

The reason for the use of plasma instead of other body fluid and cells is that * ______________________.

- - - - - - - - - - - - - - - - - -

it is easier to obtain, or a similar response

---

126 You may have a cardiac patient who has edema and yet his serum sodium concentration is reduced. How do you think this can occur?

* ______________________

______________________.

- - - - - - - - - - - - - - - - - -

He may be on a low sodium diet and taking a diuretic

---

127 A 24-hour urine sodium test is helpful for determining sodium retention or loss in the body. A normal range for 24-hour urine sodium is 40–220 mEq/L.

If a patient's 24-hour urine sodium is 32 mEq/L, serum sodium level is 133 mEq/L, and the patient has symptoms of heart failure, do you think the patient could be retaining sodium? ________. Explain. * ______________________

______________________.

- - - - - - - - - - - - - - - - - -

Yes, Very good, for I asked what you thought. A low urine sodium indicates sodium retention in the body, especially with symptoms of overhydration. A low serum sodium level can be misleading. Hyponatremia could be due to fluid volume excess (hypervolemia), which causes the sodium to be diluted

---

128 If your patient is vomiting following a surgical intervention and is receiving dextrose and water intravenously, one may expect a sodium (excess/deficit) if vomiting persists.

With a patient having severe vomiting without water replacement, one may expect a sodium (excess/deficit). Why? *______________________________

____.

— — — — — — — — — — — — — — — — —

deficit; excess. The loss of water would be greater than the loss of sodium

---

129 Excessive intravenous administration dextrose and water can cause sodium dilution. Dextrose is utilized rapidly.

Explain how sodium can be diluted.*______________________________

____.

— — — — — — — — — — — — — — — — —

Following the utilization of dextrose, the remaining water dilutes sodium and other electrolytes

---

130 If a feeble or debilitated patient receives numerous tap-water enemas for the purpose of cleaning the bowel, the enemas can cause a sodium ______________.

— — — — — — — — — — — — — — — — —

loss

---

131 Diarrhea can cause either sodium deficit or sodium excess. Babies having diarrhea can lose more ____________________ than the ion, sodium; therefore, a sodium ____________________ would result.

— — — — — — — — — — — — — — — — —

water; excess

---

132 With congestive heart failure, there would be a sodium (retention/excretion).

Why? *____

____.

— — — — — — — — — — — — — — — — — —

retention. Poor circulation reduces glomerular filtration; therefore, sodium is retained

---

133 If hyponatremia is due to SIADH (syndrome of inappropriate antidiuretic hormone) secretion or congestive heart failure, which do you think would be the treatment of choice?

( ) a. 5% Saline intravenously
( ) b. Restrict water intake

— — — — — — — — — — — — — — — — — —

a. –; b. X. Concentrated saline should be administered cautiously because pulmonary edema (lung congestion) could occur.

DRUGS AND THEIR EFFECT ON SODIUM BALANCE

Diuretics, certain antipsychotics, antineoplastics, and barbiturates can cause sodium deficit. Corticosteroids, ingestion and infusion of sodium are the major causes of sodium excess. Table 15 lists the drugs that affect sodium balance.

TABLE 15. DRUGS AFFECTING SODIUM BALANCE

Sodium Imbalance	Drugs	Rationale
Hyponatremia (serum sodium deficit)	Diuretics	Diuretics, K-wasting, and K-sparing cause sodium excretion.
	Lithium	Lithium promotes urinary sodium loss.
	Antineoplastics: Vincristine Cyclophosphamide/Cytoxan	These group of agents stimulate ADH release and cause hemodilution and decreased sodium level.
	Barbiturates Morphine Nicotine Ibuprofen/Motrin Antipsychotics: Amitrystyline/Elavil Thioridazine/Mellaril Thiothixene/Navane Tranylcypromine/Parnate	
	Clonidine/Catapres Carbamazepine/Tegretol Tolbutamide/Orinase	

TABLE 15. (Continued)

Sodium Imbalance	Drugs	Rationale
Hypernatremia (serum sodium excess)	Corticosteroids Cortisone Prednisone	Steroids promote sodium retention and potassium excretion.
	Hypertonic saline Sodium penicillin Sodium phosphate Sodium bicarbonate Cough medicines	Administration of sodium salts in excess.
	Cholestyramine/Questran	Ion exchange.
	Amphotericin B Propoxyphene/Darvon Demeclocycline/Declomycin	These agents promote urinary water loss without sodium.
	Lactulose	Water loss in excess of sodium via gut.

Adapted with permission from A. Nanji: Drug-induced electrolyte disorders. *Drug Intelligence and Clinical Pharmacy*, 17:175–177, March 1983.

134 Enter SD for sodium deficit/hyponatremia and SE for sodium excess/hypernatremia beside drugs that affect sodium balance. Refer to Table 15 as needed.

_______ a. Lithium

_______ b. Cortisone

_______ c. Diuretics

_______ d. Sodium penicillin

_______ e. Antipsychotic agents

_______ f. Ibuprofen/Motrin

_______ g. Amphoterin B

_______ h. Lactulose

_______ i. Barbiturates

_______ j. Cyclophosphamide/Cytoxan

_______ k. Tolbutamide/Orinase

— — — — — — — — — — — — — — — — —

a. SD; b. SE; c. SD; d. SE; e. SD; f. SD; g. SE; h. SE; i. SD; j. SD; k. SD

---

135 Patients who are receiving steroids, such as cortisone and prednisone, should be cautioned in the use of excess salt. Why? *____.

\- - - - - - - - - - - - - - - - -

Steroids promote sodium retention (sodium-retaining effect).

---

136 Hyponatremia enhances the action of quinidine and hypernatremia reduces or decreases it.

With a serum sodium of 156 mEq/L would the action of quinidine be (increased/decreased)? ________________________.

\- - - - - - - - - - - - - - - - -

decreased

---

137 Cough medicines, antibiotics, and sulfonamides contain sodium. Name a drug that has a sodium-retaining effect. *____________________________________.

\- - - - - - - - - - - - - - - - -

cortisone and cortisonelike products

SODIUM CORRECTION

---

138 The majority of Americans consume 8–15 g of sodium per day, which is 4 to 5 times the amount of sodium required by the body. Daily sodium requirements are 2–4 g. A teaspoon of salt has 2.3 g of sodium.

With an increase in sodium intake, what happens to water intake and to body fluids? *____.

\- - - - - - - - - - - - - - - - -

Sodium holds water. Extracellular fluid (ECF) would be increased

---

139 To restore sodium balance due to sodium deficit, either normal saline solution (0.9% NaCl) or 3% or 5% salt solution is recommended. Several physicians suggest that the serum sodium be below 130 mEq/L before giving saline and below 120 mEq/L before giving a concentrated salt solution, i.e., 3% or 5% saline.

Remember, rapid infusion of concentrated salt solutions can result in pulmonary edema. Can you explain the reason? *______________________________

____

____.

- - - - - - - - - - - - - - - -

Sodium retains fluid. A high concentration of sodium pulls out intracellular fluid, thus overexpanding the vascular compartment. Fluid collects in the lungs

CASE REVIEW

Mrs. Unger has a high temperature and diaphoresis. She has been nauseated and has taken only ginger ale for the last several days. Her serum sodium is 129 mEq/L.

1. What type of sodium imbalance does Mrs. Unger have? *____________________

   ____

2. Give the "normal" serum sodium range. *____________________.
3. Give some of the reasons for Mrs. Unger's imbalance.

   a. *____________________.

   b. *____________________.
4. Name some of the clinical signs and symptoms the nurse might observe.

   a. *____________________.

   b. *____________________.

   c. *____________________.

   d. *____________________.
5. When testing Mrs. Unger's urine, would the specific gravity be:

   ( ) a. 1.010 or below

   ( ) b. 1.015

   ( ) c. 1.020 or above

Mrs. Unger was given 3% sodium chloride solution. Her serum sodium level rose to 152 mEq/L. She was given quinidine for her irregular pulse rate.

6. Do you think her serum potassium should have been evaluated? ____________.

   Why? *______________________________.

7. Name some of the clinical signs and symptoms the nurse will observe for hypernatremia.

   a. *______________________.

   b. *______________________.

   c. *______________________.

8. Explain the effect of hypernatremia on quinidine. ______________________

   ______________________.

9. If Mrs. Unger were to receive cortisone and antibiotics, what would this do to her hypernatremic state? *______________________

   ______________________.

10. Sodium is most plentiful in the extracellular compartment. Explain why sodium might enter the cells. *______________________

    ______________________.

---

1. Sodium deficit or hyponatremia
2. 135-146 mEq/L
3. Fever; diaphoresis; ginger ale intake for several days (lack of food)
4. Abdominal cramps; muscular weakness; headaches; nausea and vomiting
5. a. 1.010 or below
6. Yes. She could have a loss of potassium from lack of food and due to illness. Arrhythmia may be a sign of hypokalemia
7. Flushed skin; elevated body temperature; rough, dry tongue; and tachycardia
8. It reduces or decreases quinidine's action
9. They increase the hypernatremic state. Cortisone would retain sodium, and antibiotics would increase sodium
10. During cell catabolism, potassium leaves the cells and sodium enters

NURSING DIAGNOSES

Potential alteration in nutrition related to excess intake in food rich in sodium or as a result of inadequate nutritional intake.

Potential alteration in electrolyte balance (sodium) related to vomiting, diarrhea, gastric suction, surgery burns, potent diuretics, SIADH, adrenal gland diseases, or diet extremely high or low in salt.

Knowledge deficit related to a lack of understanding of the disease process that is causing sodium imbalance.

Potential for noncompliance to treatment regime for sodium deficit or excess related to inadequate explanation.

Fluid volume excess related to excess sodium and water retention.

Potential impairment of skin integrity related to peripheral edema secondary to body sodium and water excess.

Potential activity intolerance related to peripheral edema secondary to body sodium and water excess.
Potential alteration in urinary elimination related to sodium and water retention.

NURSING ACTIONS

1. Recognize the clinical conditions and problems associated with hyponatremia, i.e., vomiting, diarrhea, excessive perspiration, excessive water (plain) intake, continuous use of 5% dextrose in water, use of potent diuretics with low sodium diet, SIADH from surgery, and with hypernatremia, i.e., severe vomiting and diarrhea, use of cortisone preparations, CHF and renal arterial insufficiency.
2. Check serum sodium levels daily or as ordered. The sodium levels < 125 mEq/L or > 160 mEq/L should be reported immediately to the physician.
3. Check electrolytes, chloride, and potassium when serum sodium levels are not within normal range.
4. Assess for signs and symptoms of hyponatremia, i.e., headache, confusion, muscular weakness, and abdominal cramps, and of hypernatremia, i.e., flushed skin, elevated body temperature, rough, dry tongue, and tachycardia.
5. Check specific gravity of urine. A specific gravity < 1.010 could indicate hyponatremia and > 1.030 could indicate hypernatremia.
6. Check the serum osmolality level. A serum osmolality level of < 280 mOsm/L could indicate hyponatremia and > 295 mOsm/L, hypernatremia. Sodium is mostly responsible for the serum osmolality value.
7. Keep an accurate intake and output record. Excess water intake can cause hyponatremia due to hemodilution.
8. Check the urine sodium level. A decreased urine sodium, < 40 mEq/L, frequently indicates sodium retention in the body, even though the serum sodium level could be within normal range (caused by hemodilution). Also check for rales in the lung and for pitting edema from sodium and fluid retention.
9. Assess for signs and symptoms of pulmonary edema when the patient is receiving several liters of normal saline (0.9% NaCl) or 3–5% saline. Sodium holds water in the blood vessels, and when administering a concentrated saline solution, overhydration can occur. Symptoms include dyspnea, cough, chest rales, neck and hand vein engorgement.
10. Restrict water when hyponatremia is due to hypervolemia (excess fluid volume).
11. Assess for signs and symptoms of water intoxication, i.e., headaches and behavioral changes (confusion, delirium, convulsions) when hyponatremia is due to hypervolemia.
12. Observe changes in vital signs, especially the pulse rate. If hyponatremia is due to hypovolemia (loss of fluid and sodium), shocklike symptoms, such as tachycardia, could occur. Frequently, hyponatremia is due to hemodilution from excess fluid volume.
13. Teach patient with hypernatremia to avoid foods rich in salt, i.e., canned foods, lunch meats, ham, pork, pickles, potato chips, etc. Also Pepsi-Cola has more sodium than Coca-Cola.
14. Identify drugs that have a sodium-retaining effect on the body, i.e., cortisone prep-

arations, cough medicines, certain laxatives. Check serum sodium levels, lungs for rales, and feet and legs for edema.

CALCIUM

---

140 Refer to Table 3 or Diagram 3 if needed. Calcium is a(n) (anion/cation) found in intracellular fluid. Which fluid has the greatest calcium concentration? ________

________________.

— — — — — — — — — — — — — — — —

cation; extracellular

---

141 Calcium is a durable chemical substance of the body that is the last element to find its place in the adult body composition and the last element to leave after death.

The element for the preservation of bony remains of dead creatures and also responsible for the x-ray photograph of bones is ________________.

— — — — — — — — — — — — — — — —

calcium

---

142 Refer to Table 3. Calcium concentration in the blood serum (plasma) is *________

________________, whereas in the cells it is *________________

________________.

— — — — — — — — — — — — — — — —

5 mEq/L; 2 mEq/L or a trace

Table 16 explains the functions of calcium. Calcium is needed for neuromuscular activity, normal cellular permeability, coagulation of blood, and bone and teeth formation. Study the table carefully, and refer to it as needed.

TABLE 16. CALCIUM AND ITS FUNCTIONS

Body System	Functions
Neuromuscular	Normal nerve and muscle activity. Calcium causes transmission of nerve impulses, and contraction of skeletal muscles.
Cardiac	Contraction of heart muscle (myocardium).
Cellular and blood	Maintenance of normal cellular permeability. ↑ calcium decreases cellular permeability and ↓ calcium increases cellular permeability. Coagulation of blood. Calcium promotes blood clotting by converting prothrombin into thrombin.
Bones and teeth	Formation of bone and teeth. Calcium and phosphorus make bones and teeth strong and durable.

---

143 Name five functions of calcium in the body. Refer to Table 16 as needed.

1. *______________________.
2. *______________________.
3. *______________________.
4. *______________________.
5. *______________________.

— — — — — — — — — — — — — — — — —

1. normal nerve and muscle activity
2. contraction of myocardium
3. maintenance of normal cellular permeability
4. coagulation of blood
5. formation of bone and teeth

---

144 A high serum concentration of calcium (increases/decreases) the permeability of membranes ______________, whereas a low serum concentration of calcium (increases/decreases) the permeability of membranes ______________.

— — — — — — — — — — — — — — — — —

decreases; increases

---

145 A calcium deficit increases cellular permeability and neuromuscular excitability (tetany symptoms). How does calcium promote blood clotting? *______________________________.

Explain how tetany occurs. *______________________________.

— — — — — — — — — — — — — — — — —

calcium converts prothrombin into thrombin; calcium deficit causes neuromuscular excitability

---

---

146 For body utilization, calcium must be in the ionized form. In body fluids, calcium is found in both ionized or nonionized (bound to plasma proteins) forms.

In an alkalotic state (body fluids are more alkaline), large amounts of the calcium become protein bound and cannot be utilized. If the body fluids were more acid (acidotic state), would calcium more likely be ionized/nonionized)?

______________________.

- - - - - - - - - - - - - - - - -

ionized. Calcium could be utilized.

---

147 To maintain "normal" cellular membrane, the calcium should be __________ in the body.

- - - - - - - - - - - - - - - - -

ionized

---

148 The parathyroid glands, which are four small oval-shaped glands located on the posterior thyroid gland, regulate the serum level of calcium. These glands secrete parathyroid hormone, which is responsible for the homeostatic regulation of calcium ion in the body fluids.

When the serum calcium level is low, the parathyroid glands will secrete more parathyroid hormone. Explain what would happen if the serum calcium level was high. *______________________

______________________.

- - - - - - - - - - - - - - - - -

It would inhibit the secretion of the parathyroid hormone, or less would be secreted

---

149 The regulation of serum calcium is maintained by the negative feedback system.

A low serum calcium stimulates the parathyroid gland to *______________.

What do you think happens when there is a high serum calcium? *__________

______________________.

- - - - - - - - - - - - - - - - -

secrete parathyroid hormone. It prevents the secretion of parathyroid hormone from the parathyroid gland

---

---

150 The negative feedback system is also responsible for the hormonal actions of most of our other body glands.

Explain the action of ADH, (antidiuretic hormone) in relation to the negative feedback system.

* ____

____.

— — — — — — — — — — — — — — — —

When serum solute concentration (serum osmolality) is high, there is more ADH secreted and more water reabsorbed from the renal tubules. The opposite occurs with a low serum osmolality

---

151 A low serum calcium level tells the parathyroid gland to secrete more parathyroid hormone. The parathyroid hormone increases serum calcium by mobilizing calcium from the bone, increasing renal absorption of calcium, and promoting calcium absorption from the intestine in the presence of vitamin D.

Parathyroid hormone increases serum calcium by:

( ) a. Mobilizing calcium from the bone
( ) b. Decreasing renal absorption of calcium
( ) c. Increasing renal absorption of calcium
( ) d. Promoting calcium absorption from the intestine with vitamin D

— — — — — — — — — — — — — — — —

a.X; b. –; c. X; d. X

---

152 *Hypocalcemia* means a calcium deficit in the extracellular fluid; therefore, another name for calcium excess would be ______________ ________.

— — — — — — — — — — — — — — — —

hypercalcemia

Table 17 gives various conditions that cause either a calcium deficit or excess. Study this table carefully and then proceed to the frames that follow. Refer to this table as needed.

TABLE 17. CONDITIONS CAUSING CALCIUM DEFICIT AND CALCIUM EXCESS

Conditions	Calcium Deficit (Hypocalcemia)	Calcium Excess (Hypercalcemia)
Inadequate protein diet	Inadequate protein intake inhibits body's utilization of calcium.	
Lack of calcium in diet and lack of Vitamin D intake	Today, calcium deficit is rare from lack of intake. Vitamin D should be present for calcium absorption from the GI tract.	
Diseases, extensive infections, and burns	Calcium is pulled from extracellular fluid and trapped in burns, diseased tissues, and infections of great body cavity.	
Hypoparathyroidism	When parathyroid glands are injured or destroyed, parathyroid hormone is reduced and there is a calcium loss.	
Diarrhea (lack of calcium absorption from GI tract)	A deficit will occur when large amounts of calcium are lost in loose bowel movements and due to lack of calcium absorption from GI tract.	
Overuse of antacids and laxatives	Constant use of antacids can decrease acidity which decreases calcium ionization. Overuse of laxatives can increase intestinal mobility, preventing calcium absorption.	
Hyperphosphatemia	An increased serum phosphate (phosphorus) level promotes calcium excretion.	
Renal failure	It causes phosphorus retention and calcium excretion.	
Hyperparathyroidism		When parathyroid glands are overactive, calcium builds up in extracellular fluid. This can result from tumors of parathyroid gland, which produce excessive quantities of parathyroid hormone.
Multiple fractures and prolonged immobilization		These conditions increase calcium in extracellular fluid by releasing calcium from bones.
Tumor of bones		Tumors of bone are composed of many abnormal bone cells. Calcium from these cells is released and absorbed into extracellular fluid.

TABLE 17. (Continued)

Conditions	Calcium Deficit (Hypocalcemia)	Calcium Excess (Hypercalcemia)
Malignancies: bronchogenic carcinoma, breast cancer, leukemia		Calcium is released from malignant cells and from immobilization.
Hypophosphatemia		A decreased serum phosphate (phosphorus) level promotes calcium retention.

---

153 What effect does an inadequate protein diet have on calcium? *______________

____.

- - - - - - - - - - - - - - - - - -

It inhibits the body's utilization of calcium

---

154 Burns, diseases, and extensive infections can cause a calcium ______________.

Why? *____.

- - - - - - - - - - - - - - - - - -

deficit. They pull calcium from the extracellular fluid

---

155 Hypoparathyroidism can cause a calcium ______________. How? *________

____.

Hyperparathyroidism can cause a calcium ______________. How? *________

____.

- - - - - - - - - - - - - - - - - -

deficit. Loss of the parathyroid hormone;
excess. Overproduction of the parathyroid hormone

---

156 Calcium and phosphorus are regulated by the parathyroid gland, found in many foods, and absorbed together. The serum values of calcium and phosphate (ionized phosphorus) are opposites. With hyperphosphatemia, would (hypocalcemia/hypercalcemia) occur? ______________. With hypophosphatemia, would (hypocalcemia/hypercalcemia) occur? ______________.

- - - - - - - - - - - - - - - - - -

hypocalcemia; hypercalcemia

---

157 What effect does prolonged immobilization have on calcium? * ______________

____.

— — — — — — — — — — — — — — — — — —

It increases serum calcium by releasing Ca from the bones

---

158 What other three conditions, not mentioned in the preceding frames, can cause a calcium excess? * ______________, ______________, and * ______________.

— — — — — — — — — — — — — — — — — —

tumors of the bone, malignancies, multiple fractures

---

159 Place a D for calcium deficit and E for calcium excess in the following conditions:

( ) a. Infections
( ) b. Lack of vitamin D
( ) c. Hyperparathyroidism
( ) d. Hypoparathyroidism
( ) e. Diarrhea
( ) f. Multiple fractures
( ) g. Prolonged immobilization
( ) h. Burns
( ) i. Tumor of the bone
( ) j. Cancer of the lung and breast
( ) k. Excessive use of laxatives and antacids

— — — — — — — — — — — — — — — — — —

a. D; b. D; c. E; d. D; e. D; f. E; g. E; h. D; i. E; j. E; k. D

Table 18 lists signs and symptoms of hypocalcemia and hypercalcemia according to the parts of the body that are affected. Know the normal range of serum calcium. Study this table carefully and refer to it as needed.

TABLE 18. SIGNS AND SYMPTOMS RELATED TO HYPOCALCEMIA AND HYPERCALCEMIA

Classes	Hypocalcemia	Hypercalcemia
CNS and muscular abnormalities	Anxiety, irritability Tetany Twitching around mouth Tingling and numbness of fingers Carpopedal spasm Spasmodic contractions Laryngeal spasm Convulsions Abdominal cramps Muscle cramps	Depression/apathy Deep pain over bony areas Muscles are flabby
Cardiac abnormalities	Weak cardiac contractions Prolonged QT interval	Shortened QT interval Signs of heart block Cardiac arrest in systole
Blood abnormalities	Blood does not clot normally, reduction of prothrombin.	
Skeletal abnormalities	Fractures occur if deficit persists due to calcium removed from bones.	Pathologic fractures and bone pain Thinning of bones is apparent since calcium is transferred from bone to ECF.
Renal abnormalities		Flank pain Calcium stones formed in the kidney. This can result when calcium leaves bones due to injury and immobilization.
Milliequivalents per liter	Below 4.5 mEq/L	Above 5.8 mEq/L
Milligrams per deciliter	Below 8.5 mg/dl	Above 10.5 mg/dl

---

160 Indicate which of the following signs and symptoms relate to *hypo*calcemia:

( ) a. Cramps in the abdomen
( ) b. Abnormalities in clotting of blood
( ) c. Flabby muscles
( ) d. Stones in the kidneys
( ) e. Tingling of the fingers
( ) f. Twitching around the mouth
( ) g. Laryngeal spasm

- - - - - - - - - - - - - - - - -

a. X; b. X; c. –; d. –; e. X; f. X; g. X

---

161 Hypocalcemia is present when the serum calcium level is below ________ mEq/L or ________ mg/dl. Hypercalcemia is present when the serum calcium level is above ________ mEq/L or ________ mg/dl.

— — — — — — — — — — — — — — — —

4.5, 8.5; 5,8, 10.5

---

162 Indicate which of the following signs and symptoms relate to *hyper*calcemia:

( ) a. Cramps in the abdomen
( ) b. Abnormalities in clotting of blood
( ) c. Flabby muscles
( ) d. Stones in the urinary tract
( ) e. Deep pain over the bony areas
( ) f. Tingling of the fingers
( ) g. Thinning of the bones

— — — — — — — — — — — — — — — —

a. –; b. –; c. X; d. X; e. X; f. –; g. X

---

163 There are cardiac abnormalities with hypocalcemia and hypercalcemia. In hypocalcemia the QT interval is (prolonged/shortened).________.
In hypercalcemia, the QT interval is (prolonged/shortened). ________.

What else can happen? *________.

— — — — — — — — — — — — — — — —

prolonged; shortened; heart block or cardiac arrest. Good.

---

164 Fractures can occur with hypocalcemia and hypercalcemia. In hypercalcemia, pathologic fractures are common. Explain how these can occur. *________

________.

— — — — — — — — — — — — — — — —

Calcium is transferred from bone to ECF, which weakens and thins the bone

---

CLINICAL APPLICATIONS

165 Calcium acts like a sedative on the central nervous system (CNS).

Let us say you are caring for a debilitated patient who becomes severely agitated. You notice the patient's hands trembling and mouth twitching. These symptoms may indicate a calcium (excess/deficit). What will your action be?

1. *______________.
2. *______________.
3. *______________.

\- - - - - - - - - - - - - - - - -

deficit

1. Check the lab report on serum calcium.
2. Notify physician of symptoms and lab report.
3. Check patient's dietary intake of Ca.

---

166 Lack of calcium will cause neuromuscular irritability. Explain. *______________

____.

What does hypocalcemia do to blood clotting? *______________

______________________________.

\- - - - - - - - - - - - - - - - -

It leads to hyperactivity of the nervous system and painful muscular contractions (symptoms of tetany).
It decreases clotting and causes bleeding.

---

167 Prolonged vomiting leads to alkalosis from loss of hydrogen and chloride ions from the stomach.

When the body fluids are alkaline, what happens to calcium? *______________

____.

\- - - - - - - - - - - - - - - - -

Calcium is ionized poorly and hypocalcemia can occur, especially with a decrease in calcium intake

---

168 Acidosis (increases/decreases) the ionization of calcium. ______________.

— — — — — — — — — — — — — — — — —

increases

---

169 Trousseau's and Chvostek's signs are helpful for diagnosing hypocalcemia.

To test for Trousseau's sign, inflate a blood pressure cuff for 3 minutes. The test is positive if carpopedal spasm is present (thumb adduct and fingers and hand develop contractions). To test for Chvostek's sign, tap the seventh cranial nerve (facial nerve) in front of the ear. It is positive if there are spasms of the cheek and corner of the lip.

The two diagnostic signs for hypocalcemia are:

1. * ______________________________;
2. * ______________________________.

— — — — — — — — — — — — — — — — —

1. Trousseau's sign; 2. Chvostek's sign

---

170 Describe a positive Trousseau's sign.

* ______________________________

______________________________.

Describe a positive Chvostek's sign.

* ______________________________

______________________________.

— — — — — — — — — — — — — — — — —

Positive Trousseau's sign—carpopedal spasm after blood pressure cuff is inflated for 3 minutes or a similar response. Positive Chvostek's sign—spasms of the cheek and corner of the lip after tapping the facial nerve or a similar response

---

171 The kidneys excrete approximately 50 to 250 mg/dl of calcium in the urine daily. If the kidneys excrete less than 50 mg/dl, what type of calcium imbalance could be occurring? ______________________.

— — — — — — — — — — — — — — — — —

hypercalcemia (most likely)

---

172 The nursing action for a patient with hypercalcemia is to prevent renal calculi. There are three ways this can be accomplished:

1. Drink at least 12 glasses of fluid a day.
2. Keep urine acid.
3. Prevent urinary tract infections.

How do you think the urine can be kept acid? *___________________________

____.

- - - - - - - - - - - - - - - - -

Eat foods that are high in acid content (meat, fish, poultry, eggs, cheese, peanuts, cereals), and/or drink at least 1 pint (2 glasses) of cranberry juice daily. Orange juice does not make the urine acid.

DRUGS AND THEIR EFFECT ON CALCIUM BALANCE

Phosphate preparations, corticosteroids, loop diuretics, aspirin, anticonvulsants, magnesium sulfate, and mithramycin are some of the groups of drugs that can lower the serum calcium level. Excess calcium salt ingestion and infusion and thiazide and chlorthalidone diuretics are drugs that can increase the serum calcium level. Table 19 list drugs that affect calcium balance.

TABLE 19. DRUGS AFFECTING CALCIUM BALANCE

Calcium Imbalance	Drugs	Rationale
Hypocalcemia (serum calcium deficit)	Magnesium sulfate Propylthiouracil/Propacil Colchicine Mithramycin Neomycin	These agents inhibit parathyroid hormone/PTH secretion and decrease the serum calcium level.
	Aspirin Anticonvulsants Glutethimide/Doriden Estrogens	These agents could alter the vitamin D metabolism that is needed for calcium absorption.
	Phosphate preparations: Oral, enema and intravenous: Sodium phosphate Potassium phosphate	Phosphates can increase the serum phosphorus level and decrease the serum calcium level.
	Corticosteroids: Cortisone Prednisone	Steroids decrease calcium mobilization and inhibit the absorption of calcium.
	Loop Diuretics: Furosemide/Lasix	Loop diuretics reduce calcium absorption from the renal tubules.
Hypercalcemia (serum calcium excess)	Calcium salts Vitamin D	Excess ingestion of calcium and Vitamin D and infusion of calcium could increase the serum Ca level.

TABLE 19. (Continued)

Calcium Imbalance	Drugs	Rationale
	IV lipids	Lipids can increase the calcium level.
	Diuretics: Thiazides Chlorthalidone/Hygroten	These agents can induce hypercalcemia.

Adapted with permission from A. Nanji: Drug-induced electrolyte disorder, *Drug Intelligence and Clinical Pharmacy*, 17:179–180, March 1983.

---

173 Enter CD for calcium deficit/hypocalcemia and CE for calcium excess/hypercalcemia opposite the following drugs. Refer to the table as needed.

( ) a. Magnesium sulfate
( ) b. Aspirin
( ) c. Anticonvulsants
( ) d. Calcium sulfates
( ) e. Thiazide diuretics
( ) f. Corticosteroids
( ) g. Loop diuretics
( ) h. Vitamin D

- - - - - - - - - - - - - - - - -

a. CD; b. CD; c. CD; d. CE; e. CE; f. CD; g. CD; h. CE

---

174 Mithramycin, an antineoplastic antibiotic, is used to treat hypercalcemia. This agent lowers the serum calcium level.

Will steroids and mithramycin (increase/decrease) serum calcium?

________________________.

- - - - - - - - - - - - - - - - -

decrease

---

175 Hypercalcemia can cause cardiac arrhythmias. An elevated serum calcium enhances the effect of digitalis and can cause digitalis toxicity.

Give three (3) signs and symptoms of digitalis toxicity:

a. *____________________.

b. *____________________.

c. ____________________.

\- - - - - - - - - - - - - - - - -

a. Bradycardia with or without arrhythmias (slow heart rate)
b. Nausea and vomiting
c. Anorexia

---

176 During a hypercalcemic state should the dose of digitalis preparations, e.g., digoxin, be (increased/decreased)? ____________________.

\- - - - - - - - - - - - - - - - -

decreased

---

177 Steroids such as cortisone tend to decrease calcium mobilization and inhibit the absorption of calcium.

Will steroids (increase/decrease) serum calcium? ____________________.

\- - - - - - - - - - - - - - - - -

decrease

---

178 A loop diuretic affects the renal tubules by reducing absorption of calcium and increasing calcium excretion. Do you recall the name of a loop diuretic?

____

Name two other electrolytes that are excreted by loop diuretics.

____________________ and ____________________.

\- - - - - - - - - - - - - - - - -

Furosemide/Lasix; potassium and sodium

CALCIUM CORRECTION

---

179 What food products should people have that are high in calcium to prevent or correct body's calcium deficit? * ______________________.

- - - - - - - - - - - - - - - - -

milk and milk products

---

180 Normally, calcium is not required for IV therapy since there is a tremendous reservoir in the bone. However, the body needs vitamin D for the utilization of dietary calcium.

What other essential composition of the diet is needed for calcium utilization? ______________________. Refer to Table 17 if needed.

- - - - - - - - - - - - - - - - -

protein

---

181 To correct acute hypocalcemia with tetany symptoms, 10 ml of 10% calcium gluconate diluted in 5% dextrose in water solution is frequently recommended. The 10% calcium gluconate is equivalent to 4.5 mEq/L of calcium. Calcium gluconate should not be diluted in normal saline since the sodium encourages calcium loss.

Calcium gluconate for intravenous administration should be diluted in the following intravenous solution(s):

( ) a. Normal saline (0.9% NaCl)
( ) b. 5% Dextrose in water

- - - - - - - - - - - - - - - - -

a. –; b. X

---

182 A suggested rate of IV flow (with calcium as an additive) is 2 to 3 ml per minute. When tetany symptoms are present, the physician may order calcium gluconate to be administered by a slow intravenous push. Infiltration of intravenous calcium can cause sloughing of the subcutaneous tissue.

A suggested rate of IV flow is ________ ml per minute. If the IV set states that 10 drops equal 1 ml, then ________ drops equal 2 ml, and ________ drops equal 3 ml. The nurse should frequently check for infiltration. Why?

* ____.

— — — — — — — — — — — — — — — — —

2 to 3 ml; 20; 30. To prevent infiltration since calcium can cause sloughing of the tissue or similar response

---

183 Frame 73 stated that calcium gluconate was used to counteract the effect of potassium excess on the heart muscle. Does calcium gluconate (increase/decrease) the cardiac effect due to hyperkalemia? ________________.

What type of ECG improvement should the nurse observe when using calcium gluconate for hyperkalemia? * ________________________________.

— — — — — — — — — — — — — — — — —

decrease
A decrease in the T wave

CASE REVIEW

Mr. Morgan, age 58, has had gastric upset for the past six weeks. He has been taking antacids and drinking several glasses of milk each day. The stomach discomfort was not relieved and so he was admitted to the hospital for further diagnostic study. A possible malignant neoplasm was to be determined. His serum calcium was 5.9 mEq/L.

1. His serum calcium level would indicate what type of calcium imbalance?

   ________________________.

2. "Normal" range for calcium balance is * ________________________________.
3. Explain what happens to body calcium when there is a decrease in gastric acidity and an increase in body alkaline fluids. * ________________________________

   ____.

4. Give four functions of calcium in the body.

   a. * ________________________________.

b. *______________________________.

c. *______________________________.

d. *______________________________.

5. If Mr. Morgan was bedridden, would you expect his serum calcium to be (elevated/decreased)? ______________. Explain. *______________________________ ______________________________.

6. Give a symptom that can occur from immobilization. *______________________________ ______________.

7. Give three nursing interventions to prevent renal calculi resulting from hypercalcemia.

   a. *______________________________.

   b. *______________________________.

   c. *______________________________.

8. Previously, Mr. Morgan had a "heart condition" and he was started on Digoxin. What effect does hypercalcemia have on Digoxin?

   *______________________________. Should his Digoxin dosage be (increased/decreased) until his hypercalcemic state is corrected? *______________.

9. If Mr. Morgan's serum calcium became 3.9 mEq/L, what type of calcium imbalance would be present? ______________________________.

10. Give five common signs and symptoms of hypocalcemia.

    ______________, ______________, ______________, ______________, and ______________.

---

1. Hypercalcemia
2. 4.5-5.8 mEq/L or 8.5-10.5 mg/dl
3. Decrease in ionized calcium for utilization. In alkaline fluids, calcium is non-ionized and protein-bound. Increase in his serum calcium can result from large amounts of milk intake and from a malignant neoplasm. Variety of neoplasms (tumors) can cause hypercalcemia
4. a. maintenance of normal cell permeability
   b. formation of bone and teeth
   c. normal clotting mechanism
   d. normal muscle and nerve activity
5. Elevated. Prolonged immobilization would increase serum calcium by releasing calcium from the bones
6. Kidney stones
7. a. Drink at least 12 glasses of fluid a day

   b. Eat foods high in acid content to keep urine acid
   c. Prevent urinary tract infections
8. Hypercalcemia enhances the action of Digoxin making it more powerful. Dosage should be decreased
9. Hypocalcemia
10. Tetany (carpopedal), twitching of the mouth, tingling of the fingers, spasm of the larynx, abdominal cramps, and muscle cramps

NURSING DIAGNOSES

Potential alteration in nutrition related to inadequate protein and calcium diet.

Potential alteration in electrolyte balance (calcium) related to lack of calcium and vitamin D intake, hypo-hyperparathyroidism, multiple fractures, prolonged immobilization, tumors of the bone, hyperhosphatemia, diarrhea, and laxative abuse.

Knowledge deficit related to a lack of understanding of the disease process that is causing calcium imbalance.

Potential for noncompliance to the treatment regime for calcium deficit or excess related to an inadequate explanation.

Potential for injury related to fractures secondary to calcium loss from the bones.

Potential alteration in cardiac output related to hypercalcemia.

Potential for injury (bleeding) related to the inability for blood to coagulate secondary to calcium loss.

NURSING ACTIONS: CALCIUM

1. Recognize clinical conditions and problems of hypocalcemia, i.e., insufficient diet in protein and calcium, lack of vitamin D intake, chronic diarrhea, extensive infections, tissue trauma, and overuse of laxatives and antacids; and of hypercalcemia, i.e., multiple fractures, prolonged immobilization, tumors of the bone, and cancer of the lung or breast.
2. Observe for signs and symptoms of hypocalcemia, i.e., tetany symptoms (twitching around mouth, carpopedal spasms, laryngospasms), abnormal cramps, and muscle cramps; and of hypercalcemia, i.e., pain over bony areas, flabby muscles, renal calculi, pathologic fractures.
3. Check for prolonged bleeding or reduced clot formation. A low serum calcium inhibits the production of prothrombin.
4. Determine the acid-base status when hypocalcemia is present. In an acidotic state, calcium will be ionized and can be utilized by the body even though there is a calcium deficit. This is not true when alkalosis occurs. Calcium is not ionized in an alkalotic state; and if calcium deficit is present, tetany symptoms occur.
5. Report abnormal serum calcium levels. With hypocalcemia, the serum calcium is < 4.5 mEq/L or < 8.5 mg/dl; and with hypercalcemia, the serum calcium is > 5.8 mEq/L or > 10.5 mg/dl.
6. Teach patients to eat foods rich in calcium, vitamin D, and protein, especially the older adult. Explain the importance of calcium in the diet to prevent osteoporosis

and to aid normal clot formation. Tell the patient that protein is needed to aid in calcium absorption. Nonfat dry milk can be used to meet calcium requirements.

7. Teach "bowel-conscious" persons that chronic use of laxatives can increase intestinal motility, which prevents calcium absorption from the intestine. Suggest fruits for bowel elimination, instead of laxatives.
8. Explain to persons using antacids that constant use of antacids can decrease calcium in the body. Antacids decrease acidity, which decreases calcium ionization.
9. Administer oral calcium supplements an hour before meals to enhance intestinal absorption.
10. Observe for symptoms of hypocalcemia in patients receiving massive transfusions of citrated blood. The serum calcium level may not be affected, but the citrates prevent calcium ionization.
11. Monitor the pulse regularly for bradycardia when the patient is receiving digitalis and calcium, either orally or intravenously. Increased serum calcium enhances the action of digitalis, and digitalis toxicity can result.
12. Regulate IV 10% calcium gluconate in a liter of 5% dextrose in water to run 20 to 30 drops per minute, or according to the physician's order. Do not administer calcium gluconate in saline (0.9% NaCl) for the sodium encourages calcium loss.
13. Assess for positive Trousseau's and Chvostek's signs of hypocalcemia. For Trousseau's sign, inflate the blood pressure cuff for 3 minutes and observe for carpopedal spasm. For Chvostek's sign, tap the facial nerve in front of the ear for spasms of the cheek and mouth.
14. Monitor ECG and note changes related to hypocalcemia, e.g., prolonged QT interval, and those related to hypercalcemia, e.g., shortened QT interval.
15. Check IV solutions containing calcium for infiltration. Calcium is irritating to the subcutaneous tissues and can cause tissue sloughing.
16. Promote active and passive exercise for bedridden patients. Immobilization promotes calcium loss from the bone.
17. Handle patients gently who have long-standing hypercalcemia and bone demineralization to prevent fractures.
18. Teach patients with hypercalcemia to avoid foods rich in calcium and to avoid taking massive amounts of vitamin D supplements.
19. Encourage patients with hypercalcemia to keep hydrated, in order to increase calcium dilution in the serum and urine, and to prevent renal calculi formation.
20. Explain to patients with hypercalcemia that the purpose for maintaining an acid urine is to increase solubility of calcium. An acid-ash diet may be ordered that includes meats, fish, poultry, eggs, cheese, cereals, nuts, cranberry juice, prune juice. Orange juice will not change the urine pH.
21. Administer prescribed loop diuretics to enhance calcium excretion. Thiazide diuretics inhibit calcium excretion and are not indicated in hypercalcemia.

MAGNESIUM

184 Magnesium is a(n) (cation/anion). Its highest concentration is found in what type of fluid? ________. Refer to Table 3 if needed.

- - - - - - - - - - - - - - - - -

cation; intracellular

---

185 What other cation has its highest concentration in the intracellular fluid?

________.

- - - - - - - - - - - - - - - - -

potassium

---

186 Magnesium is widely distributed throughout the body. Half of the body magnesium is in the bone. What other ion is found plentifully in the bone? ________.

- - - - - - - - - - - - - - - - -

calcium

---

187 Magnesium has a higher concentration in the cerebrospinal fluid, also known as spinal fluid, than in the blood plasma. The serum concentration of magnesium is *________. Refer to Table 3 if needed.

- - - - - - - - - - - - - - - - -

2 mEq/L (actually the range is 1.5–2.5 mEq/L)

---

188 One-third of magnesium is protein bound and approximately two-thirds is ionized, which can be utilized by the body. From dietary magnesium, 60% of Mg is excreted via the feces (mainly the magnesium that was not absorbed) and 40% via the kidneys.

Magnesium is excreted via the ________ and ________.

- - - - - - - - - - - - - - - - -

feces; kidneys

Table 20 describes the various functions of magnesium. Study the table and refer to it as needed.

TABLE 20. MAGNESIUM AND ITS FUNCTIONS

Body System	Functions
Neuromuscular	Transmits neuromuscular activity. Important mediator of neural transmission in CNS.
Cardiac	Contracts the heart muscle (myocardium).
Cellular	Activates many enzymes for proper carbohydrate and protein metabolism. Responsible for the transportation of sodium and potassium across cell membranes. Influences utilization of potassium, calcium, and protein. When there is a magnesium deficit, there is frequently a potassium and/or calcium deficit.

189 Magnesium plays an important role in enzyme activity. An *enzyme* is a catalyst capable of inducing chemical changes in other substances. Magnesium acts as a coenzyme in the metabolism of carbohydrates and protein.

Magnesium is also involved in maintaining neuromuscular stability. What other ion has this similar function? ________________.

calcium

190 Indicate which of the following are functions of magnesium:

( ) a. Neuromuscular activity
( ) b. Contraction of the myocardium
( ) c. Exchange of CO_2 and O_2
( ) d. Enzyme activity
( ) e. Responsibility (partial) for Na and K crossing cell membranes

a. X; b. X; c. –; d. X; e. X

191 When there is a magnesium deficit, two other cations may also be decreased. Name them. ________________ and ________________.

potassium and calcium

192 *Hypomagnesemia* means a low serum concentration of magnesium. Therefore, *hypermagnesemia* would mean * ____.

- - - - - - - - - - - - - - - - -

high serum concentration of magnesium

Table 21 lists the conditions causing hypomagnesemia and hypermagnesemia, each with an explanation. Study this table carefully and refer to it as needed. Refer to the glossary for unknown words.

TABLE 21. CONDITIONS CAUSING MAGNESIUM DEFICIT AND MAGNESIUM EXCESS

Conditions	Magnesium Deficit (Hypomagnesemia)	Magnesium Excess (Hypermagnesemia)
Prolonged inadequate intake of food	Magnesium is found in various foods, particularly green vegetables. Severe anorexia with minimum to no food can cause deficit.	
	Continuous IV therapy without magnesium supplement.	
Chronic alcoholism, malnutrition	An inadequate intake of food over long period of time causes deficit.	
Prolonged diuresis	It will cause magnesium loss via urine.	
Severe diarrhea	Majority of dietary Mg is excreted in feces. Diarrhea inhibits intestinal absorption of Mg and promotes excretion.	
Congestive heart failure (CHF)	Prolonged diuretic therapy can cause Mg deficit.	
Severe dehydration (especially patients with diabetic acidosis)	Loss of magnesium from intracellular fluid and from body	
Renal insufficiency, uremia		Retention of magnesium in blood plasma.
Laxatives, i.e., magnesium sulfate (Epsom salt), milk of magnesia, and magnesium citrate solution		They will increase concentration of magnesium in plasma if given in excessive amounts over prolonged period of time.

TABLE 21. (Continued)

Conditions	Magnesium Deficit (Hypomagnesemia)	Magnesium Excess (Hypermagnesemia)
Antacids, i.e., Maalox, Mylanta, Aludrox, and DiGel		These antacids contain magnesium hydroxide and aluminum hydroxide. Normally, magnesium salts are poorly absorbed, but if renal insufficiency is present, magnesium excess could occur.
Severe dehydration		If oliguria or anuria results, magnesium retention could occur.

193 Magnesium is found in various foods, but a prolonged inadequate intake can cause a * ______________________.

— — — — — — — — — — — — — — — — —

magnesium deficit

194 Chronic alcoholism and malnutrition are conditions that can cause magnesium ______________.

— — — — — — — — — — — — — — — — —

deficit

195 Two other situations that can cause hypomagnesemia are * ______________________ and * ______________________.

— — — — — — — — — — — — — — — — —

continuous IV therapy; prolonged diuresis; severe dehydration

196 Prolonged renal insufficiency can cause * ______________________.

Name a laxative that, if used constantly, could cause magnesium excess, especially if there is renal impairment. * ______________________.

Name an antacid that, if used constantly, could cause magnesium excess, especially if there is renal impairment. * ______________________.

— — — — — — — — — — — — — — — — —

magnesium excess; Epsom salt or milk of magnesia (MOM); Maalox or mylanta

197 Place a D for magnesium deficits and an E for magnesium excess in the following:

( ) a. Renal insufficiency
( ) b. Prolonged diuresis
( ) c. Constant use of Epsom salt or milk of magnesia
( ) d. Chronic alcoholism
( ) e. Malnutrition
( ) f. Prolonged inadequate intake
( ) g. Severe diarrhea
( ) h. Constant use of antacids with magnesium hydroxide

- - - - - - - - - - - - - - - - -

a. E; b. D; c. E; d. D; e. D; f. D; g. D; h. E

Study Table 22 carefully and refer to this table as needed.

TABLE 22. SIGNS AND SYMPTOMS RELATED TO HYPOMAGNESEMIA AND HYPERMAGNESEMIA

Classes	Hypomagnesemia	Hypermagnesemia
Neuromuscular abnormalities	Hyperirritability; bizarre, involuntary muscular activity appears as tremors. Twitching of face. Spasticity Convulsion	CNS depression. Decreased respiration. Inhibition of neuromuscular transmission. Lethargy Coma
Cardiac abnormalities	Cardiac arrhythmia	Bradycardia
Milliequivalents per liter	↓ 1.5 mEq/L	↑ 2.5 mEq/L

---

198 Magnesium influences the nervous system, and so too much or not enough Mg will affect the neuromuscular function.

Hyperirritability, bizarre muscular activity, and twitching of the face are signs and symptoms of ______________________.

- - - - - - - - - - - - - - - - -

hypomagnesemia

---

199 Tetany symptoms, i.e., tremors, twitching, carpopedal spasm, and generalized spasticity are associated with hypomagnesemia and alkalosis.

Name another electrolyte (cation) imbalance that is associated with tetany symptoms. ______________________________.

— — — — — — — — — — — — — — — — —

hypocalcemia or calcium deficit

---

200 Central nervous system depression, inhibited neuromuscular transmission, decreased respiration, and lethargic state are signs and symptoms of __________.

— — — — — — — — — — — — — — — — —

hypermagnesemia

---

201 Match the following cardiac signs and symptoms relating to a magnesium deficit or excess:

__________	1. Cardiac arrhythmia	a. Hypomagnesemia
		b. Hypermagnesemia
__________	2. Bradycardia	

— — — — — — — — — — — — — — — — —

1. a; 2. b

---

202 Place D for hypomagnesemia and E for hypermagnesemia beside the following signs and symptoms:

( ) a. Hyperirritability
( ) b. CNS depression
( ) c. Lethargy
( ) d. Bizarre muscular activity resembling tremors
( ) e. Twitching of the face
( ) f. Convulsion
( ) g. Decreased respiration
( ) h. Bradycardia
( ) i. Cardiac arrhythmia

— — — — — — — — — — — — — — — — —

a. D; b. E; c. E; d. D; e. D; f. D; g. E; h. E; i. D

CLINICAL APPLICATIONS

203 Magnesium is needed by the heart for myocardial contraction. It is said that magnesium will slow the rate of the atrium and will correct atrial flutter.

ECG changes due to magnesium imbalances are similar to potassium imbalances. With hypomagnesemia, the T wave is flat or inverted and the ST segment is depressed. With hypermagnesemia, what T wave change do you think takes place? ____

____.

- - - - - - - - - - - - - - - - -

peaked T wave. Good. The T wave change is similar to the one occurring in hyperkalemia.

---

204 When a patient is being treated for hypokalemia and is not responding to therapy, the serum magnesium should be checked. If a magnesium deficit is present, hypokalemia will not be completely corrected.

Another name for hypokalemia is *________________________________.

- - - - - - - - - - - - - - - - -

potassium deficit

---

205 In diabetic acidosis, magnesium will leave the cells. When insulin and dextrose are given intravenously, magnesium will return to the cells.

If the diabetic condition is corrected too fast, then (hypomagnesemia/hypermagnesemia) would occur. Why? *________________________________

____.

- - - - - - - - - - - - - - - - -

hypomagnesemia. Mg leaves the ECF rapidly and returns to the cells

---

206 The kidneys regulate the concentration of magnesium in the body. When there is a slight increase in magnesium concentration, the kidneys will excrete the excess. If there is a decreased serum magnesium level, what do you think the kidneys will do? ______________________.

If a patient has renal insufficiency and is receiving magnesium sulfate, what type of magnesium imbalance could occur? ______________________

______________________.

Why? * ______________________.

— — — — — — — — — — — — — — — — —

Kidneys will conserve Mg *or* Mg will be reabsorbed from the kidney tubules—not excreted; hypermagnesemia. Kidneys regulate Mg balance—cannot excrete it

---

207 For patients on prolonged hyperalimentation (TPN), the serum magnesium level should be checked.

What type of magnesium imbalance can occur when magnesium is not included in the solutions for TPN? ______________________.

— — — — — — — — — — — — — — — — —

hypomagnesemia

DRUGS AND THEIR EFFECT ON MAGNESIUM BALANCE

---

208 Long-term administration of saline infusions may result in magnesium and calcium loss.

Can you explain why long-term or excessive use of saline infusions can cause magnesium and calcium deficits?

* ______________________

— — — — — — — — — — — — — — — — —

It expands the extracellular fluid (ECF), causes dilution, and inhibits tubular absorption of Mg and Ca.

Diuretics, antibiotics, laxatives, and digitalis are groups of drugs that promote magnesium loss (hypomagnesemia). Excess intake of magnesium salts is the major cause of serum magnesium excess (hypermagnesemia). Table 23 list drugs that affect magnesium balance.

TABLE 23. DRUGS AFFECTING MAGNESIUM BALANCE

Magnesium Imbalance	Drugs	Rationale
Hypomagnesemia (serum magnesium deficit)	Diuretics: Furosemide/Lasix Ethacrynic acid/Edecrin Mannitol	Diuretics promote urinary loss of magnesium.
	Antibiotics: Gentamicin Tobramycin Carbenicillin Capreomycin Neomycin Polymyxin B Amphotericin B	These agents can cause magnesium loss via kidney.
	Digitalis	
	Laxatives	Laxative abuse causes magnesium loss via GI.
	Corticosteroids: Cortisone Prednisone	Steroids can decrease serum magnesium level.
Hypermagnesemia (serum magnesium excess)	Magnesium Salts Oral and Enema: Magnesium hydroxide/MOM Magnesium sulfate/Epsom salt Magnesium citrate	Excess use of magnesium salts could increase serum magnesium level.
	Magnesium sulfate (maternity)	Use of excess $MgSO_4$ in treatment of toxemia could cause hypermagnesemia.
	Lithium	Hypermagnesemia associated with lithium.

Adapted with permission from A. Nanji: Drug-induced electrolyte disorders. *Drug Intelligence and Clinical Pharmacy*, 17:180–181, March 1983.

209 Place MD for magnesium deficit/hypomagnesemia and ME for magnesium excess/hypermagnesemia beside the following drugs:

_______ a. Furosemide/Lasix

_______ b. Tobramycin

_______ c. Magnesium hydroxide/MOM

_______ d. Digitalis

_______ e. Magnesium sulfate for toxemia

_______ f. Laxatives

_______ g. Cortisone

_______ h. Lithium

---

a. MD; b. MD; c. ME; d. MD; e. ME; f. MD; g. MD; h. ME

---

210 Excessive use of steroids (corticosteroids) can cause hypomagnesemia.
A decrease in adrenal cortical hormone can cause (hypomagnesemia/hypermagnesemia). ______________.

- - - - - - - - - - - - - - - - -

hypermagnesemia. GOOD.

---

211 Hypomagnesemia enhances the action of digitalis and causes digitalis toxicity. Magnesium sulfate will correct hypomagnesemia and symptoms of digitalis toxicity.
Give at least three (3) symptoms of digitalis toxicity.

______________, ______________, and ______________.

What other electrolyte (cation) deficit will cause digitalis toxicity?

______________.

- - - - - - - - - - - - - - - - -

nausea and vomiting; anorexia; bradycardia. Potassium.

MAGNESIUM CORRECTION

---

212 For the management of hypomagnesemia, establishing adequate renal flow is first indicated and then rehydration and administering the ion, ______________.
Why would renal sufficiency be needed when administering magnesium?

* ______________________________

______________________________.

- - - - - - - - - - - - - - - - -

magnesium
It is necessary in excreting excess magnesium if too much magnesium was administered

213 Magnesium sulfate is the parenteral replacement for hypomagnesemia and can be administered intramuscularly or intravenously. The drug is available in strengths of 10, 12.5, and 50%. Many physicians will order 10 ml of a 50% solution for adults.

For intramuscular injections the dosage is divided and for intravenous infusion the dosage is diluted into 1 liter of solution. The two injectable routes in which magnesium sulfate can be delivered to the body are

______________________ and ______________________.

- - - - - - - - - - - - - - - - - -

intramuscular; intravenous.

---

214 Oral magnesium comes as sulfate, gluconate, chloride, citrate, and hydroxide in liquid, tablet, and powder form.

For magnesium supplement (maintenance or replacement) magnesium gluconate/Magonate and magnesium-protein complex/Mg-PLUS may be ordered by the physician.

For severe hypomagnesemia do you think the physician would order magnesium replacement to be administered (orally/intramuscularly/intravenously)?

______________________. Why? ______________________

______________________.

- - - - - - - - - - - - - - - - - -

intravenously. A direct and quick method for replacing serum magnesium deficit.

CASE REVIEW

Mrs. Landis has had diuresis for several days. In the hospital her diagnoses were prolonged diuresis, severe dehydration, and malnutrition. She received 3 liters of 5% dextrose in 1/2 of normal saline (0.45% NaCl). Her serum magnesium was 1.3 mEq/L.

1. What is the "normal" serum magnesium range? * ______________________.
2. Name the type of magnesium imbalance present. ______________________.
3. Mrs. Landis received fluids intravenously. Explain the relationship of IV fluids to magnesium deficit. * ______________________ ______________________.
4. Name two clinical causes of hypermagnesemia. * ______________________ and * ______________________.

Mrs. Landis's pulse was irregular. She developed tremors and twitching of the face.

Physician ordered 10 ml of magnesium sulfate IV to be diluted in 1 liter of solution. Other drugs that she was receiving included Digoxin and Lasix.

5. Name Mrs. Landis's clinical signs and symptoms of hypomagnesemia. *________ ________, ________________ and *________________.
6. What cation, in a *hypo* state, causes CNS abnormalities similar to hypomagnessemia? ________________.
7. Name at least two symptoms of hypermagnesemia. *________________ and *________________.
8. The physician ordered IV magnesium sulfate diluted in 1 liter of IV fluids. The nursing implication is first to check Mrs. Landis's urinary output. Explain the rationale. *____.
9. The nurse should be assessing digitalis toxicity while Mrs. Landis's serum magnesium is low. Explain. *____.
10. Lasix can cause (hypomagnesemia/hypermagnesemia) ________________ ________________.

- - - - - - - - - - - - - - - - - -

1. 1.7–2.3 mEq/l
2. hypomagnesemia
3. It causes dilution of magnesium in the ECF
4. renal insufficiency; use of Epsom salt ($MgSO_4$) as a laxative
5. irregular pulse (arrhythmia); tremors; twitching of the face
6. calcium
7. CNS depression; decrease in respiration; bradycardia
8. Kidneys excrete excess magnesium and kidney impairment can cause hypermagnesemia
9. Hypomagnesemia enhances the action of digitalis
10. hypomagnesemia

NURSING DIAGNOSES

Potential alteration in nutrition related to prolonged inadequate intake of food rich in magnesium and malnutrition.

Potential alteration in electrolyte balance (magnesium) related to chronic alcoholism, prolonged diuresis, severe diarrhea, severe dehydration, and magnesium-containing laxatives and antacids.

Knowledge deficit related to a lack of understanding of the disease process that causes magnesium imbalance.

Potential for noncompliance to the treatment regime for magnesium deficit or excess related to an inadequate explanation.
Potential alteration in cardiac output related to magnesium imbalance (hypo-hypermagnesemia).
Potential for ineffective breathing patterns related to hypermagnesemia that causes respiratory depression.

NURSING ACTIONS

1. Recognize clinical conditions and problems of hypomagnesemia, i.e., chronic alcoholism, malnutrition, severe diarrhea, prolonged use of diuretic therapy, or parenteral fluids without magnesium salts; and of hypermagnesemia, i.e., renal insufficiency, chronic use of laxatives and antacids containing magnesium salts.
2. Observe for signs and symptoms of hypomagnesemia, i.e., hyperirritability, tetany symptoms (tremors, twitching, laryngospasm), cardiac arrhythmias; and of hypermagnesemia, i.e., lethargy, decreased neuromuscular activity, decreased respiration, bradycardia.
3. Check the serum magnesium for imbalance. With hypomagnesemia, the serum magnesium level is < 1.5 mEq/L; and with hypermagnesemia, the serum magnesium level is > 2.5 mEq/L. If magnesium and potassium deficits are present, magnesium correction is necessary. When correcting potassium deficit, potassium will not be replaced in the cells until magnesium is replaced.
4. Assess for positive Trousseau's and Chvostek's signs of severe hypomagnesemia. Tetany symptoms occur in both magnesium and calcium deficits. (See Frame 169 for Trousseau's and Chvostek's signs.)
5. Monitor vital signs and report cardiac arrhythmias and bradycardia.
6. Teach patient having a magnesium deficit to eat foods rich in magnesium, i.e., green vegetables, fish, seafoods, whole grains, and nuts.
7. Check patients with hypomagnesemia who are taking Digoxin for digitalis toxicity, e.g., nausea and vomiting, bradycardia. Magnesium deficit enhances the action of Digoxin (digitalis preparations).
8. Report urine output of less than 25 ml per hour or 600 ml per day when the patient is receiving magnesium supplements. Magnesium excess is excreted by the kidneys. With poor urine output, hypermagnesemia could occur.
9. Report to physicians when patients receive continuous magnesium-free intravenous fluids. Solutions for hyperalimentation should contain some magnesium.
10. Administer intravenous magnesium sulfate diluted in solution slowly unless the patient has a severe deficit. Rapid infusion can cause a hot or flushed feeling.
11. Have IV calcium gluconate available for emergency to reverse hypermagnesemia from overcorrection of a magnesium deficit.
12. Monitor ECG changes during a magnesium imbalance. A flat or inverted T wave is associated with magnesium deficit, and a peaked T wave is associated with magnesium excess.
13. Teach patients to avoid constant use of laxatives and antacids containing magnesium, especially if urinary output is poor.

14. Observe for respiratory distress. Serum magnesium excess can cause respiratory depression.
15. Provide adequate fluids for hydration and for improving kidney function.

CHLORIDE

215 Chloride is a(n) (anion/cation) ______________________.

The chloride ion frequently appears in combination with the sodium ion. Which fluid has the greatest concentration of chloride—intracellular or extracellular? ______________________.

- - - - - - - - - - - - - - - - -

anion; extracellular fluid

---

216 Refer to Table 3 as needed. The concentration of chloride in plasma is *______________________ and in intracellular fluid is *______________________.
Highest concentration is in the cerebrospinal fluid, 125 mEq/L.

- - - - - - - - - - - - - - - - -

104 mEq/L with a range of 98-108 mEq/L; 1 mEq/L

Table 24 explains the four functions of the chloride ion. Study the table and refer to it as needed.

TABLE 24. CHLORIDE AND ITS FUNCTIONS

Body Involvement	Functions
Osmolality (tonicity) of ECF	Chloride, like sodium, changes the serum osmolality. When serum osmolality is increased (> 295 mOsm/L), there are more sodium and chloride ions in proportion to water. With a decreased serum osmolality (< 280 mOsM/L), there are less sodium and chloride ions.
Body water balance	When sodium is retained, chloride is frequently retained, causing an increase in water retention.
Acid-base balance	The kidneys will excrete the anion, chloride or bicarbonate, and sodium will reabsorb either chloride or bicarbonate to maintain acid-base balance.
Acidity of gastric juice	Chloride combines with hydrogen ion in the stomach to form hydrochloric acid (HCl).

---

217 A deficiency of chloride can lead to a deficiency of potassium and vice versa. Usually chloride losses will follow those of sodium. Most chloride ingestion is in combination with sodium.

The chloride ion is mostly found in combination with a ________ ion.

What is the name for sodium chloride? ________.

— — — — — — — — — — — — — — — — —

sodium; salt

---

218 Chloride, like sodium, influences the serum osmolality.

What two ions are usually increased when the serum osmolality is elevated?

________ and ________.

— — — — — — — — — — — — — — — — —

sodium and chloride

---

219 When there is a body water deficit, what occurs to:

a. serum sodium and serum chloride? ________.

b. serum osmolality? ________.

c. body water? ________.

— — — — — — — — — — — — — — — — —

a. increases; b. increases; c. reabsorbed

---

220 For every sodium ion absorbed from the renal tubules, a chloride or bicarbonate ion is also absorbed; thus the loss of sodium and chloride in proportion can differ.

The organ responsible for electrolyte homeostasis by the excretion and absorption of ions is the ________.

— — — — — — — — — — — — — — — — —

kidneys

---

221 If metabolic alkalosis is present, the kidneys excrete the bicarbonate ion and sodium will be reabsorbed with (bicarbonate/chloride) ion. ______________.

If metabolic acidosis is present, the kidneys excrete (bicarbonate/chloride) ion ______________, and the sodium will be reabsorbed with which ion? ______________.

- - - - - - - - - - - - - - - - -

chloride; chloride; bicarbonate

---

222 Chloride plays a part in oxygen and carbon dioxide exchange in the red blood cells. When the red blood cells are oxygenated, chloride travels from the red blood cells to the plasma, and bicarbonate leaves the plasma to the red blood cells. This is called the *chloride shift.*

This chloride shift is necessary in maintaining * ______________ ______________.

- - - - - - - - - - - - - - - - -

homeostasis or equilibrium

---

223 The four functions of chloride are for maintenance of:

a. * ______________.

b. * ______________.

c. * ______________.

d. * ______________.

- - - - - - - - - - - - - - - - -

a. osmolality of extracellular fluid (serum osmolality)
b. body water balance
c. acid-base balance
d. acidity of the gastric juice (HCl)

---

224 The name for chloride deficit is hypochloremia; therefore, the name for chloride excess is ______________.

- - - - - - - - - - - - - - - - -

hyperchloremia

Table 25 lists conditions that cause either a chloride deficit or excess. Study the table carefully and refer to the glossary for unknown words. Refer back to this table as needed.

TABLE 25. CONDITIONS CAUSING CHLORIDE DEFICIT AND CHLORIDE EXCESS

Condition	Chloride Deficit (Hypochloremia)	Chloride Excess (Hyperchloremia)
Continuous vomiting; gastric suction	Acidity of gastric juice is composed of hydrogen and chloride. Loss of gastric juice via vomiting or through gastric suction will cause chloride deficit.	
Diarrhea	Loss of gastrointestinal salty secretions will cause chloride deficit.	
Loss of potassium	Loss of potassium is accompanied by loss of chloride.	
Prolonged use of diuretics	Most diuretics interfere with absorption of chloride ions from renal tubules.	
Excessive sweating	Chloride, combined with sodium, is lost via skin. This can result from increased temperature, fever, or muscular exercise.	
Prolonged use of IV dextrose in water	Dextrose in water will dilute serum chloride as well as sodium and potassium.	
Low sodium diet	Decrease in sodium can cause a decrease in chloride.	
Acid-base imbalance	Metabloic alkalosis. Increased HCO_3 ion is associated with a decreased Cl ion.	Metabolic acidosis. Increased Cl ion is associated with a decreased HCO_3 ion.
Dehydration		Serum solutes, including chloride, become concentrated when there is a body water (ECF) deficit.
Head injury		Chloride ions are frequently retained with sodium.
Excessive adrenocortical hormone production		Excessive adrenal cortical hormone will cause excess of sodium in body. Sodium combines with chloride, increasing chloride ion in blood serum.

---

225 Name the two ions that cause the acidity of the gastric juice.______________

______________ and ______________.

What gastrointestinal conditions can cause a chloride deficit? ______________,

______________, ______________, and ______________.

— — — — — — — — — — — — — — — —

hydrogen; chloride
continuous vomiting; gastric suction; diarrhea

---

226 The loss of chloride is frequently accompanied by the loss of what other ion?

______________.

— — — — — — — — — — — — — — — —

potassium (Yes, sodium could be an answer too)

---

227 Chloride is frequently combined with the ______________ ion.
Name at least two causes for sodium and chloride loss or salt loss through the

skin. ______________ and *______________.

— — — — — — — — — — — — — — — —

sodium (Yes, hydrogen could be an answer too)
fever; muscular exercise, and increased environmental temperature

---

228 What type of chloride imbalance occurs with excessive use of diuretics?

______________. Why? *______________.
Name the chloride imbalance occurring with continuous use of IV dextrose

in water? ______________. Why? *______________

______________.

— — — — — — — — — — — — — — — —

hypochloremia (chloride deficit). Diuretics interfere with chloride absorption
hypochloremia. Dextrose in water dilutes the serum chloride level

---

---

229 Excessive adrenal cortical hormone can cause a sodium excess and a chloride ____________? Why? *____.

- - - - - - - - - - - - - - - - -

excess. Cortisone has a sodium-retaining effect.
Sodium combines with chloride, increasing the chloride ion in the blood

---

230 With severe dehydration, what happens to the chloride level? ____________.
Explain *____.

- - - - - - - - - - - - - - - - -

increases. ECF deficit increases chloride concentration, or hemoconcentration causes chloride excess, or a similar response

---

231 Place D for chloride deficit and E for chloride excess in the following:
( ) a. Vomiting
( ) b. Diarrhea
( ) c. Gastric suction
( ) d. Excessive adrenal cortical hormone production
( ) e. Sweating
( ) f. Diuretics
( ) g. Potassium loss
( ) h. Severe dehydration
( ) i. Head injury

- - - - - - - - - - - - - - - - -

a. D; b. D; c. D; d. E; e. D; f. D; g. D; h. E; i. E

Table 26 describes the signs and symptoms of chloride imbalances. Hypochloremic symptoms are similar to metabolic alkalosis and hyperchloremic symptoms are similar to metabolic acidosis. Study the table carefully and refer to it as needed.

TABLE 26. SIGNS AND SYMPTOMS RELATED TO HYPOCHLOREMIA AND HYPERCHLOREMIA

Classes	Hypochloremia	Hyperchloremia
Neuromuscular abnormalities	Hyperirritability Tetany— muscular excitability	Weakness Lethargic Unconsciousness (later)
Respiratory abnormalities	Slow and shallow breathing	Deep, rapid, vigorous breathing
Cardiac abnormalities	↓ Blood pressure with severe Cl and ECF losses	
Milliequivalent per liter	↓ 98 mEq/L	↑ 108 mEq/L

---

232 In hypochloremia, the respiratory symptom is similar to metabolic alkalosis. Indicate which type of breathing occurs with a chloride deficit.

( ) a. slow, shallow breathing
( ) b. deep, rapid, vigorous breathing

Do you know why? * ______________________________

______________________________.

- - - - - - - - - - - - - - - -

a. X; b. –. The lungs conserve carbon dioxide ($CO_2 + H_2O = H_2CO_3$ or carbonic acid to increase acid, and restore pH. Very good.

---

233 In hyperchloremia, the respiratory symptom is similar to metabolic acidosis. Indicate which type of breathing occurs with a chloride excess.

( ) a. slow, shallow breathing
( ) b. deep, rapid, vigorous breathing

Do you know why? * ______________________________

______________________________.

- - - - - - - - - - - - - - - -

a. –; b. X. The lungs blow off carbon dioxide to prevent the formation of H_2CO_3—carbonic acid. Very good.

CLINICAL APPLICATIONS

---

234 Hypochloremia usually indicates alkalosis (hypochloremic alkalosis) due to increased levels of bicarbonate.

Persistent vomiting and gastric suction cause a loss of hydrogen and chloride ions. A loss in hydrogen and chloride results in *____________________ ____________________.

— — — — — — — — — — — — — — — —

hypochloremic alkalosis

---

235 With vomiting, a hypokalemic state also can occur.

Name four clinical symptoms of hypokalemia. ________________, ________________, *________________, and *________________.

— — — — — — — — — — — — — — — —

dizziness; arrhythmia; muscular weakness; abdominal distention

---

236 A potassium deficit cannot be fully corrected until a chloride deficit is corrected.

With vomiting, what type of potassium supplement (Kaon/K-Lyte/potassium chloride) can replace ion deficits. ____________________.

Explain *____________________ ____________________.

— — — — — — — — — — — — — — — —

potassium chloride
Both chloride and potassium are lost due to vomiting

---

237 Hypochloremia can result from excessive IV administration of 5% dextrose in water. How can hypochloremia be prevented?

*____________________ ____________________.

— — — — — — — — — — — — — — — —

Administering 5% dextrose in normal saline or in 1/2 normal saline

---

---

238 One liter of normal saline contains 154 mEq/L of chloride, which exceeds the daily need; however, with severe hypochloremic alkalosis, normal saline solution is suggested.

What is the "normal" serum chloride range? *____________________.

\- - - - - - - - - - - - - - - - - -

98–108 mEq/L

---

239 A normal urine chloride level in 24 hours is 150–250 mEq/L. The amount of chloride excreted depends on the amount of salt intake, body fluid imbalance, and acid-base imbalance.

With a body fluid deficit, the serum chloride and sodium levels are increased due to hemoconcentration. Would you expect the urine chloride level to be (increased/decreased)? ____________________.

\- - - - - - - - - - - - - - - - - -

decreased.

---

240 Nursing responsibility for hypochloremia includes:

a. *____________________.

b. *____________________.

\- - - - - - - - - - - - - - - - - -

a. assessing the serum chloride level
b. suggesting that normal saline or 1/2 NSS (half normal saline solution) be included with the IV orders

---

CASE REVIEW

Mr. Reynolds, 68 years old, has been vomiting for several days. In the hospital, a nasogastric tube was inserted. His serum electrolytes were as follows: serum chloride, 94 mEq/L; serum sodium, 132 mEq/L; and serum potassium, 3.2 mEq/L. Dextrose, 5% in 0.45% normal saline, with 40 mEq/L of KCl was started.

1. What are the four main functions of the chloride ion?

   a. *____________________.

   b. *____________________.

   c. *____________________.

   d. *____________________.

2. According to Mr. Reynolds's serum electrolytes, what three imbalances were present? ______________________, ______________________, and ______________________.
3. The acidity of Mr. Reynolds's gastric juice was decreased because of vomiting and nasogastric suctioning. Name the clinical condition that occurs due to the loss of hydrogen and chloride ions. *______________________.
4. What is the "normal" range of serum chloride? *______________________.
5. Most of the chloride ingested is in combination with the ion ______________________.
6. With hypokalemia and hypochloremia, explain what can occur if only potassium is replaced. *______________________

______________________.
7. Mr. Reynolds received 5% dextrose in 0.45% NaCl or 1/2 normal saline and KCl as the IV fluid. How many mEq/L does 1 liter of 1/2 normal saline contain?

______________________.
8. Nursing implications in caring for Mr. Reynolds include:
   a. *______________________.
   b. *______________________.
   c. *______________________.

- - - - - - - - - - - - - - - - -

1. a. Serum osmolality
   b. Acid-base balance
   c. Body water balance
   d. Acidity of the gastric juice
2. hypochloremia; hyponatremia; hypokalemia
3. hypochloremic alkalosis
4. 98-108 mEq/L
5. sodium
6. A potassium deficit cannot be fully corrected without a correction of chloride deficit
7. 77 mEq/L
8. a. assessing Mr. Reynolds's serum electrolyte values
   b. assessing for signs and symptoms of hypokalemia and hyponatremia
   c. checking the pH for acid-base imbalance
   d. others—measure GI loss (vomiting and gastric drainage for suction)

NURSING DIAGNOSES

Potential alteration in electrolyte balance (chloride) related to continuous vomiting, gastric suction, diarrhea, sodium and potassium loss, prolonged use of diuretics, salt-free solutions, alkalosis, head injury, and excessive adrenocortical hormone production.

Knowledge deficit related to a lack of understanding of the conditions that cause chloride imbalance.

Potential for noncompliance to the treatment regime for chloride deficit or excess related to an inadequate explanation.

NURSING ACTIONS: CHLORIDE

1. Recognize clinical conditions and problems associated with hypochloremia, i.e., vomiting, diarrhea, gastric suction, prolonged use of diuretics and IV 5% dextrose in water, excessive perspiration, low sodium diet; and with hyperchloremia, i.e., severe dehydration, head injury, and excessive use of steroids.
2. Observe for signs and symptoms of hypochloremia, i.e., hyperirritability, tetany symptoms, slow and shallow breathing, and later a decrease in blood pressure; and of hyperchloremia, i.e., weakness, lethargy, and deep, rapid, vigorous breathing.
3. Check serum chloride results. A serum chloride < 98 mEq/L is indicative of hypochloremia, and > 108 mEq/L is indicative of hyperchloremia.
4. Report serum potassium loss as well as serum chloride loss. Potassium deficit cannot be fully corrected until chloride deficit is corrected.
5. Check serum CO_2 or arterial HCO_3. An increased CO_2 or HCO_3 (serum CO_2 > 32 mEq/L or arterial HCO_3 > 28 mEq/L) could indicate metabolic alkalosis and hypochloremia, and a decreased CO_2 or HCO_3 (serum CO_2 < 22 mEq/L or arterial HCO_3 < 24 mEq/L) could indicate metabolic acidosis.
6. Observe for respiratory difficulties, i.e., slow and shallow breathing (hypochloremia), and deep, rapid, vigorous breathing (hyperchloremia).
7. Interpret the types of intravenous fluids ordered; and if continuous IV dextrose in water is administered, notify the physician.
8. Record the amount of gastric secretions from gastric suction, and report continuous and excessive losses.
9. Check serum chloride level and the drugs patient is receiving. If the patient is taking potassium bicarbonate (Kaon without Cl or Klyte), and has a chloride deficit, the nurse should notify the physician. Giving bicarbonate will increase the chloride deficit.
10. Teach the patient with hyperchloremia to avoid foods rich in sodium (salt), e.g., ham, bacon, pickles, potato chips, pretzels, etc. Most of the chloride ion in food is combined with sodium.
11. Check the 24-hour urine chloride value and compare to the 24-hour urine sodium value. In acidosis, the kidneys excrete the chloride ion and conserve the bicarbonate ion. Sodium would be reabsorbed and combined with bicarbonate to correct acidosis.

PHOSPHORUS

241 Phosphorus is a major anion found in high concentration in (extracellular/intracellular) ______________ fluid. Refer to Table 3.

The ions phosphorus (P) and phosphate (PO_4) are used interchangeably. Phosphorus is measured in the serum; in the cells it is a form of phosphate.

— — — — — — — — — — — — — — — — —

intracellular (highest concentration).

---

242 The normal serum phosphorus level is 1.7–2.6 mEq/L or 2.5–4.5 mg/dl.

A serum phosphorus level < 1.7 mEq/L or < 2.5 mg/dl is identified as hypophosphatemia. A serum level > 2.6 mEq/L or > 4.5 mg/dl is labeled

______________.

— — — — — — — — — — — — — — — — —

hyperphosphatemia

---

243 The normal serum phosphorus range in adults is ______________ mEq/L or ______________ mg/dl.

The serum phosphorus level is usually higher in children: 4.0–7.0 mg/dl.

— — — — — — — — — — — — — — — — —

1.7–2.6 mEq.L; 2.5–4.5 mg/dl.

Table 27 explains the functions of phosphorus by body system and structure. Study the table carefully and refer to the table as needed.

TABLE 27. PHOSPHORUS AND ITS FUNCTIONS

Body System and Structure	Function
Neuromuscular	Normal nerve and muscle activity.
Bones and teeth	Bone and teeth formation, strength, and durability.
Cellular	Formation of high-energy compounds (ATP, ADP). Phosphorus is the backbone of nucleic acids and stores metabolic energy.
	Formation of the red-blood-cell enzyme 2,3-diphosphoglycerate (2,3-DPG) responsible for oxygen delivery to tissues.
	Utilization of B vitamins.
	Transmission of hereditary traits.
	Metabolism of carbohydrates, proteins, and fats.
	Maintenance of acid-base balance in body fluids.

---

244 An important function of Phosphorus is in neuromuscular activity. Name at least two cations that play an important role in neuromuscular activity.

____________________ and ____________________.

\- - - - - - - - - - - - - - - - - -

potassium and sodium. Answers could also include calcium and magnesium.

---

245 Phosphorus, like calcium, is needed for strong and durable teeth and

____________________.

\- - - - - - - - - - - - - - - - - -

bones.

---

246 Intracellular ATP is needed for cellular energy.

The red-blood-cell enzyme 2,3-DPG is responsible for * ____________________

____________________________________.

\- - - - - - - - - - - - - - - - - -

oxygen delivery to tissues.

---

247 Other functions of phosphorus include the following:

( ) a. Utilization of vitamin A.
( ) b. Utilization of B vitamins.
( ) c. Metabolism of carbohydrates, proteins, and fats.
( ) d. Maintenance of acid-base balance in body fluids.
( ) e. Transmission of hereditary traits.

\- - - - - - - - - - - - - - - - - -

a. –; b. X; c. X; d. X; X.

Table 28 lists the conditions that cause phosphorus deficit or excess. Study this table carefully before proceeding to the frames that follow. Refer to this table as needed.

TABLE 28. CONDITIONS CAUSING PHOSPHORUS DEFICIT AND EXCESS

Conditions	Phosphorus Deficit (Hypophosphatemia)	Phosphorus Excess (Hyperphosphatemia)
Poor dietary intake	Inadequate phosphorus intake.	
Malnutrition	Poor nutrition; lack of protein.	
Alcoholism	Phosphorus loss caused by poor diet (malnutrition), diuresis, or aluminum antacids.	
Intestinal Malabsorption	Vitamin D deficit. Phosphorus is absorbed in jejunum in the presence of Vitamin D.	
Vomiting, anorexia, and chronic diarrhea	Loss of phosphorus through the GI tract. Decrease in cellular ATP (energy) stores.	
Acid-base disorders	Respiratory alkalosis from prolonged hyperventilation decreases serum phosphorus by causing an intracellular shift of phosphorus. Metabolic alkalosis is another cause of this shift.	
Parathyroid disorders	Hyperparathyroidism. Parathyroid hormone/PTH enhances renal phosphate excretion and calcium reabsorption.	Hypoparathyroidism. Lack of PTH causes calcium loss and phosphorus excess.
Drugs:		
Aluminum antacids	Phosphate binds with aluminum to decrease phosphorus level.	
Laxatives		Frequent use of phosphate laxatives increases serum phosphorus level.
Diuretics	Thiazides decrease phosphorus level.	
Oral and IV Phosphate		Excess administration can cause phosphorus excess.
Hyperalimentation/TPN	Phosphorus-poor or -free solution. IV concentrated glucose and protein, given rapidly, shifts phosphorus into cells, thus causing a serum phosphorus deficit.	
Renal failure	Renal wasting of phosphorus. Hemodialysis: Dialyzed with phosphate-poor dialysate. Low protein and phosphorus diet.	Renal shutdown decreases phosphorus excretion.

TABLE 28. (Continued)

Conditions	Phosphorus Deficit (Hypophosphatemia)	Phosphorus Excess (Hyperphosphatemia)
Diabetic ketoacidosis	Glycosuria and polyuria increase phosphate excretion. Dextrose infusion with insulin causes a phosphorus shift into cells; decreasing serum phosphorus level.	
Burns	Phosphorus loss due to utilization in tissue building.	

---

248 A decreased serum phosphorus level is known as phosphorus deficit or

________________.

An elevated serum phosphorus level is known as phosphorus excess or

________________.

— — — — — — — — — — — — — — — —

hypophosphatemia; hyperphosphatemia

---

249 Alcoholism can cause severe hypophosphatemia. Potential phosphorus loss could be the result of ________________ and

________________.

— — — — — — — — — — — — — — — —

poor diet (malnutrition); diuresis. Also use of aluminum antacids.

---

250 Name the vitamin that is necessary to phosphorus absorption in the small intestine. ________________.

— — — — — — — — — — — — — — — —

Vitamin D

---

251 Vomiting, anorexia, and chronic diarrhea can produce (hypophosphatemia/ hyperphosphatemia) ________________.

— — — — — — — — — — — — — — — —

hypophosphatemia

---

252 Explain how hypophosphatemia can occur as a result of prolonged hyperventilation. * ____.

What type of acid-base imbalance could result from prolonged hyperventilation * ______________________________.

- - - - - - - - - - - - - - - - -

phosphorus shift to cells (intracellular fluid). Respiratory alkalosis. Good.

---

253 In parathyroid disorders the parathyroid hormone/PTH influences phosphorus balance.

An increase in PTH secretion causes a phosphorus (loss/excess) ____________________; a decrease in PTH secretion causes a phosphorus (loss/excess) ____________________.

- - - - - - - - - - - - - - - - -

loss; excess

---

254 Certain groups of drugs affect phosphorus balance. Enter PD for phosphorus deficit and PE for phosphorus excess against the drug groups that could cause phosphorus imbalance.

________ a. Aluminum antacids

________ b. Phosphate-containing laxatives

________ c. Thiazide diuretics

________ d. Oral phosphate ingestion

________ e. Intravenous phosphate administration

- - - - - - - - - - - - - - - - -

a. PD; b. PE; c. PD; d. PE; e. PE

---

255 Severe hypophosphatemia could result from hyperalimentation/TPN. Two (2) reasons for serum phosphorus deficit related to hyperalimentation are

* ____;

* ____.

- - - - - - - - - - - - - - - - -

phosphate-poor or -free solution; concentrated glucose and/or protein, given too rapidly, causes phosphorus shift from serum into cells.

---

256 Renal failure could produce a phosphorus deficit or excess. Hypophosphatemia results from renal wasting of phosphorus and hyperphosphatemia, from

* ______________________.

— — — — — — — — — — — — — — — —

renal shutdown.

---

257 A patient in diabetic ketoacidosis may have severe hypophosphatemia. Give two (2) ways by which phosphorus deficit can be detected:

1. ______________________

2. ______________________

— — — — — — — — — — — — — — — —

1. Glycosuria and polyuria increase phosphorus excretion.
2. Dextrose infusion with insulin causes a phosphorus shift from serum into cells.

Table 29 lists signs and symptoms of hypo-hyperphosphatemia. Learn the normal ranges of serum phosphorus. Study this table carefully and refer to it as needed.

TABLE 29. SIGNS AND SYMPTOMS RELATED TO HYPOPHOSPHATEMIA AND HYPERPHOSPHATEMIA

Classes and values	Hypophosphatemia	Hyperphosphatemia
CNS and muscular abnormalities	Muscle weakness Tremors Paresthesia Bone pain Seizures	Tetany (with decreased calcium) Hyperreflexia Flaccid paralysis Muscular weakness
Hematologic abnormalities	Tissue hypoxia (decreased oxygen-containing hemoglobin and hemolysis) Possible bleeding (platelet dysfunction) Possible infection (leukocyte dysfunction)	
Cardiopulmonary	Weak pulse (myocardial dysfunction) Hyperventilation	Tachycardia
GI Abnormalities		Nausea, diarrhea Abdominal cramps
Milliequivalents per liter	Below 1.7 mEq/L	Above 2.6 mEq/L
Milligrams per deciliter	Below 2.5 mg/dl	Above 4.5 mg/dl

---

258 Indicate which of the following signs and symptoms relate to hypophosphatemia:

( ) a. Muscle weakness
( ) b. Paresthesia
( ) c. Bone pain
( ) d. Flaccid paralysis
( ) e. Tissue hypoxia
( ) f. Tachycardia

- - - - - - - - - - - - - - - - - -

a. X (also could occur with hyperphosphatemia); b. X; c. X; d. –; e. X; f. –.

---

259 Indicate which of the following signs and symptoms relate to hyperphosphatemia:

( ) a. Muscle weakness
( ) b. Paresthesia
( ) c. Hyperreflexia
( ) d. Flaccid paralysis
( ) e. Tachycardia
( ) f. Abdominal cramps
( ) g. Bone pain

- - - - - - - - - - - - - - - - - -

a. X (more common with hypophosphatemia; b. –; c. X; d. X; e. X; f. X; g. –.

---

260 Symptoms of phosphorus imbalance are very often vague; therefore serum values are needed.

Hypophosphatemia is present when the serum phosphorus level is less than ________________ mEq/L or ________________ mg/dl. Hyperphosphatemia is present when the serum phosphorus level is greater than ________________ mEq/L or ________________ mg/dl.

- - - - - - - - - - - - - - - - - -

1.7; 2.5; 2.6; 4.5

CLINICAL APPLICATIONS

---

261 If a severely malnourished patient is receiving 25% dextrose solution (hyperalimentation) should the infusion rate be (fast/slow) when first administered?

__________________. What type of serum phosphorus imbalance could occur if the infusion rate was faster than 80 ml/hr when started?

__________________________________. Why? ______________

____________________________.

- - - - - - - - - - - - - - - - - -

slow. Hypophosphatemia. Concentrated glucose tends to shift phosphorus into cells; the result is serum phosphorus deficit.

---

262 Any carbohydrate-loading diet can cause a phosphorus shift from serum into cells.

During tissue repair following trauma phosphorus shifts into cells. The serum phosphorus imbalance that could occur is called ______________________.

- - - - - - - - - - - - - - - - - -

hypophosphatemia

DRUGS AND THEIR EFFECT ON PHOSPHORUS BALANCE

The major drug group that causes hypophosphatemia is aluminum antacid; the drug groups responsible for hyperphosphatemia are phosphate laxatives, phosphate enemas, and the administration of oral and intravenous phosphates. Table 30 lists the names and rationales of drugs that affect phosphorus balance.

TABLE 30. DRUGS THAT AFFECT PHOSPHORUS BALANCE

Phosphorus Imbalance	Drugs	Rationale
Hypophosphatemia (Serum Phosphorus Deficit)	Aluminum Antacids: Amphojel Basaljel Aluminum/Magnesium Antacid: Di-Gel Gelusil Maalox Maalox Plus Mylanta Mylanta II	Aluminum-containing antacids bind with phosphorus; therefore the serum phosphorus level is decreased.

TABLE 30. (Continued)

Phosphorus Imbalance	Drugs	Rationale
	Thiazides: Chlorothiazide Hydrochlorothiazide	Phosphorus can be lost in diuresis when thiazides are used.
	Androgens Corticosteroids: Cortisone Prednisone Glucagon Gastrin Epinephrine Mannitol	These agents have a mild to moderate effect on phosphorus loss.
Hyperphosphatemia (serum phosphorus excess)	Oral Phosphates: Sodium phosphate/ Phospho-Soda Potassium phosphate/ Neutra-Phos K Intravenous Phosphates: Sodium phosphate Potassium phosphate	Excess oral ingestion and IV infusion can increase the serum phosphorus level.
	Phosphate Laxatives: Sodium phosphate Sodium biphosphate/ Phospho-Soda Phosphate Enema: Fleet sodium phosphate	Continuous use of phosphate laxatives and enemas can increase the serum phosphorus level.

Adapted with permission from A. Nanji: Drug-induced electrolyte disorders. *Drug Intelligence and Clinical Pharmacy*, 17:181, March 1983.

---

263 Prolonged intake of aluminum antacids, with or without magnesium, decreases the serum phosphorus level. Why? * ____. The phosphorus imbalance that results is ________________.

\- - - - - - - - - - - - - - - - - -

Aluminum-containing antacids bind with phosphorus; hypophosphatemia.

---

264 Aluminum antacids may be ordered for hyperphosphatemia. Do you know why?
* ____.

\- - - - - - - - - - - - - - - - - -

Aluminum binds with phosphorus to decrease the serum phosphorus level.

265 Enter PD for phosphorus deficit and PE for phosphorus excess beside the drugs that can cause phosphorus imbalance:

_______ a. Amphojel

_______ b. Cortisone

_______ c. Phospho-Soda

_______ d. Fleet's sodium phosphate

_______ e. Epinephrine/adrenalin

_______ f. Glucagon

_______ g. IV potassium phosphate

— — — — — — — — — — — — — — — — —

a. PD; b. PD; c. PE; d. PE; e. PD; f. PD; g. PE

PHOSPHORUS CORRECTION

When the serum phosphorus level falls below 1.5 mEq/L or 2 mg/dl oral and/or intravenous phosphate-containing solutions are usually ordered.

If the serum phosphorus level falls below 0.5 mEq/L or 1 mg/dl severe hypophosphatemia will appear. Intravenous phosphate-containing solutions are indicated.

---

266 Name two (2) drugs, oral or intravenous, that are administered to replace phosphorus deficit (refer to Table 30). *_______________________________ and *_______________________________.

— — — — — — — — — — — — — — — — —

Sodium phosphate/Phospho-Soda; potassium phosphate/Neutra-Phos K.

---

267 Concentrated IV phosphates are hyperosmolar and must be diluted. If IV potassium phosphate (K PO_4) is given in intravenous solution the IV rate should be no more than 10 mEq/hr to avoid phlebitis.

If intravenous potassium phosphate solution infiltrates do you know what could happen to the tissue? *____

____.

— — — — — — — — — — — — — — — — —

necrosis or sloughing of tissue. Potassium is extremely irritating to subcutaneous tissue.

---

268 Foods rich in phosphorus include milk (especially skim milk), milk products, meat (beef and pork), whole grain cereals, and dried beans.

Phosphorus-rich foods would be indicated if the serum phosphorus level were

( ) a. 0.3 mEq/L or 1 mg/dl
( ) b. 0.9 mEq/L or 1.5 mg/dl
( ) c. 1.6 mEq/L or 2.4 mg/dl

– – – – – – – – – – – – – – – – –

a. –; b. –; c. X

CASE REVIEW

Mrs. Peterson had a history of alcohol abuse. She was admitted for GI bleeding. In her own words she had not eaten a balanced diet for two months and had been taking Amphojel to relieve an "upset stomach." Mrs. Peterson complained of hand paresthesias and "overall" muscle weakness.

1. What type of phosphorus imbalance could Mrs. Peterson have?

   __________________________.

2. Give the "normal" serum phosphorus range. ____________ mEq/L;

   ________________ mg/dl.

3. Give two (2) reasons for Mrs. Peterson's imbalance:

   a. * ________________________________.

   b. * ________________________________.

4. What among Mrs. Peterson's signs and symptoms indicated a phosphorus deficit?

   a. * ________________________________.

   b. * ________________________________.

5. Name other clinical signs and symptoms of hypophosphatemia:

   a. * ________________________________.

   b. * ________________________________.

   c. * ________________________________.

6. Explain how aluminum hydroxide lowers the serum phosphorus level.

   * ___________________________________.

Mrs. Peterson was given a 10% dextrose solution intravenously. Her serum phosphorus level was 1.5 mg/dl and her potassium level, 3.0 mEq/L. Several hours later potassium phosphate was added to her intravenous solution.

7. What effect does concentrated dextrose (glucose) solution have on serum phos-

   phosphorus? * __________________________________.

8. What is the responsibility of the nurse who is attending Mrs. Peterson while she is receiving IV potassium phosphate diluted in solution?

   *______________________________________.

9. Name two (2) oral phosphate drugs:

   a. *______________________________________.

   b. *______________________________________.

- - - - - - - - - - - - - - - - - -

1. hypophosphatemia
2. 1.7–2.6 mEq/L; 2.5–4.5 mg/dl
3. poor diet (possible malnutrition); ingestion of aluminum hydroxide antacid, Amphojel
4. hand paresthesias; muscle weakness
5. bone pain, tissue hypoxia, weak pulse, hyperventilation.
6. phosphorus binds with aluminum, thus, lowering the serum phosphorus level
7. concentrated glucose cause a shift of phosphorus from serum into the cells.
8. monitor IV rate so Mrs. Peterson receives approximately 10 mEq/hr of KPO_4; check IV solution with KPO_4 for infiltration.
9. sodium phosphate/Phospho-Soda; potassium phosphate/NeutraPhos K

NURSING DIAGNOSES

Potential alteration in nutrition related to inadequate phosphorus intake, lack of vitamin D, and malnutrition.

Potential alteration in electrolyte balance (phosphorus) related to chronic alcoholism, parathyroid disorders (hypo-hyperparathyroidism), vomiting, chronic diarrhea, phosphorus-poor or -free solutions, renal failure, diabetic ketoacidosis, burns and drugs, aluminum antacids, phosphate laxatives, and thiazides.

Knowledge deficit related to a lack of understanding of the disease process that causes phosphorus imbalance.

Potential for noncompliance to the treatment regime for phosphorus deficit or excess related to an inadequate explanation.

Potential impairment of urinary elimination related to hyperphosphatemia.

NURSING ACTIONS

1. Explain the clinical conditions and problems that contribute to hypo-hyperphosphatemia; i.e., alcoholism with poor diet, vitamin D deficit, hyperalimentation/TPN, renal failure, and diabetic ketoacidosis.
2. Teach the patient/client to eat an adequate phosphorus-containing diet; i.e., milk (especially skim milk), meats (especially beef, pork), dried beans, and whole grain cereals.
3. Observe the patient for signs and symptoms of phosphorus imbalance: hypophosphatemia, i.e., muscle weakness, paresthesia, bone pain, weak pulse, and over-

breathing, and hyperphosphatemia, i.e., tachycardia, abdominal cramps, nausea, and hyperreflexia.

4. Check laboratory findings and report abnormal electrolyte results, including serum phosphorus, to the physician.
5. Observe the patient for signs and symptoms of hypocalcemia (e.g. tetany) when phosphate supplements are being administered. An increase in the serum phosphorus level decreases the calcium level.
6. Explain treatment and nursing care to the patient if possible and answer all questions or refer them to the physician.
7. Assess intake and output. If the urinary output is greatly decreased notify the physician. Phosphorus is excreted by the kidneys and poor renal function can cause hyperphosphatemia.
8. Monitor infusion rates closely while the patient is receiving IV fluids that contain potassium phosphate (KPO_4). The suggested amount of KPO_4 per hour is 10 mEq; KPO_4 is irritating to the blood vessels and can cause phlebitis.
9. Check for signs of infiltration at the IV site; KPO_4 is extremely irritating to subcutaneous tissue and could cause sloughing of tissue and necrosis.
10. Inform the physician if your patient is receiving phosphorus-poor or -free solution for hyperalimentation.

FOODS RICH IN ELECTROLYTES

Table 31 lists by name the foods that are rich in electrolyte content, according to class. If no foods are listed all in that class are low in that electrolyte. You will, no doubt, have to refer to this table. It is important, however, to memorize the foods rich in potassium. Because of the potassium deficit that frequently occurs, you must remember the foods rich in this electrolyte. Again, study this table carefully before proceeding to the next frames.

---

269 What classes of food are rich in potassium? ____________,

____________, ____________, ____________,

and * ____________.

- - - - - - - - - - - - - - - - -

fish; nuts; vegetables; fruits; fruit juice (see Table 31)

TABLE 31. FOODS RICH IN POTASSIUM, SODIUM, CALCIUM, MAGNESIUM, CHLORIDE AND PHOSPHORUS

Classes	Potassium	Sodium	Calcium	Magnesium	Chloride	Phosphorus
Daily requirements	3-4 g	2-4 g	800 mg	300 mg	3-9 g	800 mg
Beverages	Cocoa, Coca Cola, coffee, wines	Pepsi-Cola, tea, decaffeinated coffee		Cocoa		
Fruit and fruit juices	Citrus fruits: Oranges, grapefruit					
	Juices: Grapefruit (canned), orange (canned), prune (canned), tomato (canned)					
	Fruits: Apricots (dry), bananas, cantaloupe, dates, raisins (dry), watermelon, prunes			Average	High only in dates and bananas	
Bread products and cereal	Average to low amount	White bread, soda crackers, and wheat flakes		Cereals with oats		Whole-grain cereal
Dairy products	Average to low Milk, buttermilk	Butter, cheese, and margarine	Milk, cheese	Milk (average)	Cheese, milk	Cheese, milk
Nuts	Almonds, Brazil nuts, cashews, and peanuts	Low, except if salted	Brazil nuts (moderate)	Almonds, Brazil nuts, peanuts, and walnuts		Peanuts

Vegetables	Baked beans, carrots (raw), celery (raw), dandelion greens, lima beans (canned), mustard greens, tomatoes, spinach NOTE: Nearly all vegetables are rich in potassium when raw, but K will be lost if water used in cooking is discarded.	Average to low Celery (high average)	Baked beans, kale, mustard and turnip greens, broccoli		Spinach, celery	Dry beans
Meat, fish, and poultry	Average—meats High average—sardines, codfish, scallops	Corned beef, bacon, ham, crab, tuna fish, sausage (pork) Low in poultry	Salmon, meats	Bacon, shrimp Low in poultry Low in meats; Egg, average	Eggs, crabs, fish (average), turkey	Beef, pork, fish, chicken, turkey
Miscellaneous	Catsup (average), spices, potato chips, and peanut butter	Catsup, mayonnaise, potato chips, pretzels, pickles, dill, olives, mustard, Worcestershire sauce, celery salt, salad dressing—French and Italian	Molasses	Chocolate and chocolate bars, chocolate syrup Molasses Table salt		

270 Foods with the highest concentration of potassium are nuts and dried fruits. Juices, such as orange juice, are a quick source of potassium. Bananas, dates, prunes, and apricots have a greater concentration of potassium than have oranges and orange juice.

A quick source of potassium is * ________________.
Name four fruits containing a high concentration of potassium.

a. ____________.

b. ____________.

c. ____________.

d. ____________.

- - - - - - - - - - - - - - - - - -

orange juice

a. bananas
b. dates
c. prunes
d. apricots

---

271 Many Americans consume 5 to 10 g of sodium chloride per day, which is more than needed.

The daily requirement for sodium per day is * ____________.

The daily requirement for potassium per day is * ____________.

- - - - - - - - - - - - - - - - - -

2 to 4 g; 3 to 4 g

---

272 Name the classes of food that are rich in sodium. * ____________, * ____________, ____________, and ____________.

- - - - - - - - - - - - - - - - - -

bread products; dairy products; meat; fish

---

273 The classes of food that are rich in calcium are:

1. *________________.

2. *________________.

3. ____________.

4. ____________.

- - - - - - - - - - - - - - - - -

1. baked beans
2. dairy products
3. fish (salmon)
4. meat

---

274 The classes of food that are rich in magnesium are oats from bread products, shrimp from fish, and ____________.

- - - - - - - - - - - - - - - - -

nuts

---

275 Give examples of foods that are rich in chloride.

1. fruits: *____________.

2. meats: ____________.

3. fish: ____________.

4. dairy products: ____________ and ____________.

5. vegetables: ____________ and ____________.

- - - - - - - - - - - - - - - - -

1. dates or bananas
2. turkey
3. crab
4. cheese; milk
5. spinach; celery

---

276 Name five (5) foods that are rich in phosphorus. ____________________,

____________________, ____________________, ____________________,

and *____________________.

— — — — — — — — — — — — — — — — —

milk; beef; pork; fish; whole-grain cereal. Also chicken, turkey, peanuts, and dried beans.

---

277 Check the classes of food that are rich in K (potassium), Na (sodium), Ca (calcium), Mg (magnesium), Cl (chloride), and P (phosphorus).

	K	Na	Ca	Mg	Cl	P	
a.	( )	( )	( )	( )	( )	( )	Fruits
b.	( )	( )	( )	( )	( )	( )	Fruit juices
c.	( )	( )	( )	( )	( )	( )	Bread products
d.	( )	( )	( )	( )	( )	( )	Dairy products
e.	( )	( )	( )	( )	( )	( )	Nuts
f.	( )	( )	( )	( )	( )	( )	Vegetables
g.	( )	( )	( )	( )	( )	( )	Meat
h.	( )	( )	( )	( )	( )	( )	Fish

— — — — — — — — — — — — — — — — —

a. K, Cl (dates)
b. K
c. Na, Mg (oats), P
d. Na, Ca, Cl, P
e. K, Mg, P
f. K, Ca, Cl
g. Na, Cl, Ca, P
h. K, Na, Ca, Mg, Cl, P

REFERENCES

Abbott Laboratories: *Fluid and Electrolytes.* North Chicago, 1970, pp 15-21.

Baker WL: hypophosphatemia. *Am J Nurs* 85(9):998–1003, September 1985.

Beeson P, McDermott W: *Textbook of Medicine,* ed 14. Philadelphia, WB Saunders Co, 1975, pp 1579-1599.

Bolte H: Treatment of acute and chronic hypokalemia. *Acta Cardiol Suppl* 17: 213-215, 1973.

Burgess A: *The Nurse's Guide to Fluid and Electrolyte Balance.* ed 2. New York, McGraw-Hill Book Co, 1979, pp 16–17, 230–231.

Burgess R: Fluids and electrolytes. *Am J Nurs* 65:92–93, 1965.

Crowell CE (ed); Programmed instruction—potassium imbalance. *Am J Nurs* 67:343–366, 1967.

Elbaum N: Detecting and correcting magnesium imbalance. *Nursing '77* 7: 34-35, August 1977.

Felver L: Understanding the electrolyte maze. *Am. J. Nurs* 80(9):1591–1595, September 1980.

Groer MW, Shekleton ME: *Basic Pathophysiology,* ed 2. St Louis, CV Mosby Co, 1983, pp 294–316.

Heidland A, Hennemann HM, Rockel A: The role of magnesium and substances promoting the transport of electrolytes. *Acta Cardiol Suppl* 17: 65-67, 1973.

Kerkovits G: Antiarrhythmics and electrolytes. *Acta Cardiol Suppl* 17: 155-157, 1973.

Krause MV, Mahan LK: *Food, Nutrition and Diet Therapy*. Philadelphia, WB Saunders Co, 1979, pp 908-912.

Lancour J: ADH and aldosterone: How to recognize their effects. *Nursing '78* 8(9): 36-41, September 1978.

MacLeod S: The rational use of potassium supplements. *Postgrad Med* 57(2): 123-127, 1975.

Makoff D: Common fluid and electrolyte disorders in the cardiac patient. *Geriatrics* 67-76, November 1972.

Menzel L: Clinical problems of fluid balance. *Nursing Clinics of North America* 15(3): 549-576, September 1980.

Metheney NM, Snively WD: *Nurses' Handbook of Fluid Balance*, ed 4. Philadelphia, JB Lippincott Co, 1983, pp 39–63.

Mitchell H: *Nutrition in Health and Disease*, ed 16. Philadelphia, JB Lippincott Co, 1976, pp 557-559.

Monitoring Fluid and Electrolytes Precisely. Horsham, PA, Nursing Skillbook, 1983, pp 79–112.

Nanji A: Drug-induced electrolyte disorders. *Drug Intelligence and Clinical Pharmacy* 17:175–185, March 1983.

Narins RG et al.: Diagnostic strategies in disorders of fluid, electrolyte and acid-base homeostasis. *Am J Med* 72: 496–518, March 1982.

O'Dorisio TM: Hypercalcemia crisis. *Heart and Lung* 7(3): 425-432, May-June 1978.

Pestana C: *Fluids and Electrolytes in the Surgical Patient.* Baltimore, Williams & Wilkins Co, 1977, pp 73-85.

Snively WD: *Sea Within.* Philadelphia, JB Lippincott Co, 1960, pp 62-81.

Spencer RT et al.: *Clinical Pharmacology*. Philadelphia, JB Lippincott Co, 1983, pp 834–838.

Statland H: *Fluid and Electrolytes in Practice*, ed 3. Philadelphia, JB Lippincott Co, 1963, pp 98-126.

Suki WN: Disposition and regulation of body potassium: An overview. *Am J Med Sci* 272:31–41, July-August 1976.

Travenol Laboratories, Inc: *Guide to Fluid Therapy*, Travenol Lab Deerfield, Ill. 1970, pp 14-55.

Tripp A: Hyper and hypocalcemia. *Am J Nurs* 76: 1142-1145, 1976.

Turner D: *Handbook of Diet Therapy*, ed 5. Chicago, University of Chicago Press, 1970.

Wiener MB, Pepper GA: *Clinical Pharmacology and Therapeutics in Nursing*, ed 2. New York, McGraw-Hill Book Co, 1985, pp 325–350.

Wilson ED: *Principles of Nutrition*, ed 3. New York, John Wiley & Sons, Inc, 1975.

Wyngaarden J, Smith, L: *Cecil Textbook of Medicine*, ed 16. Philadelphia, WB Saunders Co, 1982, pp 481–486, 1131–1134, 1318–1332.

CHAPTER THREE

Acid-Base Balance and Imbalance

BEHAVIORAL OBJECTIVES

Upon completion of this chapter, the student will be able to:

Explain the influence of the hydrogen ion (H^+) on body fluids.
Give the pH ranges for acidosis and alkalosis.
Explain the four regulatory mechanisms for pH control and how the regulatory mechanisms can maintain acid-base balance.
Explain how various clinical conditions can cause metabolic acidosis and alkalosis, and respiratory acidosis and alkalosis.
Observe clinically for symptoms of metabolic acidosis and alkalosis, and respiratory acidosis and alkalosis. (You may need some assistance in recognizing all these symptoms.)
Explain the body's defense action and the clinical management for acid-base balance and be able to apply this with assistance to various clinical situations.

INTRODUCTION

Our body fluid must maintain a balance between acidity and alkalinity in order for life to be maintained. *Acid* comes from the Latin word meaning "sharp," and acid is frequently referred to as being sour. On the other hand, alkaline is referred to as being sweet. According to the Bronsted-Lowry concept of acids and bases, an "*acid* is any molecule or ion that can *donate a proton* to any other substance, whereas a *base* is any molecule or ion that can *accept a proton.*" Also, the more readily an acid gives up its protons, the stronger it is as an acid. Acids and bases are not synonymous with anions and cations.

Other theories state that the concentration of the hydrogen ion (plus or minus) will determine either the acidity or the alkalinity of a solution. The amount of ionized hydrogen in the extracellular fluid is extremely small; around 0.0000001 gram per liter. Instead of using this cumbersome figure, the symbol pH is used, which stands for the negative logarithm (exponent) of the hydrogen ion concentration. Mathematically, it would be expressed as 10^{-7}, the base being 10 and the power –7 being the logarithm of the number. The minus sign is dropped and the symbol used to designate the hydrogen ion concentration would then be pH 7. As the hydrogen ion concentration rises (in solution), the pH value falls, indicating a decreased negative logarithm of the hydrogen ion concentration, thus indicating acidity. As the hydrogen concentration falls, the pH rises, thus indicating alkalinity.

The hydroxyl ions (OH^-) are base ions and when in excess will cause the alkalinity of the solution. A solution of pH 7 is neutral since at that concentration the number of hydrogen ions (H^+) is exactly balanced by the number of hydroxyl ions (OH^-).

The above information will help you in the basic understanding of acidity and alkalinity. This will aid in understanding the material presented in this chapter: the regulatory mechanism for pH control; the clinical conditions causing metabolic acidosis and alkalosis and respiratory acidosis and alkalosis; symptoms and clinical management of acidosis and alkalosis; and clinical applications.

There are two case reviews and two sets of nursing diagnoses and nursing actions.

Refer to the Introduction as needed to answer the first six frames. An asterisk (*) on an answer line indicates a multiple-word answer. The meanings for the following symbols are: ↑ increased, ↓ decreased, > greater than, < less than.

---

1 According to the Bronsted-Lowery concept of acids and bases, an acid is a proton (donor/acceptor) __________ and a base is a proton (donor/acceptor) ______________.

\- - - - - - - - - - - - - - - - -

donor; acceptor

---

2 For our purposes, we shall consider that the acidity or alkalinity of a solution depends on the concentration of the *________________.

An increase in concentration of the hydrogen ions makes a solution more __________ and a decrease makes it more ____________.

\- - - - - - - - - - - - - - - - -

hydrogen ions; acid; alkaline

---

3 Explain the meaning of the pH symbol.* ______________________________

______________________________.

What effect does this have on a solution? * ______________________________

______________________________.

\- - - - - - - - - - - - - - - - - -

negative logarithm of the hydrogen ion
It determines the acidity or alkalinity of a solution

---

4 As the hydrogen ion concentration increases, the pH value * ______________

______________. What does this indicate? * ______________

______________________________.

As the hydrogen ion concentration falls, the pH value * ______________

______________. What does this indicate? * ______________

______________________________.

\- - - - - - - - - - - - - - - - - -

falls or decreases; indicates acidity;
rises or increases; indicates alkalinity

---

5 A pH of 7 represents 10 times the number of hydrogen ions as does a pH of 8.
Which of the two pH's would be considered alkaline? ______________.

\- - - - - - - - - - - - - - - - - -

pH 8

---

6 A solution at pH 7 is neutral. Why? * ______________________________

______________________________.

The symbol for the hydrogen ion is ______ and for the hydroxyl ion, ______.

A hydroxyl ion and CO_2 would yield * ______________________________.

\- - - - - - - - - - - - - - - - - -

The number of hydrogen ions is balanced by the number of hydroxyl ions
H^+, OH^-; HCO_3, known as bicarbonate

---

7 The pH of extracellular fluid in health is maintained at a level between 7.35 and 7.45. The body fluid is slightly (acid/alkaline) ______________________.

- - - - - - - - - - - - - - - -

alkaline

---

8 With a pH higher than this range (7.35-7.45), the body is considered to be in a state of alkalosis.

What do you think we would call a state in which the pH is below 7.35? ______________________.

- - - - - - - - - - - - - - - -

acidosis

---

9 The pH norm of blood serum is 7.4; a variation of 0.4 of a pH unit in either direction can be fatal.

In a healthy individual, the pH range of blood serum is * ______________________.

- - - - - - - - - - - - - - - -

7.35-7.45

---

10 Within our bodies, the pH of the different body fluids varies. The normal pH for urine is 6.0; for gastric juice, 1.0-2.0; for bile, 5.0-6.0; and for intracellular fluid, 6.9-7.2. These body fluids are (acid/alkaline) ______________________.

Why? * ______________________ ______________________.

The normal pH for intestinal juice is 6.5-7.6. The fluids from the intestinal tract can be:

( ) a. acid
( ) b. alkaline
( ) c. neutral

- - - - - - - - - - - - - - - -

acid. The pH is below 7.35
a. X; b. X; c. X

---

11 Whether a substance is acid, neutral, or alkaline depends on the number of________________ ions present in a given weight or volume.

- - - - - - - - - - - - - - - - - -

hydrogen

---

12 When the number of hydrogen ions increases in the body fluid, the body fluid becomes ________________.

When the number of hydrogen ions decreases, the body fluid becomes ________________.

- - - - - - - - - - - - - - - - - -

acid; alkaline

---

13 In health, there are 1 1/3 mEq/L of acid to each 27 mEq of alkali for each liter of extracellular fluid, which represents a ratio of 1 part of acid to 20 parts of alkali.

Why do you think the measurement of acid and alkali is based on the extracellular fluid? *________________.

If the ratio of 1:20 is maintained, the patient is said to be in acid-base (balance/imbalance) ________________.

- - - - - - - - - - - - - - - - -

more available for analysis; balance

Diagram 7 demonstrates by the arrow that the body is in acid-base balance when there is 1 part acid to 20 parts alkali. The pH would be 7.4. If the arrow tilts left due to an alkali deficit or acid excess, then acidosis occurs, and if the arrow tilts right due to an alkali excess or acid deficit, then alkalosis occurs. Carbonic acid is H_2CO_3.

Study this diagram carefully. Know what will happen when the arrow tilts either left or right. Refer to the diagram when needed.

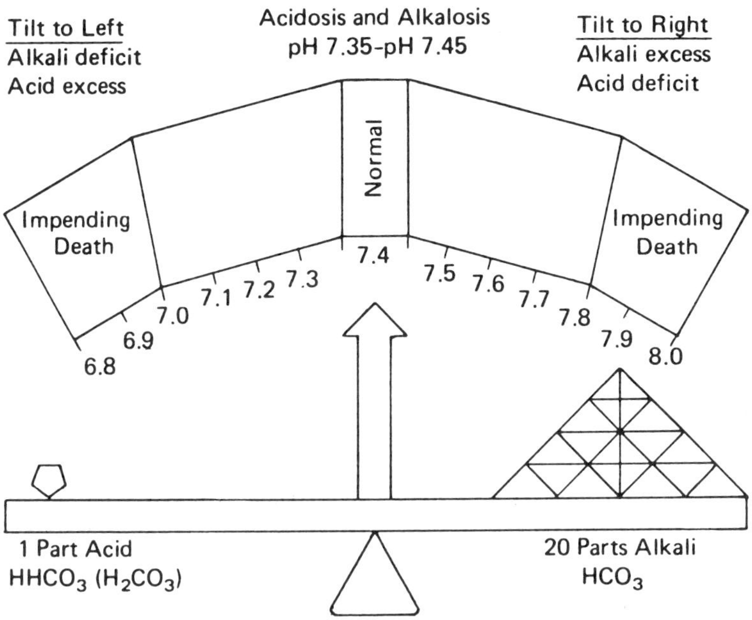

Diagram 7. Acidosis and alkalosis.

---

14 Your patient's serum pH is 7.1. Tell everything you can about the patient's condition on the basis of Diagram 7. * ______________________________

____

______________________________.

— — — — — — — — — — — — — — — —

An acidotic condition has occurred. It is due to either too much acid or too little alkali in the extracellular fluid

---

15 Another patient's serum pH is 7.8. Tell everything you can about the patient's condition on the basis of the diagram.* ______________________________

____

______________________________.

- - - - - - - - - - - - - - - - -

Disturbance causing alkalosis. It is due to too much alkali or to too little acid. Borderline on impending death

---

16 If the ratio 1:20 of the extracellular fluid is no longer present and the acid is increased or the alkali is decreased, then we say the patient suffers from ________

____________________.

If the alkaline reserve is increased or the acid decreases, then he suffers from

____________________.

- - - - - - - - - - - - - - - - -

acidosis; alkalosis

---

17 Which of the following will occur as the result of acidosis? The balance is tilted

____________________.

There is/are:

( ) a. alkali deficit
( ) b. alkali excess
( ) c. acid deficit
( ) d. acid excess

The pH is below ____________________.

- - - - - - - - - - - - - - - - -

left
a. X; b. –; c. –; d. X
7.35

---

18 Which of the following will occur as the result of alkalosis? The balance is tilted

________________________.

There is/are:

( ) a. alkali deficit
( ) b. alkali excess
( ) c. acid deficit
( ) d. acid excess

The pH is above ____________________.

— — — — — — — — — — — — — — — — — —

right
a. –; b. X; c. X; d. –
7.45

---

19 The body is provided with several mechanisms for controlling its pH even though considerable amounts of acid or alkali enter the body.

The pH of blood serum in a healthy individual is between *______________.

The pH norm of blood serum is ____________________.

— — — — — — — — — — — — — — — — — —

7.35 and 7.45; 7.4

REGULATORY MECHANISM FOR PH CONTROL

20 The main regulatory mechanisms for pH control are:

1. the buffer system
2. the ion exchange
3. the respiratory regulation
4. the renal regulation

Name the four regulatory mechanisms for pH control.

1. *__________________________.
2. *__________________________.
3. *__________________________.
4. *__________________________.

— — — — — — — — — — — — — — — — — —

1. buffer system
2. ion exchange
3. respiratory regulation
4. renal regulation

Table 32 is divided into three parts. Part A explains the buffer system. Buffers maintain the acid-base balance of body fluids by protecting the fluids against changes in pH. They act like chemical sponges, for they can either soak up surplus hydrogen ions or release them.

There are four main examples of the buffer system, including bicarbonate-carbonic acid; phosphate; hemoglobin-oxyhemoglobin; and protein. They are shown in Table 32A along with their interventions—their action as a buffer, and the rationale—the reason for their action. The bicarbonate-carbonic acid buffer system is more readily available and acts within a fraction of a second to prevent excessive changes in H^+ concentration, and therefore it is the principal buffer system of the body. Strong acids when added will combine with the bicarbonate ion to form carbonic acid, which is a weak acid. This prevents the fluids from becoming strongly acid.

Remember that bicarbonates and phosphates are anions. Refer to the glossary for unknown words. Study the table carefully and refer to it as needed.

TABLE 32A. REGULATORY MECHANISM FOR pH CONTROL.
A. BUFFER SYSTEMS

Regulatory Mechanism	Intervention	Rationale
a. Bicarbonate-carbonic acid buffer system (principal buffer system of body)	Acids combine with bicarbonates in blood to form neutral salts (bicarbonate salt) and carbonic acid (weak acid). Carbonic acid (H_2CO_3) is weak and unstable acid, changing to water and carbon dioxide in fluid ($H_2CO_3 \leftrightharpoons H_2O + CO_2$). A strong base will combine with weak acid, e.g., H_2CO_3	When a strong acid such as HCl enters body, H^+ ions combine with HCO_3^- ions of $NaHCO_3$, yielding carbonic acid and neutral salt. ($HCl + Na\underline{HCO_3} \rightarrow \underline{H_2CO_3} + NaCl$) Weak acid, e.g., H_2CO_3, does not release H^+ as readily as does a strong acid, e.g., HCl, and thus ionizes less effectively. When strong base such as NaOH is added to system, it is neutralized by carbonic acid, yielding water and HCO_3^- from a salt ($Na\underline{OH} + \underline{H_2}CO_3 \rightarrow \underline{H_2O} + Na\underline{HCO_3}$)
b. Phosphate buffer system	The phosphate buffer system increases amount of sodium bicarbonate ($NaHCO_3$) in extracellular fluids, making extracellular fluids more alkaline. The H^+ is excreted as NaH_2PO_4 and Na and bicarbonate ions combine.	Excess H^+ combine in renal tubules with Na_2HPO_4 (disodium phosphate), forming NaH_2PO_4 ($H^+ + Na_2HPO_4 \rightarrow Na^+ + NaH_2PO_4$); Na^+ is reabsorbed ($Na^+ + HCO_3^- \rightarrow NaHCO_3$) and H^+ is passed into urine.
c. Hemoglobin-oxyhemoglobin buffer system	Maintains same pH level in venous blood as in arterial blood.	Venous blood has higher CO_2 content and bicarbonate ion concentration than arterial blood, but pH is the same since oxyhemoglobin (acid) in erythrocyte has taken over some anion function which

TABLE 32A. (Continued)

Regulatory Mechanism	Intervention	Rationale
		was provided by excess bicarbonate in venous plasma.
d. Protein buffer system	Proteins can exist in form of acids (H protein) or alkaline salts (B protein) and in this way are able to bind or release excess hydrogen as required.	Proteins are amphoteric, carrying both acidic and basic charge.

---

21 (Refer to the Introduction of Table 32A if necessary.)

Explain the purpose of the buffer system. *______________________________

____.

How do the buffer systems accomplish their purpose? *____________________

____.

\- - - - - - - - - - - - - - - - - -

They protect fluids against changes in pH
They soak up surplus H^+ and release them as needed

---

22 What is the principal buffer system of the body? *________________________

____.

\- - - - - - - - - - - - - - - - - -

bicarbonate-carbonic acid buffer system

---

23 When acid enters the body, the H^+ is picked up by the bicarbonate, changing it to *_______________________________.

A base added to the body is neutralized by carbonic acid to form ____________ and ____________________.

\- - - - - - - - - - - - - - - - - -

carbonic acid; water; bicarbonate

---

24 The bicarbonate–carbonic acid buffer system is the most important buffer system in the body. It maintains acid-base balance 55% of the time. Give an example with a formula of how a strong acid combines with a bicarbonate to yield a weak acid. * ____

____.

Give an example with a formula of how a strong base is neutralized by a weak acid to yield water and bicarbonate from a weak salt. * ______________________

____.

— — — — — — — — — — — — — — — — — —

$HCl + NaHCO_3 \rightarrow H_2CO_3 + NaCl$
$NaOH + H_2CO_3 \rightarrow H_2O + NaHCO_3$

---

25 The phosphate buffer system maintains the acid-base balance by combining the excess H^+ with sodium salts, forming * ______________________________.

The H^+ is excreted in ______________.

This system (excretes/retains)______________ excess acid in the body.

— — — — — — — — — — — — — — — — — —

NaH_2PO_4; urine; excretes

---

26 What is the function of the hemoglobin–oxyhemoglobin buffer system? * ______

____.

— — — — — — — — — — — — — — — — — —

It maintains the same pH level in the venous blood and arterial blood

---

27 What is unique about the protein buffer system? * ______________________

____.

This system carries a(n) ______________ and ______________ charge.

— — — — — — — — — — — — — — — — — —

It is amphoteric—it has the ability to bind or release excess H^+; acidic; basic

28 The four buffer systems in the body are:

1. *______________________________.

2. ______________________________.

3. *______________________________.

4. ______________________________.

- - - - - - - - - - - - - - - -

1. bicarbonate-carbonic acid
2. phosphate
3. hemoglobin-oxyhemoglobin
4. protein

Table 32B gives explanations of the ion exchange and the respiratory regulation as regulatory mechanisms for pH control. The ion exchange is frequently referred to as the *chloride shift.* Refer to Chapter 2 for clarification of the chloride shift if needed. The respiratory regulation depends on the lungs in exhaling CO_2 or retaining the CO_2 for the control of pH. Again, the inverventions refer to the action, and rationale refers to the reason.

Study this table carefully and refer to it as needed.

TABLE 32B. REGULATORY MECHANISM FOR pH CONTROL. B. ION EXCHANGE AND RESPIRATORY REGULATION

Regulatory Mechanism	Intervention	Rationale
Ion exchange	Ion exchange of HCO_3 and Cl occurs in red blood cell as result of O_2 and CO_2 exchange. There is redistribution of anions in response to increase in CO_2. Chloride ion enters red blood cell (RBC) as bicarbonate ion diffuses into plasma in order to restore ionic balance.	Increase in serum carbon dioxide causes CO_2 to diffuse into red blood cells combining with H_2O to form H_2CO_3. This weak acid dissociates, forming acid and base ions, $H_2CO_3 \rightarrow H^+ + HCO_3^-$. Hydrogen ion is buffered by hemoglobin, and HCO_3^- ion moves into plasma as chloride ion (Cl^-) shifts into cell to replace it.

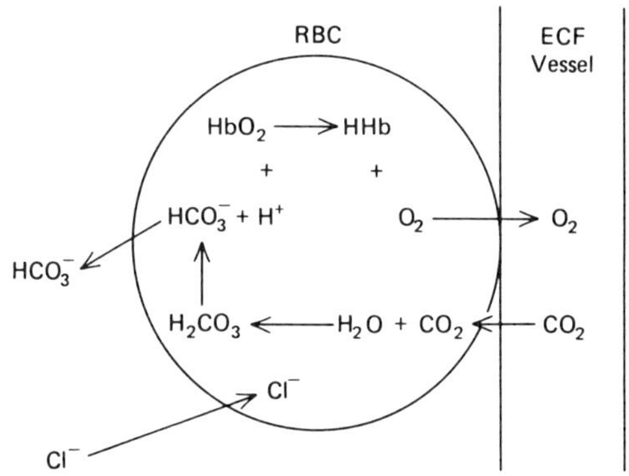

Respiratory regulation (acts quickly in case of emergency)	For regulation of acid, lungs will blow off more CO_2 and for regulation of alkaline, respiratory center will be depressed in order to retain CO_2. It takes 1 to 3 minutes for respiratory system to readjust H^+ concentration.	Respiratory center in medulla controis rate and depth of respiration and is sensitive to changes in blood pH or CO_2 concentration. When pH is decreased, carbonic acid is exhaled in form of carbon dioxide and moist air ($H_2CO_3 \rightarrow H_2O + CO_2$); thus acid is eliminated.

Note: Illustration adapted with permission from JL Keyes: Blood-gases and blood-gas transport. *Heart and Lung,* 1974, pp 945–954.

---

29 When carbon dioxide enters the red blood cell, what happens to the bicarbonate and chloride anions? * ____.

This ion exchange is frequently called the * ________________________________.

\- - - - - - - - - - - - - - - -

Bicarbonate diffuses out the cell and the chloride ion enters the cell; chloride shift

---

30 When the red blood cells are oxygenated, what anion is commonly present?

______________________________.

When carbon dioxide enters the red blood cells, what anion also enters?

______________________________.

What anion leaves as carbon dioxide enters? ______________________________.

- - - - - - - - - - - - - - - - -

bicarbonate (HCO_3); chloride (Cl); bicarbonate (HCO_3)

---

31 How does the respiratory regulatory mechanism control the serum pH?

* ______________________________

______________________________.

- - - - - - - - - - - - - - - - -

When the serum pH is decreased, the lungs will blow off CO_2. With an increased pH, the lungs retain CO_2

---

32 Where is the respiratory center located? ______________________________. What does the respiratory center control? * ______________________________.

- - - - - - - - - - - - - - - - -

medulla; rate and depth of the respiration

---

33 How do the two mechanisms in Table 32B deal with an increased CO_2?

1. * ______________________________ ______________________________.
2. * ______________________________.

- - - - - - - - - - - - - - - - -

1. As the CO_2 enters the red cells, HCO_3 leaves the red cell and Cl enters
2. The lungs will blow off more CO_2

Table 32C describes how the kidneys regulate pH in the body. The kidneys compensate for an excess production of acid by excreting the acid and returning the bicarbonate to the extracellular fluid. The acid occurs as the result of normal metabolism.

Study the table carefully, noting how in each instance the excess H^+ are neutralized. When you think you understand the renal regulatory mechanism described in the table, answer the frames that follow. Refer to the table when needed.

TABLE 32C. REGULATORY MECHANISM FOR pH CONTROL.
C. RENAL REGULATION

Regulatory Mechanism	Intervention	Rationale
a. Acidification of phosphate buffer salts	Exchange mechanism occurs between H^+ of renal tubular cells and disodium salt (Na_2HPO_4) in tubular urine.	Sodium salt (Na_2HPO_4) dissociates into Na^+ and $NaHPO_4^-$; Na^+ moves into tubular cell and hydrogen unites with $NaHPO_4$, forming dihydrogen phosphate salt, NaH_2PO_4, which is excreted.

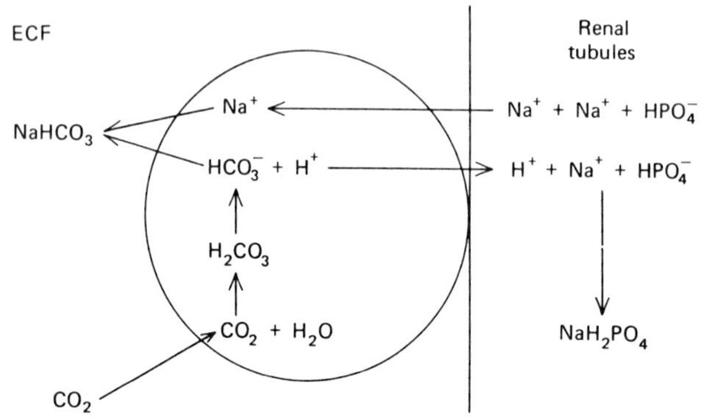

Regulatory Mechanism	Intervention	Rationale
b. Reabsorption of bicarbonate	Carbon dioxide is absorbed by tubular cells from blood and combines with water present in cells to form carbonic acid, which in turn ionizes, forming H^+ and HCO_3^--. Na^+ of tubular urine exchanges with H^+ of tubular cells and combines with HCO_3^- to form sodium bicarbonate and is reabsorbed into blood.	The enzyme, carbonic anhydrase, is responsible for formation of carbonic acid H_2CO_3. Ionization of $H_2CO_3 \rightarrow H^+ + HCO_3^-$. Free H^+ will exchange with Na^+ Exchange of H^+ and Na^+ permits reabsorption of bicarbonate with sodium ($NaHCO_3$) and excretion of an acid or H^+.

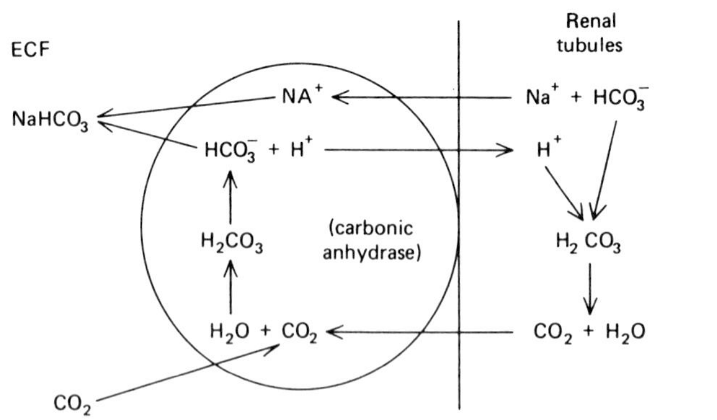

Note: Adapted with permission, from AC Guyton: *Textbook of Medical Physiology*, ed 6. Philadelphia, WB Saunders Co, 1981.

TABLE 32C. (Continued)

Regulatory Mechanism	Intervention	Rationale
c. Secretion of ammonia	Ammonia (NH_3) unites with HCl in renal tubules and H^+ is excreted as NH_4Cl (ammonium chloride).	Almost half of H^+ excretion is from this method—$HCl + NH_3 \rightarrow NH_4Cl$. Ammonia is formed in renal tubular cells by oxidative breakdown of amino acid glutamine in presence of enzyme, glutanimase. Ammonia can also be converted into urea by liver and excreted as urea by kidneys.

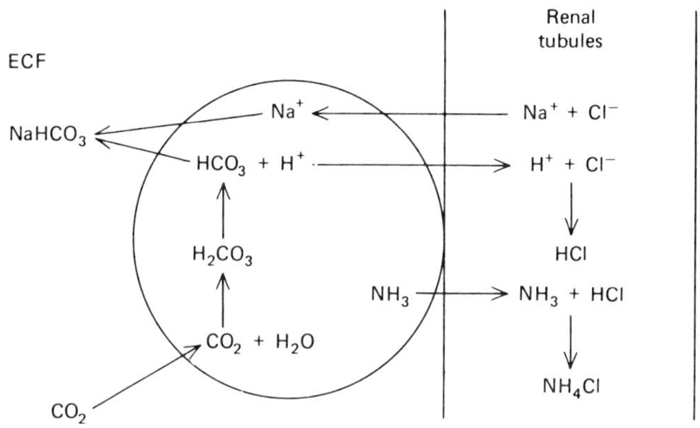

Note: Adapted with permission, from AC Guyton: *Textbook of Medical Physiology*, ed 6. Philadelphia, WB Saunders Co, 1981.

---

34 Name the three renal regulatory mechanisms for pH control.

1. *______________________________.
2. *______________________________.
3. *______________________________.

- - - - - - - - - - - - - - - -

1. acidification of phosphtae buffer salts
2. reabsorption of bicarbonate
3. secretion of ammonia

---

35 The hydrogen ions exist largely in the renal tubules in the buffered state. They are excreted indirectly, replacing a cation in excretion.

Explain how the hydrogen ion is excreted after it combines with the sodium salt—Na_2HPO_4. * ____

____.

— — — — — — — — — — — — — — — — —

The H^+ replaces a Na^+, forming a dihydrogen phosphate salt—NaH_2PO_4. This salt is excreted.

---

36 The kidneys regulate the H^+ levels in the body by varying the excretion and reabsorption of H^+ and HCO_3^-.

Explain how the H^+ is derived from carbonic acid. * ____________________

____________________.

Explain how the bicarbonate ion is reabsorbed.* ____________________

____________________.

— — — — — — — — — — — — — — — — —

The ionization of $H_2CO_3 \rightarrow H^+ + HCO_3^-$

The Na^+ exchanges with the H^+ in renal tubules and the Na^+ combines with HCO_3^- and is reabsorbed into the blood

---

37 Explain the formation of ammonia in the renal tubular cells. * ____________________

____.

How does the ammonia aid in the excretion of the hydrogen ion?

* ____.

— — — — — — — — — — — — — — — — —

It is formed from the breakdown of amino acid glutamine

NH_3 unites with HCl and is excreted as NH_4Cl (ammonium chloride)

---

38 Which one of the three methods in Table 32C is responsible for nearly half of the hydrogen ion excretion? * ____________________.

— — — — — — — — — — — — — — — — —

secretion of ammonia

---

39 The kidneys take hours to days to adjust acid-base balance. The lungs act quickly in emergencies for it takes *__________ minutes to readjust the balance.

- - - - - - - - - - - - - - - - -

1-3

---

40 Hydrogen ions circulate throughout the body fluids in two forms, namely, *volatile acid* and *nonvolatile acid.*

A volatile acid (carbonic acid—H_2CO_3) circulates as CO_2 and H_2O and is excreted as a gas. The gas excreted is ______________, which helps in maintaining acid-base balance.

A nonvolatile acid (fixed acid) is produced as the result of various organic acids within the body. It must be excreted from the body in water, e.g., urine. What regulatory mechanism may be responsible for excreting nonvolatile acids?

*______________________________.

- - - - - - - - - - - - - - - - -

CO_2; kidney or renal

---

41 The lungs excrete (volatile/nonvolatile) ______________ acids and the kidneys excrete (volatile/nonvolatile) ______________ acids.

Renal Regulation	*Respiratory Regulation*
$H^+ + HCO_3^- \leftrightharpoons \; \vert H_2CO_3 \vert$	$\leftrightharpoons H_2O + CO_2$
(excreted as nonvolatile acid)	(excreted as volatile acid)

- - - - - - - - - - - - - - - - -

volatile; nonvolatile

---

42 The kidneys and lungs aid in acid-base balance. Label the chemical formula as to which organ is responsible for acid-base regulation.

__________ $H^+ + HCO_3^- \leftrightharpoons \vert H_2CO_3 \vert \leftrightharpoons H_2O + CO_2$ __________
(Organ) (Organ)

- - - - - - - - - - - - - - - - -

kidneys; lungs

Diagram 8 demonstrates the normal acid-base balance or equilibrium and the cause and results of imbalance. The top scale shows acid-base balance with 1 part acid and 20 parts base. To determine the acid-base imbalance, three variables should be known: the pH, P_{CO_2} (respiratory component), and HCO_3 or serum CO_2 (metabolic component).

Respiratory acidosis has a pH↓ and P_{CO_2}↑. The person is hypoventilating and CO_2 is being retained. The CO_2 combines with H_2O to form carbonic acid, H_2CO_3.

Respiratory alkalosis has pH↑ and P_{CO_2}↓. The person is hyperventilating—blowing off CO_2—and therefore decreasing the body's carbonic acid.

Metabolic acidosis has a pH↓ and HCO_3↓ or *serum CO_2↓. There is a gain of acid or a loss of bicarbonate from the extracellular fluid, which can result from an increase of body acid.

Metabolic alkalosis has pH↑ and HCO_3↑ or *serum CO_2↑. There is an increase in the bicarbonate or a loss in a strong acid.

To determine the acidotic or alkalotic states, (1) assess the pH of the blood; (2) assess the P_{CO_2} (respiratory component: range 36-44 mm Hg) or arterial blood; (3) assess the HCO_3 of arterial blood (metabolic component [renal] : range 24-28 mEq/L) or serum CO_2 (range 22-32 mEq/L).

*Serum CO_2 is a serum bicarbonate determinant and is frequently called *CO_2 combining power.* It refers to the amount of cations, e.g., H^+, Na^+, K^+ and so on, available to combine with HCO_3. The level of HCO_3 in the blood is determined by the amount of CO_2 dissolved in the blood.

Study the diagram carefully and refer to the diagram or this introduction as needed. Refer to Diagram 7 for further clarification if necessary.

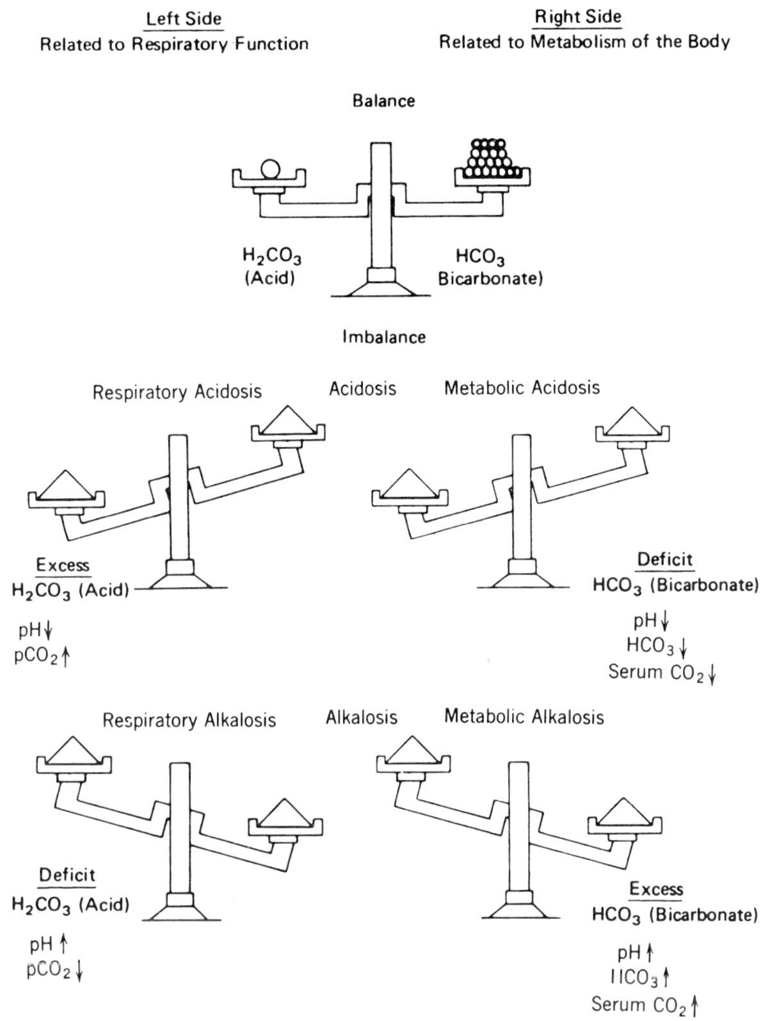

Diagram 8. Acid-Base Balance and Imbalance.

---

43 In acid-base balance the left side or the carbonic acid side is controlled by:
( ) a. respiratory function
( ) b. body metabolism
The right side or the base bicarbonate is controlled by:
( ) a. respiratory function
( ) b. body metabolism

— — — — — — — — — — — — — — — —

a. X; b. –; a. –; b. X

---

44 With respiratory acidosis, there is * ____. With respiratory alkalosis, there is * ____. With metabolic acidosis, there is * ____. With metabolic alkalosis, there is * ____.

\- - - - - - - - - - - - - - - - - -

carbonic acid excess, also pH↓, P_{CO_2}↑;
carbonic acid deficit, also pH↑, P_{CO_2}↓;
base bicarbonate deficit, also pH↓, HCO_3↓, serum CO_2↓;
base bicarbonate excess, also pH↑, HCO_3↑, serum CO_2↑

---

45 Refer to Introduction to Diagram 8 if necessary.

To determine acidotic and alkalotic states, the nurse must first assess ________; second, __________ of arterial blood; and third, ____________ of arterial blood or ____________ of venous blood.

\- - - - - - - - - - - - - - - - - -

pH; P_{CO_2}; HCO_3; serum CO_2

---

46 Refer to Introduction to Diagram 8 if necessary.

The normal range for pH in blood is * ____________________. The normal range for P_{CO_2} in arterial blood is * ____________________. The normal range of HCO_3 in arterial blood is * ____________________. The normal range of serum CO_2 in venous blood is * ____________________.

\- - - - - - - - - - - - - - - - - -

7.35–7.45; 36–44 mm Hg; 24–28 mEq/L; 22–32 mEq/l

---

47 Respiratory acidosis and alkalosis are determined by two of the following:
( ) a. pH
( ) b. P_{CO_2}
( ) c. HCO_3

\- - - - - - - - - - - - - - - - - -

a. X; b. X; c. –

---

48 Metabolic acidosis and alkalosis are determined by two of the following:

( ) a. pH
( ) b. P_{CO_2}
( ) c. HCO_3

\- - - - - - - - - - - - - - - - -

a. X; b. –; c. X

---

49 Place R. Ac for respiratory acidosis, R. Al for respiratory alkalosis, M. Ac for metabolic acidosis, and M. Al for metabolic alkalosis beside the following laboratory determinants.

__________, __________ a. pH↑

__________, __________ b. pH↓

__________ c. P_{CO_2}↑

__________ d. HCO_3↑

__________ e. P_{CO_2}↓

__________ f. HCO_3↓

\- - - - - - - - - - - - - - - - -

a. R. Al, M. Al;
b. R. Ac, M. Ac;
c. R. Ac;
d. M. Al;
e. R. Al;
f. M. Ac

---

50 There are compensatory reactions in response to metabolic acidosis and alkalosis, and respiratory acidosis and alkalosis. The pH returns to normal or close to normal by changing the component, e.g., P_{CO_2} or HCO_3, that originally was not affected.

The respiratory system can compensate for metabolic acidosis and alkalosis.

For metabolic acidosis, the lungs (stimulated by the respiratory center) will hyperventilate to decrease CO_2.

A pH of 7.34, P_{CO_2} of 24, and HCO_3 of 15 indicate metabolic acidosis, since the pH is slightly acid and the HCO_3 is definitely low (acidosis). The P_{CO_2} should be normal (36-44); however, it is low since the respiratory center compensates for the acidotic state by "blowing off" CO_2 (hyperventilating). Without compensation, the pH would be extremely low, e.g., pH 7.2.

For metabolic alkalosis, the lungs will ____________________ to conserve ____________.

With a pH of 7.48, P_{CO_2} of 45, and HCO_3 of 39, the pH and HCO_3 indicate

* ______________________________________.

The P_{CO_2} indicates respiratory compensation. Explain. * ____________________

____.

— — — — — — — — — — — — — — — — —

hypoventilate; CO_2;
metabolic alkalosis;
The lungs compensate for the alkalotic state by conserving CO_2

---

51 With respiratory acidosis, the kidneys will excrete more acid, H^+, and conserve HCO_3^-.

With pH of 7.35, P_{CO_2} of 68, and HCO_3 of 35, the pH is low normal, borderline on acidosis, and the P_{CO_2} is highly elevated, indicating CO_2 retention—respiratory acidosis. The HCO_3 indicates * ____________________

____.

— — — — — — — — — — — — — — — — —

kidney (renal) compensation. Without compensation the pH would be very low

---

---

52 With respiratory alkalosis, the kidneys will excrete ______________ and conserve ______________.

With a pH of 7.46, P_{CO_2} of 20, and HCO_3 of 22, the pH and P_{CO_2} indicate * ______________________________. The HCO_3 indicates renal compensation. Explain how. * ____

____.

– – – – – – – – – – – – – – – – –

bicarbonate (HCO_3^-); acid (H^+);
respiratory alkalosis;
The kidneys compensate for the alkalotic state by excreting HCO_3

---

53 When the body is in a state of metabolic acidosis, an excess of nonvolatile acid is retained in body fluids.

What do you suppose occurs in metabolic alkalosis * ______________________________

____.

How is a nonvolatile acid excreted from the body? ______________.

– – – – – – – – – – – – – – – – –

a decrease in nonvolatile acid in body fluids
urine (water)

---

54 With respiratory acidosis, there is an excess of volatile acid. In what form do you think a volatile acid is excreted from the body? ______________________.

With respiratory alkalosis, would there be a(n) (increase/decrease) in volatile acid? ______________.

– – – – – – – – – – – – – – – – –

gas; decrease

CASE REVIEW

Mr. Swift, age 46, has a history of respiratory problems. His latest problem was pneumonia. He smokes 2 to 3 packs of cigarettes a day. His condition indicates an acid-base imbalance.

1. The acidity or alkalinity of a solution depends on the concentration of the
   * ____.

2. With buffering, a strong acid is replaced by a weak acid and a neutral salt; therefore, fewer ________________ ions will be released.
3. With respiratory regulation, for acid-base balance the lungs blow off or conserve CO_2 by ________________ or ________________.
4. The kidney maintains acid-base balance by excreting ________________ or ________________ and by retaining ________________ or ________________.
5. Which acts faster in regulating or correcting acid-base imbalance (kidneys/lungs)?
6. Respiratory acidosis has a carbonic acid (excess/deficit), whereas respiratory alkalosis has a carbonic acid (excess/deficit).

   Mr. Swift's blood gases had a pH of 7.29, P_{CO_2} of 54, and HCO_3 of 25.

7. Mr. Swift's pH and P_{CO_2} indicate *________________.
8. Is there any renal (kidney) compensation? ________________.

   Two days later Mr. Swift's blood gases had a pH of 7.34, P_{CO_2} of 62, and HCO_3 of 29.

9. Mr. Swift has been (hyperventilating/hypoventilating)? ________________.
10. The nurse assesses Mr.Swift's acid-base imbalance as *________________ ________________. Explain why. *________________ ________________.
11. Would the kidneys be compensating for the imbalance? ________________ Explain how. *________________.

- - - - - - - - - - - - - - - - -

1. hydrogen ion
2. hydrogen
3. hyperventilating; hypoventilating
4. H^+; HCO_3^-; HCO_3^-; H^+
5. lungs
6. excess; deficit
7. respiratory acidosis
8. No
9. hypoventilating. It causes CO_2 retention—respiratory acidosis
10. respiratory acidosis. The pH is slightly acidotic and the P_{CO_2} is elevated, indicating respiratory imbalance and carbonic acid excess
11. Yes. Kidneys are conserving bicarbonate

NURSING DIAGNOSES

Potential for ineffective airway clearance related to thick bronchial secretion secondary to respiratory disorders; e.g., pneumonia.
Potential for ineffective breathing patterns related to inadequate gas exchange.
Potential activity intolerance related to breathlessness.
Potential for self-care deficit (hygiene and grooming) related to breathing difficulty.

NURSING ACTIONS

1. Provide adequate fluid intake to decrease tenacious secretions.
2. Encourage patient to deep-breathe and cough. This helps to eliminate bronchial secretions and improve gas exchange.
3. Assist patient with self-care.
4. Recognize the type of acid-base imbalance occurring in the patient. A pH < 7.35 is indicative of acidosis and a pH > 7.45 is indicative of alkalosis.
5. Determine the type of acidosis. With respiratory acidosis, the pH is decreased and the PCO_2 is increased. With metabolic acidosis, the pH is decreased and the HCO_3 is decreased.
6. Determine the type of alkalosis. With respiratory alkalosis, the pH is increased and the PCO_3 is decreased. With metabolic alkalosis, the pH is increased and the HCO_3 is increased.
7. Check the serum CO_2 to determine which metabolic acid-base imbalance is present. Normal serum CO_2 is 22–32 mEq/L. With metabolic acidosis, the serum CO_2 is < 22 mEq/L and with metabolic alkalosis, the serum $CO_2 > 32$ mEq/L.
8. Interpret the arterial blood gas report and determine if there is respiratory or renal compensation. The pH determines acidosis or alkalosis. The PCO_2 and HCO_3 relate to the pH. If pH is decreased, PCO_2 increased, and HCO_3 slightly increased, the acid-base imbalance is respiratory acidosis with metabolic (renal) compensation.

METABOLIC ACIDOSIS AND ALKALOSIS

Table 33 gives various clinical conditions that can cause metabolic acidosis and metabolic alkalosis. Study this table carefully until you understand and can state the reasons these conditions can cause acidosis or alkalosis, then proceed to the frames. Refer to this table as needed.

TABLE 33. CONDITIONS CAUSING METABOLIC ACIDOSIS AND METABOLIC ALKALOSIS

Conditions	Metabolic Acidosis	Metabolic Alkalosis
Uncontrolled diabetes mellitus (diabetic acidosis)	With failure to metabolize adequate quantities of glucose, liver will increase metabolism of fatty acids. Oxidation of fatty acids produces ketone bodies. H^+ circulate with ketones, which make blood more acid. Every ketone requires a base for excretion.	

TABLE 33. (Continued)

Conditions	Metabolic Acidosis	Metabolic Alkalosis
Severe diarrhea	Loss of sodium ions exceeds that of chloride ions. Cl^- combines with H^+, producing strong acid. Kidney mechanisms for conserving sodium and water and for excreting H^+ fail.	
Starvation, kidney failure, excessive exercise, severe infections	Acid metabolites, e.g., lactic acids and others, occur as result of abnormal accumulation of acid products from metabolism or cellular breakdown (from starvation). These acids neutralize alkali, and excess acid then causes balance to tilt left.	
Stomach ulcer (peptic ulcer)		Excess of alkali in ECF occurs when patient takes excessive amounts of baking soda, $NaHCO_3$, or acid neutralizers to ease ulcer pain.
Vomiting, gastric suction, mercurial diuretics		Large quantities of chloride are lost. Bicarbonate anions increase to compensate for chloride loss. Number of anions must be equal to number of cations.
Loss of potassium		Loss of potassium from body is accompanied by loss of chloride (as stated in Chapter 2 on chlorides).

55 With uncontrolled diabetes mellitus, glucose cannot be metabolized; therefore, what will occur? * ______________________________

______________________________.

- - - - - - - - - - - - - - - - -

The liver will produce fatty acids, and later ketone bodies will be produced

56 What makes the blood more acid in diabetic acidosis? * ______________________________

______________________________.

Why is there a base deficit? * ______________________________

______________________________.

- - - - - - - - - - - - - - - - -

H^+ circulating with the ketone bodies, which are also acid
The base bicarbonate is excreted with the ketone bodies

57 In severe diarrhea, explain which ion is excreted in excess? * ______________.

Explain how this can cause an acidotic state. * ______________________________

____.

— — — — — — — — — — — — — — — — —

sodium ion
The Cl^- is retained and combines with the H^+, producing HCl

58 How can starvation contribute to metabolic acidosis? * ______________________

_________________________________.

— — — — — — — — — — — — — — — — —

Acid metabolites resulting from cellular breakdown

59 How can excessive exercise, kidney failure, and severe infections contribute to metabolic acidosis? * ____

_________________________________.

— — — — — — — — — — — — — — — — —

An abnormal accumulation of acid products from metabolism

60 Indicate which of the following conditions can cause metabolic acidosis—a deficit of alkali or base bicarbonate.

( ) a. Starvation
( ) b. Gastric suction
( ) c. Excessive exercise
( ) d. Mercurial diuretics
( ) e. Severe infections
( ) f. Uncontrolled diabetes mellitus

— — — — — — — — — — — — — — — — —

a. X; b. –; c. X; d. –; e. X; f. X

---

61 *Anion gap* is a useful indicator for determining the presence or absence of metabolic acidosis.

Anion gap can be obtained by the following:

1. Adding the serum chloride/Cl and serum CO_2 values;
2. Subtracting the sum of serums Cl and CO_2 from the serum sodium/Na value. The difference is the anion gap.

If the anion gap is >16 mEq/L metabolic acidosis is suspected. The acid in the body would be stronger than the carbonic acid.

What type of acid-base imbalance is indicated by an anion gap >16 mEq/L:

( ) a. Metabolic acidosis
( ) b. Metabolic alkalosis
( ) c. Respiratory acidosis

\- - - - - - - - - - - - - - - - -

a. X; b. –; c. –

---

62 The patient's serum values are Na 142 mEq/L, Cl 102 mEq/L, and CO_2 18 mEq/L.

The anion gap is ______________________.

Is metabolic acidosis present? ______________ Why? * ______________

____.

\- - - - - - - - - - - - - - - - -

142 – 120 = 22 mEq/L anion gap. Yes. The anion gap is greater than 16 mEq/L

---

63 Conditions associated with an anion gap >16 mEq/L are diabetic ketoacidosis, lactic acidosis, poisoning, and renal failure.

Indicate the conditions that might apply to an anion gap of 25 mEq/L.

( ) a. diabetic ketoacidosis
( ) b. COPD
( ) c. respiratory failure
( ) d. renal failure
( ) e. poisoning
( ) f. lactic acidosis

\- - - - - - - - - - - - - - - - -

a. X; b. –; c. –; d. X; e. X; f. X

---

64 When a patient takes excessive amounts of baking soda or commercially prepared acid neutralizers to ease indigestion or stomach ulcer pain, what will probably occur? * ____________________. Why? * ______________________________
____.

— — — — — — — — — — — — — — — — —

metabolic alkalosis
There is excess alkali in the extracellular fluid

---

65 Name the anion that is lost in great quantities due to vomiting, gastric suction, or mercurial diuretics. ______________________.

— — — — — — — — — — — — — — — — —

chloride

---

66 Name the conditions that cause metabolic alkalosis.

1. * ______________________________________.
2. ______________________________________.
3. * ______________________________________.
4. * ______________________________________.
5. * ______________________________________.

— — — — — — — — — — — — — — — — —

1. treated stomach ulcer
2. vomiting
3. gastric suction
4. mercurial diuretics
5. loss of potassium

67 Place BE for base bicarbonate excess or metabolic alkalosis and BD for base bicarbonate deficit or metabolic acidosis beside the following:

____ a. Uncontrolled diabetes mellitus

____ b. Treated stomach ulcer

____ c. Severe diarrhea

____ d. Severe infection

____ e. Vomiting, gastric suction

____ f. Mercurial diuretics

____ g. Fever, severe infections

____ h. Excessive exercise

- - - - - - - - - - - - - - - - -

a. BD; b. BE; c. BD; d. BD; e. BE; f. BE; g. BD; h. BD

RESPIRATORY ACIDOSIS AND ALKALOSIS

Table 34 gives the clinical conditions and causes of respiratory acidosis and respiratory alkalosis. Study this table carefully and be prepared to explain why respiratory acidosis or alkalosis would occur with these conditions or situations. Proceed to the frames that follow and refer to the table whenever necessary.

TABLE 34. CONDITIONS CAUSING RESPIRATORY ACIDOSIS AND RESPIRATORY ALKALOSIS

Conditions	Respiratory Acidosis	Respiratory Alkalosis
Central nervous system depressants 1. Narcotics*	These drugs depress respiratory center in medulla causing retention of CO_2 (carbon dioxide) in the blood.	
a. Morphine	These produce increase in carbonic acid in blood plasma.	
b. Meperidine hydrochloride (trade name is Demerol) 2. Anesthetics* 3. Barbiturates* (in large doses)		
Pulmonary dysfunctions 1. Emphysema 2. Bronchiectasis 3. Pneumonia 4. Poliomyelitis	Inadequate exchange of gases in lungs due to decrease in aeration surface causes retention of CO_2 in the blood.	

TABLE 34. (Continued)

Conditions	Respiratory Acidosis	Respiratory Alkalosis
Hyperventilation 1. Anxiety, hysteria 2. Drug toxicity, aspirin 3. Diseases, tetany 4. Fever 5. Swimming 6. Deliberate overbreathing		Excessive blowing off of CO_2 through lungs reduces amount of carbonic acid in extracellular fluid.

Note: *Existing when taken in large quantities, or over an extended time, or both.

---

68 Explain how an inadequate exchange of gases in the lungs can cause respiratory acidosis? *____________________

____________________.

- - - - - - - - - - - - - - - - - -

It can cause a retention of CO_2 in the blood. H_2O and $CO_2 \rightarrow H_2CO_3$

---

69 Which of the following will depress the respiratory center, causing carbon dioxide retention?

( ) a. Morphine
( ) b. Demerol
( ) c. Hyperventilation
( ) d. Anesthesia
( ) e. Barbiturates
( ) f. Aspirin

- - - - - - - - - - - - - - - - - -

a. X; b. X; c. –; d. X; e. X; f. –

---

70 What four conditions can lead to respiratory acidosis due to a decrease in aeration surface in the lung?

1. ____________________.
2. ____________________.
3. ____________________.
4. ____________________.

- - - - - - - - - - - - - - - - - -

1. emphysema; 2. pneumonia; 3. bronchiectasis; 4. poliomyelitis

---

71 What is respiratory alkalosis? * ______________________________

____

______________________________.

- - - - - - - - - - - - - - - - - -

Blowing off CO_2, or carbonic acid deficit, or both; $PCO_2\downarrow$ due to increased breathing (hyperventilating)

---

72 Hyperventilation can result from:

1. * ______________________________.
2. * ______________________________.
3. * ______________________________.
4. ______________________________.
5. ______________________________.
6. * ______________________________.

- - - - - - - - - - - - - - - - - -

1. anxiety, hysteria
2. drug toxicity, aspirin
3. diseases, tetany
4. fever
5. swimming
6. deliberate overbreathing

---

73 Explain the difference between respiratory alkalosis and respiratory acidosis.

* ______________________________

______________________________.

- - - - - - - - - - - - - - - - - -

Respiratory alkalosis $\downarrow PCO_2$, carbonic acid deficit
Respiratory acidosis $\uparrow PCO_2$, carbonic acid excess

---

74 List the conditions or situations that may cause respiratory acidosis and respiratory alkalosis.

Respiratory Acidosis	*Respiratory Alkalosis*
1. ________.	1. ________.
2. ________.	2. ________.
3. ________.	3. ________.
4. ________.	4. ________.
5. ________.	5. ________.
6. ________.	6. ________.
7. ________.	

- - - - - - - - - - - - - - - - -

Respiratory Acidosis	*Respiratory Alkalosis*
1. narcotics	1. anxiety
2. anesthetics	2. drug toxicity
3. barbiturates	3. tetany
4. emphysema	4. fever
5. bronchiectasis	5. swimming
6. pneumonia	6. overbreathing
7. poliomyelitis	

CLINICAL APPLICATIONS

Table 35 lists selected symptoms of metabolic acidosis and alkalosis, and respiratory acidosis and alkalosis. Study these symptoms carefully. Refer to the glossary for any unknown words. Refer to this table as you find it necessary.

You may find it necessary to refer to Diagram 7 for clarification of pH changes, Diagram 8 for clarification of serum CO_2, and Table 23 for the regulatory mechanisms for compensation of acid-base imbalance.

TABLE 35. SELECTED SYMPTOMS OF METABOLIC ACIDOSIS AND ALKALOSIS, RESPIRATORY ACIDOSIS AND ALKALOSIS

Metabolic Acidosis	Metabolic Alkalosis
Kussmaul breathing—hyperactive, abnormally vigorous breathing	Shallow breathing
Flushing of skin—capillary in dilation due to CO_2 accumulation in tissues	Vomiting—loss of Cl^- and loss of K^+
Dehydration—water loss via kidney and later through vomiting	pH↑ (increased)—when compensatory mechanisms fail
Restlessness	HCO_3 and serum CO_2↑

TABLE 35. (Continued)

Metabolic Acidosis	Metabolic Alkalosis
pH↓ (decreased)—when compensatory mechanisms fail	
HCO_3 and serum CO_2↓	
Death—from renal and respiratory failure	
Respiratory Acidosis	**Respiratory Alkalosis**
Dyspnea—labored or difficult breathing (hypoventilation)	Overbreathing (hyperventilation) with rapid and shallow breathing due to H^+ or CO_2 deficit
Disorientation	Vertigo (dizziness)—overbreathing
pH↓ decreased—when compensatory mechanisms fail	Tetany spasms
PCO_2↑	Unconsciousness—later
	pH↑ increased—when compensatory mechanisms fail
	PCO_2↓

75 Metabolic acidosis results from a * ______________________.

In metabolic acidosis, the HCO_3 would be (decreased/increased) ____________

and the serum CO_2 would be (decreased/increased)? ______________.

— — — — — — — — — — — — — — — —

base bicarbonate deficit; decreased; decreased

---

76 The common symptoms of metabolic acidosis are:

1. * ______________________.
2. ______________________.
3. * ______________________.
4. ______________________.

— — — — — — — — — — — — — — — —

1. hyperactive, vigorous breathing
2. dehydration
3. flushing of the skin
4. restlessness

---

77 With metabolic acidosis, the renal and respiratory mechanisms try to reestablish balance.

Explain how the renal mechanism works to reestablish balance.

* ____.

Explain how the respiratory mechanism works to reestablish balance.

* ____.

When these two mechanisms fail, what happens to the plasma pH?

* ________________________________.

- - - - - - - - - - - - - - - - -

The H^+ exchanges for the Na^+ and thus H^+ is excreted
As the result of the Kussmaul breathing, CO_2 is blown off. $H_2O + CO_2 \rightarrow H_2CO_3$
It decreases

---

78 Metabolic alkalosis results from a * ______________________________.

In metabolic alkalosis, the HCO_3 would be (decreased/increased) ____________

and the serum CO_2 would be (decreased/increased) ______________________.

- - - - - - - - - - - - - - - - -

base bicarbonate excess; increased; increased

---

79 The common symptoms of metabolic alkalosis are:

1. * ____________________.
2. ____________________.

- - - - - - - - - - - - - - - - -

1. shallow breathing
2. vomiting

80 With metabolic alkalosis, the buffer, renal, and respiratory mechanisms will try to reestablish balance. With the buffer mechanism, the excess bicarbonate reacts with buffer acid salts; thus, there will be a decrease of bicarbonate in the extracellular fluid and an increase of carbonic acid.

Explain how the renal mechanism works to reestablish balance.

*____.

Explain how the respiratory mechanism works to reestablish balance.

*____.

When these three mechanisms fail, what happens to the plasma pH?

*____________________.

- - - - - - - - - - - - - - - - - -

H^+ is conserved and Na^+ and K^+ are excreted with HCO_3
Pulmonary ventilation is decreased; therefore, CO_2 is retained, producing H_2CO_3
It increases

---

81 Respiratory acidosis results from a *____________________. What happens to the P_{CO_2}? *____________________.

- - - - - - - - - - - - - - - - - -

carbonic acid excess. It increases

---

82 The common symptoms of respiratory acidosis are:

1. *____.
2. ____________________.

- - - - - - - - - - - - - - - - - -

1. dyspnea or hypoventilation
2. disorientation

---

83 With respiratory acidosis, the buffer, renal, and respiratory mechanisms will try to reestablish balance. As the result of the chloride shift, bicarbonate ions are released to neutralize carbonic acid excess.

With an increased CO_2, explain how the respiratory mechanism works to compensate for this imbalance. *________________________________

________________________________.

Explain how the renal mechanism works to compensate for this imbalance.

a. *________________________________;

b. *________________________________.

When these three mechanisms fail, what happens to the plasma pH?

*________________________________.

- - - - - - - - - - - - - - - - - -

CO_2 will stimulate the respiratory center to increase the rate and depth of respiration. CO_2 is blown off with water. $H_2O + CO_2 \rightarrow H_2CO_3$

a. The H^+ exchanges with the Na^+, and the Na^+ is reabosrbed with the HCO_3

b. An increased secretion of ammonium chloride

It is decreased

---

84 Respiratory alkalosis results from a *________________________________.

What happens to the Pco_2? *________________________________.

- - - - - - - - - - - - - - - - - -

carbonic acid deficit. It decreases

---

85 The common symptoms of respiratory alkalosis are:

1. *________________________________.

2. ________________________________.

3. *________________________________.

4. ________________________________.

- - - - - - - - - - - - - - - - - -

1. rapid and shallow breathing
2. vertigo
3. tetany spasms
4. unconsciousness

---

86 Frequently, with respiratory alkalosis, you will notice that sufferers are very apprehensive and anxious. They will hyperventilate to overcome their anxiety. Many times this occurs for a psychological reason, e.g., giving a speech for the first time, or fear of failing an exam. How do you think you might help the respiratory mechanism compensate for this imbalance? * ______________________________.

- - - - - - - - - - - - - - - - - -

There is a lack of CO_2; so giving CO_2, e.g., rebreathing CO_2 from a paper bag, would stimulate the lungs to breathe more deeply and then more slowly

---

87 The buffer mechanism produces more organic acids, in respiratory alkalosis, which react with the excess bicarbonate ion.

How do you think the renal mechanism works to compensate for this imbalance? * ______________________________.

When these three mechanisms fail, what happens to the plasma pH?

* ______________________________.

- - - - - - - - - - - - - - - - - - -

An increased HCO_3 excretion and a H^+ retention. It increases

85 Place M. Ac for metabolic acidosis, M. Al for metabolic alkalosis, R. Ac for respiratory acidosis, and R. Al for respiratory alkalosis beside the following symptoms:

_______ a. Hyperactive, vigorous breathing or Kussmaul breathing

_______ b. Dyspnea

_______ c. Shallow breathing

_______ d. Rapid, shallow breathing (overbreathing)

_______ e. Flushing of the skin

_______ f. Dehydration

_______ g. Vomiting

_______ h. Vertigo

_______ i. Tetany spasms

_______ j. Disorientation

- - - - - - - - - - - - - - - - -

a. M. Ac	f. M. Ac
b. R. Ac	g. M. Al
c. M. Al	h. R. Al
d. R. Al	i. R. Al
e. M. Ac	j. R. Ac

CLINICAL MANAGEMENT

Diagram 9 gives the four acid-base imbalances, explains the body's defense action against these imbalances, and gives various methods of treatment for restoring balance. This diagram describes briefly the respiratory and renal mechanisms for restoring balance in respect to the treatment given. The diagram should aid in reinforcing your understanding of these two mechanisms used for our body's defense action.

Study this diagram carefully, with particular attention to the cause of each imbalance, the body's defense action, the pH of the urine as to whether it is acid or alkaline, and the treatment for these imbalances. Refer to the diagram whenever you find it necessary.

Metabolic Acidosis (Deficit of bicarbonate in the extracellular fluid)

Lungs | Kidney

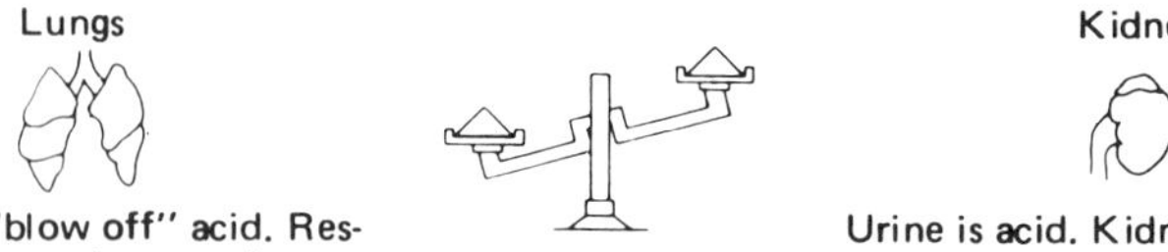

Lungs "blow off" acid. Respirations are increased.

Urine is acid. Kidneys conserve alkali and excrete acid.

Treatment: Remove the cause.* Administer an IV alkali solution, e.g., sodium bicarbonate or sodium lactate. Restore water, electrolytes, and nutrients.

Metabolic Alkalosis (Excess of bicarbonate in the extracellular fluid)

Lungs | Kidney

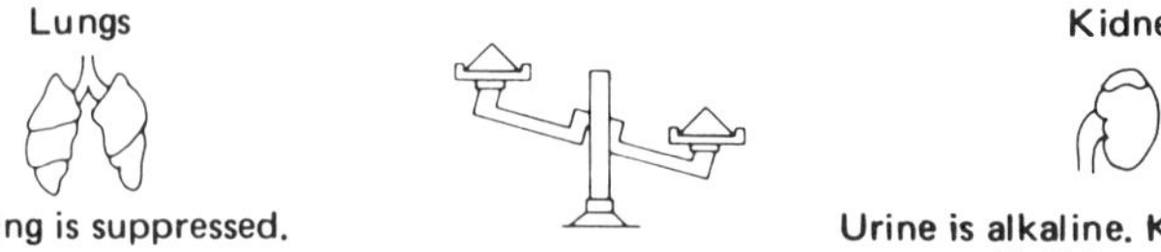

Breathing is suppressed.

Urine is alkaline. Kidneys excrete alkali ions, and retain hydrogen ions and nonbicarbonate ions.

Treatment: Remove the cause.* Administer an IV solution of chloride, e.g., sodium chloride. Replace potassium deficit.

Respiratory Acidosis (Excess of carbonic acid in the extracellular fluid)

Kidney

Lungs are affected. However, H_2CO_3 will stimulate the lungs to "blow off" acid, by breathing deeper.

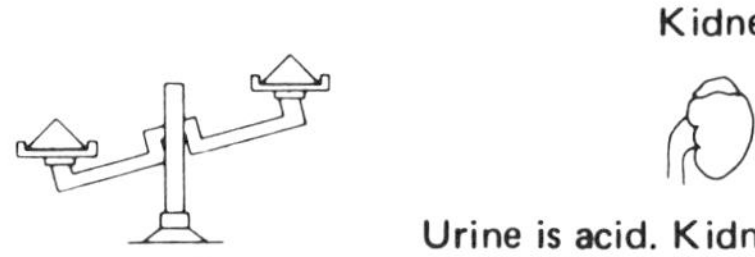

Urine is acid. Kidneys conserve alkali; excrete acid.

Treatment: Remove the cause.** Administer an IV alkali solution. Deep breathing exercise or use of a ventilator, e.g., MAI (Bennett).

Respiratory Alkalosis (Deficit of carbonic acid in the extracellular fluid)

Kidney

Lungs are affected. Treatment would be recommended.

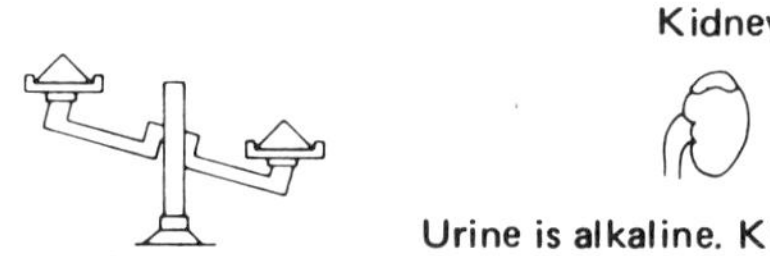

Urine is alkaline. Kidneys excrete alkali; retain acid.

Treatment: Remove the cause.** Rebreathe expired air, e.g., CO_2, from a paper bag. Anti-anxiety drugs, e.g., Valium (diazepam), Librium (chlordiazepoxide)

Diagram 9. Body's Defense Action and Treatment for Acid-Base Imbalance.

*Examples on Table 33.

**Examples on Table 34.

---

89 What is metabolic acidosis? * ______________________________.

Would the urine be (acid/alkaline)? ____________________.

What are the body's defense actions against it?

1. * ______________________________.
2. * ______________________________.

- - - - - - - - - - - - - - - - -

base bicarbonate deficit; acid

1. lungs blow off CO_2 or acid
2. kidneys excrete acid or H^+ and conserve alkali

---

90 How is metabolic acidosis treated?

1. * ____________________.
2. * ______________________________________.
3. * ______________________________.

- - - - - - - - - - - - - - - - -

1. remove cause
2. administer IV alkali solution, e.g., $NaHCO_3$
3. restore H_2O and electrolyte

---

91 What is metabolic alkalosis? * ______________________________.

Would the urine be (acid/alkaline)? ____________________.

What are the body's defense actions against it?

1. * ______________________________.
2. * ______________________________________

______________________________.

- - - - - - - - - - - - - - - - -

base bicarbonate excess; alkaline

1. breathing is suppressed
2. kidneys excrete alkali ions, e.g., HCO_3, and retain H^+ and nonbicarbonate ions

---

92 How is metabolic alkalosis treated?

1. *____________________.
2. *__________________________________.
3. *____________________________.

\- - - - - - - - - - - - - - - - - -

1. remove cause
2. administer IV chloride solution, e.g., NaCl
3. replace K deficit

---

93 What is respiratory acidosis? *________________________

Would the urine be (acid/alkaline)? ________________.
What are the body's defense actions against it?

1. *__________________________________.
2. *__________________________________.

\- - - - - - - - - - - - - - - - - -

carbonic acid excess; acid
1. Excess CO_2 stimulates the lung to blow off CO_2 or acid
2. Kidneys conserve alkali and excrete H^+ or acid urine

---

94 How is respiratory acidosis treated?

1. *____________________.
2. *__________________________________.
3. *__________________________________.

\- - - - - - - - - - - - - - - - - -

1. remove cause
2. deep breathing exercise
3. ventilator, e.g., MA 1 (Bennett)

95 What is respiratory alkalosis? * ______________________________.

Would the urine be (acid/alkaline)? ______________________.

What is the body's defense action against it? * ______________________

____.

- - - - - - - - - - - - - - - - -

carbonic acid deficit; alkaline. Kidneys excrete alkaline HCO_3, and retain acid, H^+

---

96 How is respiratory alkalosis treated?

1. * ______________________________.
2. * ____.
3. * ____.

- - - - - - - - - - - - - - - - -

1. remove cause
2. rebreathe expired air
3. antianxietydrugs, e.g., chlordiazepoxide (Librium) and diazepam (Valium)

CASE REVIEW

Mrs. Brush, age 56, had chronic renal disease. Her respirations were rapid and vigorous. She was restless. Her urine pH was 4.5 and urine output was decreased. Her laboratory results were pH of 7.2, P_{CO_2} of 38, and HCO_3 of 14.

1. The "normal" extracellular level of pH is * ____________. The average range of P_{CO_2} is * ____________, and HCO_3 is * ____________________.
Refer to Introduction to Diagram 8 if needed.)
2. According to Mrs. Brush's pH and HCO_3, her acid-base imbalance was
* ______________________________.
3. Was there effective respiratory compensation? ____________.
4 Give two of her symptoms that related to her acid-base imbalance.
a. * ______________________________________.
b. ______________________________________.
5 Would the imbalance be the result of:
( ) a. base bicarbonate excess
( ) b. base bicarbonate deficit

( ) c. carbonic acid excess
( ) d. carbonic acid deficit

6. How were Mrs. Brush's lungs and kidneys compensating for the acid-base imbalance? *____________________ and *____________________.
7. Her chronic renal disease (failure) could cause the acid-base imbalance due to *____________________.

Later Mrs. Brush's pH was 7.34, P_{CO_2} was 31, and HCO_3 was 20. Fluid with sodium bicarbonate was given IV. As a nurse, you would reassess her laboratory findings.

8. Her pH and HCO_3 indicated *____________________.
9. Was there effective respiratory compensation? __________. Explain how. *____________________.
10. Why was IV administration of fluid with sodium bicarbonate performed? *____________________.
11. Complete the following chart on acid-base imbalance as to pH, P_{CO_2}, and HCO_3. Use the arrow pointed upward for increase, the arrow pointed downward for decrease, and – for not involved (except with compensation).

Metabolic Acidosis

pH
P_{CO_2}
HCO_3 or serum CO_2

Metabolic Alkalosis

pH
P_{CO_2}
HCO_3 or serum CO_2

Respiratory Acidosis

pH
P_{CO_2}
HCO_3 or serum CO_2

Respiratory Alkalosis

pH
P_{CO_2}
HCO_3 or serum CO_2

– – – – – – – – – – – – – – – – –

1. 7.35–7.45; 36–44 mmHg; 24–28 mEq/L
2. metabolic acidosis
3. no
4. a. rapid, vigorous breathing
   b. restlessness
5. b. base bicarbonate deficit
6. rapid, vigorous breathing; excreting acid urine
7. retention or build-up of acid metabolites
8. metabolic acidosis
9. Yes. Lungs were blowing off CO_2 ($CO_2 + H_2O \rightarrow H_2CO_3$ [acid])

10. Sodium bicarbonate will restore the body's base bicarbonate
11. *Metabolic Acidosis*
    pH↓
    P_{CO_2} —
    HCO_3 or serum CO_2 ↓

    Metabolic Alkalosis
    pH↑
    P_{CO_2} —
    HCO_3 or serum CO_2 ↑

    Respiratory Acidosis
    pH↓
    P_{CO_2} ↑
    HCO_3 or serum CO_2 —

    Respiratory Alkalosis
    pH↑
    P_{CO_2} ↓
    HCO_3 or serum CO_2 —

NURSING DIAGNOSES

Alteration in acid-base balance related to inadequate gas exchange or increase in serum ketones secondary to chronic obstructive lung disease/COLD, respiratory failure, or diabetic ketoacidosis.

Impaired gas exchange related to inadequate ventilation secondary to COLD.

Potential for ineffective airway clearance related to thick bronchial secretions.

Potential activity intolerance related to breathlessness.

Potential for noncompliance to the treatment regime for acid-base imbalance related to inadequate explanation.

Potential for impaired home maintenance management related to inadequate support systems.

NURSING ACTIONS

1. Recognize clinical conditions and problems of metabolic acidosis, i.e., severe diarrhea, uncontrolled diabetes mellitus (diabetic acidosis), kidney failure, severe infections, starvation, and excessive exercise; and of metabolic alkalosis, i.e., peptic ulcer, vomiting, gastric suction, and loss of potassium.
2. Observe for signs and symptoms of metabolic acidosis, i.e., rapid, deep, vigorous breathing (Kussmaul breathing), flushed skin, and restlessness; and of metabolic alkalosis, i.e., shallow breathing, and vomiting.
3. Check the arterial bicarbonate and serum CO_2 levels for metabolic acid-base imbalance. Decreased HCO_3 (< 24 mEq/L) and serum CO_2 (< 22 mEq/L) are indicative of metabolic acidosis, and increased HCO_3 (> 28 mEq/L) and serum CO_2 (> 32 mEq/L) are indicative of metabolic alkalosis.
4. Recognize clinical conditions and problems of respiratory acidosis, i.e., emphysema, bronchiectasis, pneumonia, and CNS depressant drugs (narcotics, anesthetics, and barbiturates), and of respiratory alkalosis, i.e., anxiety, hysteria, fever, aspirin toxicity, deliberate overbreathing.
5. Observe for signs and symptoms of respiratory acidosis, i.e., dyspnea and disorientation; and of respiratory alkalosis, i.e., overbreathing (hyperventilation), dizziness, and later unconsciousness.
6. Check the P_{CO_2} for respiratory acid-base imbalance. An increased P_{CO_2} (> 45 mm Hg) is indicative of respiratory acidosis and a decreased P_{CO_2} (< 35 mm Hg) is indicative of respiratory alkalosis.

7. Teach breathing exercises and postural drainage to patients with chronic obstructive lung disease (COLD). Mucous secretions are trapped in overextended alveoli (air sacs), and breathing exercises and postural drainage help to remove secretions and restore gas exchange (ventilation).
8. Administer chest clapping on COLD patients to break up mucous plugs and secretions in the alveoli.
9. Encourage the patient who is overanxious and hyperventilating to take deep breaths and breathe slowly. Proper breathing will prevent respiratory alkalosis.
10. Assess the patient with diabetes mellitus for signs and symptoms of diabetic acidosis (metabolic acidosis): Kussmaul breathing, sweet-smelling breath, apprehension, urine positive for glucose and ketone bodies, serum $CO_2 < 22$ mEq/L.
11. Assist the patient and family members with home-plan management and call on health professionals for assistance as needed

REFERENCES

Abbott Laboratories: *Fluid and Electrolytes.* North Chicago, 1970, pp 22–25.
Best CH, Taylor NB: *The Physiological Basis of Medical Practice,* ed 9. Baltimore, Williams & Wilkins Co, 1973, pp 4–112–125.
Burgess R: Fluids and electrolytes. *Am J Nurs* 65: 92, 1965.
Cardin S: Acid-base balance in the patient with respiratory disease. *Nursing Clinics of North America* 15(3): 593–601, September 1980.
Cohen S, Boyce B, King TKC: Blood-gas and acid base concepts in respiratory care—programmed instruction. *Am J Nurs* 76(6): 1–30, 1976.
Fischbach F: *A Manual of Laboratory Diagnostic Tests.* Philadelphia, JB Lippincott Co, 1980, pp 659–675.
Guyton AC: *Textbook of Medical Physiology,* ed 6. Philadelphia, WB Saunders Co, 1981.
Jacob SW, Francone CA: *Structure and Function in Man,* ed 3. Philadelphia, WB Saunders Co,
Keyes JL: Blood-gas and blood-gas transport. *Heart and Lung* 3(6): 945–954, 1974.
Keyes JL: Blood-gas analysis and the assessment of acid-base status. *Heart and Lung* 5(2): 247–255, 1976.
Leaf A, Cotran R: *Renal Pathophysiology,* ed 3. New York, Oxford University Press, 1985, pp 108–109.
Luckmann J, Sorensen KC: *Medical-Surgical Nursing,* ed 2. Philadelphia, WB Saunders Co, 1980, pp 171–198.
Metheney NM, Snively WD: *Nurses' Handbook of Fluid Balance,* ed 4. Philadelphia, JB Lippincott Co, 1983 pp 67–86.
Ruch TC, Patton H (ed): *Physiology and Biophysics.* Philadelphia, WB Saunders Co, 1965, pp 899–916.
Sharer JE: Reviewing acid-base balance. *Am J Nurs* 75(6): 980–983, 1975.
Statland H: *Electrolytes in Practice,* ed 3. Philadelphia, JB Lippincott Co, 1963, pp 127–154.
Travenol Laboratories, Inc.: *Guide to Fluid Therapy.* Deerfield, Il., 1970, pp 56–76.
Widmann FK: *Clinical Interpretation of Laboratory Tests,* ed 9. Philadelphia, FA Davis Co, 1983.

CHAPTER FOUR

Intravenous Therapy

BEHAVIORAL OBJECTIVES

After studying this chapter the student will be able to complete the following:

Give the three basic objectives of intravenous therapy.

Write the four main classifications of intravenous (IV) solution and explain the rationale for the use of selected IV fluids.

State the methods applied in the administration of intravenous fluids.

State the uses of selected IV fluids and give examples of each.

State the uses of hyperalimentation and explain the ways of preventing major complications.

Observe clinically the physical changes in a patient when IV fluids are administered.

State the rate of flow of IV fluids according to the type of solution and therapy and the physical condition of the patient.

Calculate the number of drops per minute of IV fluid to be absorbed by the body in a specific time.

Describe the nursing assessment of intravenous therapy in regard to solutions, tubing, needles and catheters, injection site, additives, flow rate, and complications.

INTRODUCTION

The basic objectives of intravenous therapy are to provide the maintenance requirements of fluid and electrolyte, to replace previous losses, and to meet concurrent losses. The healthy person can normally preserve fluid and electrolyte balance, whereas a sick person cannot. When a patient cannot maintain this balance intravenous therapy is required.

The uses of intravenous therapy include the replacement of fluid and electrolytes, the administration of blood and nutrition, and the provision of a medium for administering drugs intravenously. Blood is a fluid made up of approximately 97% intracellular and extracellular fluids. The plasma portion is typically extracellular fluid and the blood cells contain the intracellular fluid. Whole blood is indicated when severe fluid loss due to hemorrhage has occurred and when the body needs the normal blood constituents, as in anemia.

This chapter discusses selected solutions commonly used and the rationale for their use; rate of flow for intravenous administration; hyperalimentation (total parenteral nutrition/TPN); groups of solutions, including blood, with clinical applications; and nursing assessment of the procedure. The assessment, nursing action, and rationale are described in detail and should be useful in monitoring intravenous therapy.

It must be understood that the nurse never administers intravenous fluids without a physician's order. The physician will determine the type and amount of fluid to be given and the time required to administer it. The nurse then computes the hourly requirement in accordance with the physician's order. His or her understanding of intravenous therapy and cognizance of physical changes that can occur can be life-saving to the patient.

Three case reviews, three sets of nursing diagnoses, and three sets of nursing actions are provided.

An asterisk (*) indicates a multiple-word answer. The meanings of the following symbols are ↑ increased, ↓ decreased, > greater than, < less than.

---

1 Refer to the Introduction as needed.

The three basic objectives of administering fluids and electrolytes by intravenous therapy are the following:

1. *______________________________________.
2. *__________________________________.
3. *__________________________________.

- - - - - - - - - - - - - - - - - -

1. to provide maintenance requirements
2. to replace previous losses
3. to meet concurrent losses

---

2 People requiring intravenous therapy depend on this route to meet daily maintenance needs for water, electrolytes, calories, vitamins, and other nutritional substances, and certain medications.

The first objective in administering IV therapy is to *____.

- - - - - - - - - - - - - - - - - -

provide maintenance requirements

---

3 It is necessary for a patient to have sound kidney function while receiving fluids and electrolytes through IV therapy. Renal dysfunction may result in electrolyte (retention/excretion) ________.

The electrolyte that can cause cardiac arrest when given in excessive amounts is ________.

- - - - - - - - - - - - - - - - -

retention; potassium

---

4 Multiple electrolyte solutions are also helpful in accomplishing the second objective, which is * ________

____

- - - - - - - - - - - - - - - - -

To replace previous (past) losses

---

5 Fluid and electrolyte losses that can occur due to diarrhea, vomiting, or gastric suction would be an indication of the third objective for administering fluids and electrolytes through IV therapy, which is * ________

________.

- - - - - - - - - - - - - - - - -

to meet concurrent losses

---

6 To meet present fluid and electrolyte losses, a *replacement solution* with an electrolyte content similar to that of the fluid being lost is given in approximately equal amounts to the volume lost.

This type of IV therapy would be given to * ________

________.

- - - - - - - - - - - - - - - - -

meet concurrent losses

---

7 Successful fluid and electrolyte therapy often depends on satisfying all three basic objectives, which are:

1. *______________________________.

2. *______________________________.

3. *______________________________.

— — — — — — — — — — — — — — — —

1. to provide maintenance requirements
2. to replace previous losses
3. to meet concurrent losses

---

8 The uses of intravenous (IV) therapy are the following:

a. *______________________________.

b. *______________________________.

c. *______________________________.

— — — — — — — — — — — — — — — — —

a. replace fluids and electrolytes
b. administer blood and nutrition
c. provide the medium for administering drugs intravenously

SOLUTIONS FOR INTRAVENOUS THERAPY

Many of the solutions in intravenous/IV therapy are produced commercially to meet specific types of needs. Table 36 is a list of commonly used solutions in intravenous therapy and the rationale for their use.

You will be expected to memorize the four major classifications of solutions: protein solutions, plasma expander solutions, hydrating solutions, and replacement solutions. You will also be expected to memorize the uses for IV fluids as stated under rationale and be able to give examples. However, you are not expected to memorize all the solutions and their caloric and electrolyte contents.

Note the tonicity (osmolality) of the solutions as hypotonic (hypo-osmolar), isotonic (iso-osmolar), or hypertonic (hyperosmolar). The concentration of solutions is determined by the number of osmols in solution; thus, hypo-osmolar, iso-osmolar, and hyperosmolar are the suggested terms. Salt solutions, or NaCl, are frequently referred to as saline. Normal saline is 0.9% NaCl (sodium chloride).

Study the table carefully because we will be referring to it throughout the chapter. Then go on to the frames that follow, referring to Table 36 only when necessary.

TABLE 36. SELECTED SOLUTIONS COMMONLY USED IN INTRAVENOUS THERAPY

Solution	Osmolality (Tonicity)	Caloric Value	mEq/L Na$^+$	mEq/L K$^+$	mEq/L Cl$^-$	Miscellaneous	Rationale for Use of Selected Intravenous Fluids
Protein Solutions							
Aminosol 5%	Iso	175	<10	17	—*	Amino acids	Provides protein and fluid for body. Prevents shock and promotes wound healing.
Aminosol 5% with dextrose 5%	Hyper	345	<10	17	—	Amino acids	Provides protein, calories, and fluid for body; especially helpful for patients who are old and malnourished and for those with hypoproteinemia due to other causes. Not to be used in severe liver damage.
Plasma Expander							
Dextran 40 10% in normal saline (0.9%) or 5% dextrose in water (500 ml bottle)	Hyper						Dextran is a colloidal solution used to increase plasma volume. Dextran 40 is a short-lived plasma volume expander (4 to 6 hours). Useful in early shock by correcting hypovolemia, and increasing arterial pressure, pulse pressure, and cardiac output. It improves microcirculatory flow by reducing red blood cell aggregation in the capillaries (increases small vessel perfusion). Caution: It should not be used for patients who are severely dehydrated, have renal disease, have thrombocytopenia, or are actively hemorrhaging.
Dextran 70 6% in normal saline (0.9%) or 5% dextrose in Water	Hyper						Dextran 70 is long-lived plasma volume expander (20 hours). Useful for shock or impending shock due to hemorrhage, surgery, burns. It can interfere with platelet function, thus cause prolonged bleeding. Blood for type and cross-match should be drawn before starting Dextran; it tends to coat RBC,

TABLE 36. (Continued)

Solution	Osmolality (Tonicity)	Caloric Value	mEq/L Na⁺	K⁺	Ca⁺⁺	Cl⁻	Lactate	Mg⁺⁺	HPO₄⁻⁻	Rationale for Use of Selected Intravenous Fluids
										and blood type is difficult to obtain. Overhydration may occur if oliguria or heart failure is present. Allergic reactions can occur, i.e., nausea, vomiting, dyspnea, wheezing, hypotension.
Hydrating Solution										
Sodium chloride 0.45%	Hypo	—	77		—	77	—			Useful for daily maintenance of body fluid, but is of less value for replacement of NaCl deficit. Helpful for establishing renal function.
*Dextrose 2½% in 0.45% saline	Iso	85	77		—	77	—			Helpful in establishing renal function–urine output.
*Dextrose 5% in 0.2% saline	Iso	170	34		—	34	—			Useful for daily maintenance of body fluids and when less Na and Cl are required.
*Dextrose 5% in 0.33% saline	Hyper	170	56		—	56	—			
*Dextrose 5% in 0.45% saline	Hyper	170	77		—	77	—			Useful for daily maintenance of body fluids and nutrition, and for rehydration.
*Dextrose 5% in water (50 g) (dextrose 10% is occasionally used)	Iso	170	—		—	—	—			Helpful in rehydration and excretory purposes. May cause some sodium loss in urine. Good vehicle for IV potassium.
Replacement Solutions										
Dextrose 5% in saline 0.9%	Hyper	170	154	—	—	154	—	—	—	Replacement of fluid, sodium, chloride, and calories.
Dextrose 10% in saline 0.9%	Hyper	340	154	—	—	154	—	—	—	Replacement of fluid, sodium, chloride, and calories.
*Lactated Ringer's	Iso	9	130	4	3	109	28	—	—	This solution resembles electrolyte structure of normal blood serum–plasma. Potassium present is not sufficient for daily potassium requirement.

Solution	Osmolality (Tonicity)	Caloric Value	mEq/L							Rationale for Use of Selected Intravenous Fluids
			Na^+	K^+	Ca^{++}	Cl^-	Lactate	Mg^{++}	HPO_4^{--}	
*Dextrose 5% in lactated Ringer's	Hyper	179	130	4	3	109	28	—	—	Same contents as lactated Ringer's plus calories.
Ringer's solution	Iso	—	147	4	5	156	—	—	—	Does not contain lactate, which can be harmful to people who cannot metabolize lactic acid.
Dextrose 5% Ringer's solution	Hyper	170	147	4	4	155	—	—	—	Same contents as Ringer's solution plus calories.
M/6 sodium lactate	Iso	55	167	—	—	—	167	—	—	Supplies sodium without chloride. Lactate has some caloric value and is metabolized to CO_2 for excretion or increases bicarbonate in alkalosis.
*Normal saline (0.9%)	Iso	—	154	—	—	154	—	—	—	Restores extracellular fluid volume and replaces sodium chloride deficit.
Hyperosmolar saline 5% NaCl	Hyper	—	855	—	—	855	—	—	—	Helpful in salt-depleted hyponatremia. It raises osmolality of the blood. Helpful in eliminating water excess from cells.
Isolyte E with dextrose 5%	Hyper	170	140	10	5	103		3	—	Replacement of fluid and electrolyte according to electrolyte imbalance.
Ionosol B with dextrose 5%	Hyper	178	57	25	—	49	25	5	13	Useful in treating patients requiring polyionic parenteral replacement, e.g., with alkalosis due to vomiting, diabetic acidosis, and fluid losses due to burns or stress, and postoperative dehydration.
Ionosol D-CM with dextrose 5%	Hyper	186	138	12	5	108	50	3	—	For replacement of electrolyte losses of duodenal fluid through intestinal suction, biliary or pancreatic drainage, and to correct mild acidosis.

Source: Selected portions from Abbott Laboratories: *Wall Chart, Intravenous and Other Solutions*, North Chicago, October 1968, H Statland: *Fluid and Electrolytes in Practice*, Philadelphia, JB Lippincott Co, 1963; McGaw Laboratories: *Guide to Parenteral Fluid Therapy*, CA Glendale, 1963; Travenol Laboratories, Inc.: *Guide to Fluid Therapy*, Deerfield, IN, 1970; N Methney, W Snively: *Nurses' Handbook of Fluid Balance*, Philadelphia, WB Saunders Co, 1983.

---

9 The four main classifications of parenteral solutions are:

1. * ______________________________.
2. * ______________________________.
3. * ______________________________.
4. * ______________________________.

- - - - - - - - - - - - - - - - -

1. protein solution
2. plasma expander
3. hydrating solutions
4. replacement solutions

---

10 Dextran is a colloidal solution that is used to expand the * ______________.

Dextran 40 remains in the body * __________ hours and Dextran 70 remains for __________ hours.

- - - - - - - - - - - - - - - - -

plasma volume; 4 to 6; 20

---

11 Give two ways that Dextran 40 is useful to correct hypovolemia in early shock.

1. * ______________________________.
2. * ______________________________.

How does Dextran 40 improve the microcirculation? * ______________________________

______________________________.

- - - - - - - - - - - - - - - - -

1. increases arterial blood pressure
2. increases cardiac output; also increase pulse pressure

It reduces red blood cell (erythrocyte) aggregation in the capillaries

---

12 Give three conditions in which Dextran 40 would be contraindicated.

1. *____.

2. *____.

3. *____.

- - - - - - - - - - - - - - - - - -

1. Severely dehydrated. Dextran 40 can increase dehydration by pulling more fluid from the cells and tissue spaces to the vascular space. If urine output is good, the vascular fluid will be excreted. Both cellular and extracellular dehydration will occur.
2. Renal disease. If oliguria is due to hypovolemia, Dextran 40 may improve urine output; but if renal damage is present, Dextran 40 may cause renal failure.
3. Thrombocytopenia. Dextran 40 tends to clot platelets, and bleeding time would be increased (prolonged).

Also, active hemorrhaging. Dextran 40 improves microcirculation, which could cause additional blood loss from the capillaries if hemorrhaging is continuous.

---

13 Why should blood be typed and cross-matched before administering Dextran 70?

*____

____.

The nurse should stay for 30 minutes with the patient receiving Dextran 70 to observe for allergic reactions.

Name two allergic reactions.

1. ______________________________.

2. ______________________________.

- - - - - - - - - - - - - - - - - -

Dextran 70 tends to coat RBC, which makes it difficult to type and crossmatch the blood specimen

1. urticaria (hives)
2. wheezing

Others, dyspnea, hypotension, nausea and vomiting. Very good.

---

14 The solution resembling the electrolyte structure of plasma is *_______________

____.

\- - - - - - - - - - - - - - - - -

lactated Ringer's or Ringer's solution

---

15 Hydrating solutions are helpful for daily maintenance of body fluid, rehydration, and in establishing effective renal output.

Indicate which of the following are hydrating solutions:

( ) a. Dextrose 2 1/2% in 0.45% saline
( ) b. Ringer's solution
( ) c. Dextrose 5% in water
( ) d. Sodium chloride 5%
( ) e. Dextrose 5% in 0.45% saline
( ) f. Sodium chloride 0.45%
( ) g. Dextrose 5% in 0.2% saline

\- - - - - - - - - - - - - - - - -

a. X; b. –; c. X; d. –; e. X; f. X; g. X

---

16 Dextrose solutions for intravenous therapy are prepared in two strengths: 5 and 10%. The 5% dextrose means that there are 5 g in 100 ml. If the bag contains 1000 ml of 5% dextrose how many grams would be in this bag? ________.

\- - - - - - - - - - - - - - - - -

Ratio: 5:100::X:1000

$$100X = 5000$$
$$X = 50 \text{ g}$$

Fraction: $\frac{5}{100} \times \frac{X}{1000}$

$$100X = 5000$$
$$X = 50 \text{ g}$$

---

17 Potassium is often administered by diluting an ampule or two of potassium chloride in solution. A good vehicle for this IV potassium is the hydrating solution, *____.

\- - - - - - - - - - - - - - - - -

1000 ml (cc) of 5% dextrose in water

---

18 Replacement solutions are useful for replacing fluid, calories, and electrolyte deficits due to injury or illness.

Indicate which of the following solutions are considered replacement solutions:

( ) a. Sodium chloride 0.45%
( ) b. Dextrose 5% in normal saline
( ) c. Lactated Ringer's
( ) d. Ringer's
( ) e. Dextrose 5% in water
( ) f. M/6 sodium lactate
( ) g. Hypertonic (3% or 5%) saline
( ) h. Dextrose 5% in 0.45% saline
( ) i. Multiple electrolyte or polyionic solutions (Ionosol B or D-CM solutions and Isolyte E).

- - - - - - - - - - - - - - - - -

a. –; b. X; c. X; d. X; e. –; f. X; g. X; h. –; i. X

---

19 The solution that contains sodium and not chloride is *____.

- - - - - - - - - - - - - - - - - -

M/6 sodium lactate

---

20 Hypertonic saline (3% or 5%) is used in treatment of (hypernatremia/hyponatremia) __________________________. Usually it is not administered until the serum Na is below 115 mEq/L.

This solution will raise the __________________________ of the blood.

- -

hyponatremia; osmolality

---

21 The multiple electrolyte solution most helpful in treating severe vomiting, diabetic acidosis, postoperative dehydration, and fluid loss due to burns is *____.

- - - - - - - - - - - - - - - - -

Ionosol B with 5% dextrose or Isolyte E with 5% dextrose (Ionosol B and Isolyte E come without dextrose)

---

22 The multiple electrolyte solution most helpful in replacement of electrolyte losses from the gastrointestinal tract is * ______________________.

Can you explain why? * ______________________

______________________.

— — — — — — — — — — — — — — — — —

Ionosol D-CM
Electrolyte content of the solution is similar to the gastrointestinal juice

---

23 Do you think lactated Ringer's solution would be administered in a case of potassium deficit? Why or why not? * ______________________

______________________.

— — — — — — — — — — — — — — — — —

No! The amount of potassium is not sufficient in replacing potassium deficits

---

24 Name three purposes for the use of hydrating solutions:

1. * ______________________.
2. * ______________________.
3. * ______________________.

— — — — — — — — — — — — — — — — —

1. daily maintenance
2. rehydration
3. effective renal output

HYPERALIMENTATION/TOTAL PARENTERAL NUTRITION (TPN) (HYPEROSMOLAR ALIMENTATION)

Hyperalimentation is the administration of nutrients in concentrated form to maintain metabolism and promote tissue synthesis. For patients who cannot take adequate nutrition by mouth, enteral (gastric or enteric tube feeding) or parenteral hyperalimentation may be required. This section discusses parenteral hyperalimentation, known as total parenteral nutrition (TPN) or intravenous hyperalimentation (IVH).

25 Nutritional solutions for hyperalimentation contains glucose, nitrogen source (protein hydrolysate or crystalline amino acids), vitamins, minerals, and electrolytes.

Would the hyperalimentation solution be (hypo-osmolar/hyperosmolar)?

______________________. Explain what effect the concentration of the solution would have on body cells. *______________________

______________________.

— — — — — — — — — — — — — — — — —

Hyperosmolar. It can pull fluid from the intracellular compartment (cells) into the extracellular compartment, thus causing cellular dehydration.

---

26 The sugar concentration is higher than other hyperosmolar solutions, and if administered via peripheral arm and leg veins, it can cause extensive damage.

Do you know what type of vein damage can occur? *______________________

______________________.

— — — — — — — — — — — — — — — — —

phlebitis or thrombophlebitis (clots). Good.

---

27 Subclavian and internal jugular veins are used when administering hyperosmolar solutions.

Which veins can be used for hyperalimentation?

( ) a. Peripheral arm veins
( ) b. Internal jugular vein
( ) c. Leg veins
( ) d. Subclavian veins

— — — — — — — — — — — — — — — — —

a. –; b. X; c. –; d. X

---

28 When patients do not receive enough protein-sparing calories in the form of amino acids, fat, and carbohydrates, the body's protein and fat will be converted to carbohydrates. To prevent this conversion, the body needs 600–800 calories daily for resting state, approximately 1600 calories daily for sitting state, and 2500–3500 calories daily following general surgical procedures.

If the patient received 2500 ml (2 1/2 liters) of 5% dextrose in water he would receive 500 calories (CHO: 1 g = 4 calories; 50 g = 5%D/1000 ml, frame 16).

1000 ml 5% D = 50 g X 4 = 200 calories
200 X 2 1/2 liters = 500 calories

Would this be a sufficient number of calories for the resting state? ______.

Explain why * ______.

No. 600–800 calories are needed in the resting state.

---

29 2500 ml of 10% dextrose in water provides 1000 calories. Concentrations of sugar higher than 10% dextrose can cause severe peripheral vein damage. A constant use of 10% dextrose in water may cause phlebitis.

Would 1000 calories daily be sufficient for a patient following surgery? ______

Explain why. * ______

______

______.

No. 2500-3500 calories are needed daily following surgical treatment. If the patient was in "reasonably good" health (wellness) before and after surgery, the body calorie and protein needs would soon be restored

---

30 The average percentage of dextrose used in hyperalimentation is between 25–30% per liter.

High sugar (glucose) concentration is mixed with a commercially prepared protein source. Vitamins and electrolytes are added. Electrolytes are frequently added immediately before the infusion according to the patient's serum electrolytes.

Since high sugar concentration is most irritating to peripheral veins, what large vein(s) is/are used in hyperalimentation? * ______

______.

Subclavian vein or internal jugular vein

31 Why is it recommended that electrolytes be added to the solution immediately before administration? * ____ ____.

To determine the amount of electrolytes needed according to the patient's serum electrolytes. Frequently, electrolytes are added with the other nutrients, 12-24 hours before use, to prevent contamination

---

Table 37 gives the indications of hyperalimentation. Study the table carefully before proceeding to the frames. Refer to the table as needed.

TABLE 37. INDICATIONS FOR HYPERALIMENTATION

Indications	Rationale
Oral or nasogastric feedings are contraindicated or not tolerated	Long-term use of intravenous glucose solutions can cause protein wasting. Hyperalimentation maintains positive nitrogen balance.
Severe malnutrition	Malnutrition can cause severe protein loss. Negative nitrogen (protein) balance occurs. Hyperalimentation restores positive nitrogen balance.
Malabsorption syndrome	Inability to absorb nutrients in the small intestine.
Dysphagia	Difficulty in masticating and swallowing due to radiation treatment required in the pharygeal area.
Gastrointestinal fistula	Fistula promotes continuous protein losses. Hyperalimentation allows intestine to rest and decreases gall bladder, pancreas, and small intestine secretions.
Major bowel resection and ulcerative colitis	They cause reduced absorptive area of small intestine. Hyperalimentation increases intestine's ability to absorb nutrients more quickly than oral feedings. Also, it puts bowel to rest.
Extensive surgical trauma and stress	Extensive surgery requires 3500-5000 calories a day to maintain protein balance. Hyperalimentation lowers chance for wound infection and provides positive nitrogen balance.
	Hyperalimentation before surgery increases nutritional status so that patient can withstand surgery and its stresses.
Extensive burns	Extensive burns require 7500-10,000 calories daily. Hyperalimentation will speed wound healing and formation of granulation tissue. It will promote skin graft take.
Metastatic cancer with anorexia and weight loss	Wasting and debilitating diseases, e.g., cancer, frequently are in negative nitrogen balance. Hyperalimentation will restore protein balance and tissue synthesis.

32 Indications for hyperalimentation, or TPN, include:

( ) a. Major bowel resection
( ) b. Minor surgical procedures
( ) c. Gastrointestinal fistula
( ) d. Severe malnutrition
( ) e. Contraindicated or intolerable oral and gastric feedings
( ) f. Severe congestive heart failure
( ) g. Extensive burns
( ) h. Metastatic cancer with weight loss
( ) i. Malabsorption syndrome

- - - - - - - - - - - - - - - - - -

a. X; b. –; c. X; d. X; e. X; f. –; g. X; h. X; i. X

---

33 With contraindicated or intolerable oral or gastric feeding and with severe malnutrition, hyperalimentation helps in restoring *______________________________.

- - - - - - - - - - - - - - - - - -

positive nitrogen balance

---

34 Following major bowel resection, what occurs to the intestinal area in relationship to nutrient absorption? *______________________________.

- - - - - - - - - - - - - - - - - -

Reduced absorptive area of the intestine

---

35 What effect does hyperalimentation have for both gastrointestinal fistula and major bowel resection? *______________________________.

- - - - - - - - - - - - - - - - - -

It allows the intestine (bowel) to rest

---

36 After extensive surgical procedure involving trauma and stress, the body's daily caloric needs are *______________________________.

Would 3000 ml (3 liters) of 10% dextrose in saline provide the patient following extensive surgical procedure with his daily caloric requirement? ______.

Explain why. *____

____.

— — — — — — — — — — — — — — — — —

3500–5000 calories

No. One liter of 10% dextrose is 400 calories; three liters is 1200 calories; thus, less than caloric need

---

37 What effect does hyperalimentation have before surgery? *________________

____.

— — — — — — — — — — — — — — — — —

It increases the patient's nutritional status so that he or she can withstand surgical procedure and its stresses

---

38 What is the caloric need for patients with extensive burns? *________________

________________________.

Give two ways that hyperalimentation aids in healing extensive burns.

a. *____.

b. *____.

— — — — — — — — — — — — — — — — —

7500–10,000 calories

a. promotes wound healing and formation of granulation tissue

b. promotes skin graft take

39 Metastic cancer with anorexia and weight loss frequently results in negative nitrogen balance.

Explain how hyperalimentation helps with this clinical problem. *________

____.

\- - - - - - - - - - - - - - - - - -

It helps in restoring positive nitrogen (protein) balance and tissue synthesis

COMPLICATIONS OF HYPERALIMENTATION

Major complications that result from hyperalimentation therapy are air embolism, infection, hyper- and hypoglycemia, and fluid overload.

Nutritional intravenous fluid is an excellent medium on which organisms, bacteria, and yeast will grow. Strict asepsis is necessary when the solution is handled at the intravenous insertion site and when dressings are changed. Most hospitals have a procedure for changing dressings in which a strict aseptic technique (i.e., gloves, masks, povidone-iodine (Betadine) ointment, iodine, and tincture of benzoin) is mandatory.

When intravenous tubing is changed at the catheter site the patient must lie flat and perform the Valsalva maneuver (take a breath, hold it, and bear down) to prevent air from being sucked into the circulation. The Valsalva maneuver increases intrathoracic pressure.

Increased blood sugar (hyperglycemia) is usually the result when hyperosmolar dextrose solutions for hyperalimentation are infused rapidly. An elevated blood glucose level may occur during early hyperalimentation until the pancreas adjusts to the concentrated sugar load. Regular insulin, in IV solution or injection, may be needed to prevent or control hyperglycemia. Other complications may include hypoglycemia from abruptly discontinuing hyperosmolar dextrose solutions, overload/overhydration from the infusion of an excessive amount of fluid, or a fluid shift from cellular to extracellular compartments.

Table 38 lists the 5 major complications associated with hyperalimentation/TPN, causes, symptoms and actions.

TABLE 38. HYPERALIMENTATION: COMPLICATIONS, SYMPTOMS, ACTIONS

Problems	Causes	Symptoms	Actions
Air embolism	IV tubing disconnected Catheter not clamped Injection port fell off Improper changing of IV tubing (no valsalva procedure)	Coughing Shortness of breath Chest pain Cyanosis	Clamp catheter Patient must lie on left side with head down Check VS Notify physician

TABLE 38. (Continued)

Problems	Causes	Symptoms	Actions
Infection	Poor aseptic technique Contaminated tubing Solutions: Conc. glucose Protein hydrolysate	Temperature above 100° F (37.7°C) Pulse increased Chills Sweating Redness, swelling, drainage at insertion site Pain in neck, arm, or shoulder Lethargy Urine: Glycosuria Bacteria Yeast growth	Notify physician Change dressing every 2 days Change solution every 12 hours Change tubing every 24 hours Check VS q4h
Hyperglycemia	Fluid infused rapidly Insufficient insulin coverage Infection	Nausea Weakness Thirst Headache Urine glucose test >1/2%	Monitor blood glucose Notify physician Decrease infusion rate Regular insulin
Hypoglycemia	Fluids stopped abruptly Too much insulin infused	Nausea Pallor Cold, clammy Increased pulse rate Shaky feeling Headache Blurred vision	Notify physician Increase infusion rate with NO insulin *or* Orange juice with 2 teaspoons of sugar if patient can tolerate fluids
Fluid overload (hypervolemia)	Increased number of IV infusions Fluid shift from cellular to vascular due to hyperosmolar solutions	Cough Dyspnea Neck vein engorgement Chest rales Weight gain	Check VS q4h Weigh daily Monitor I & O Check neck veins for engorgement Check chest sounds

Adapted with permission from L. Wilhelm: Helping your patient settle in with TPN. *Nursing 85*, 15(4):63, April 1985.

---

40 To prevent air embolism when changing IV tubing the Valsalva maneuver is performed.

Explain how it is done (see introduction to hyperalimentation).

* ______________________________

- - - - - - - - - - - - - - - - -

Take a breath, hold it, and bear down

41 What are the two immediate actions if air embolism is suspected? *________________________________ and *________________________________.

— — — — — — — — — — — — — — — — —

Clamp catheter; position patient on left side with head down

42 Hyperosmolar dextrose in a protein hydrolysate solution promotes yeast and bacteria growth. It has been reported that these organisms do not grow so rapidly in a solution that contains crystalline amino acid as they do in a solution of protein hydrolysate.

Name three (3) symptoms indicative of infection ________________, ________________ and *________________.

— — — — — — — — — — — — — — — — —

Elevated temperature; chills; redness, swelling, and drainage at insertion site; sweating and pain in arm or shoulder

43 Usually 2 1/2 to 3 liters of hyperosmolar dextrose solution for hyperalimentation is administered over 24 hours. A strict infusion rate is important because the "make-up" rate is not suggested.

To prevent and control hyperglycemia what test is suggested? *________________________________.

How often should the test be performed? *________________________________.

— — — — — — — — — — — — — — — — —

Finger stick for blood sugar level (Chemistrip bG) or urine test for glycosuria. Every 4 to 6 hours.

---

44 It is suggested that an iso-osmolar dextrose solution be administered for 12 hours after hyperalimentation is discontinued.

If hyperosmolar dextrose solution is discontinued abruptly what complication may develop? ______________________.

What signs and symptoms should the nurse be aware of? ______________,

______________, ______________, * ______________

______________ and * ______________________.

- - - - - - - - - - - - - - - - -

Hypoglycemia. Pallor; cold and clammy skin; increased pulse; shakiness. Also nausea and blurred vision.

---

45 Certain medications are compatible with hyperalimentation solutions when administered through a supplemental line. These drugs are the following: Antibiotics: cephalothin/Keflin, cefazolin/Ancef, carbenicillin/Geopen, gentamicin/Garamycin, kanamycin/Kantrex, methicillin, and tobramycin/Nebcin. Antineoplastic agents: cyclophosphamide/Cytoxan, 5-fluorouracil/5-FU, and methotrexate. Others: cimetidine/Tagamet, heparin, regular insulin, furosemide/Lasix, lidocaine/Xylocaine, aqua-Mephyton, Solu-Medrol, dopamine/Intropin, and metaraminol/Aramine.

The Hickman multilumen catheter can be used to administer intravenous fluids, hyperalimentation, blood products, and medications.

When in doubt about drug compatibility with hyperalimentation solution the nurse should * ______________________

______________________.

- - - - - - - - - - - - - - - - -

check with the physician or the pharmacist in charge of TPN.

---

46 Fat-emulsion-supplement therapy provides an increased number of calories and is a carrier of fat-soluble vitamins.

The two solutions that can provide nutrients, calories, and vitamins for hyperalimentation are * ______________________, and

* ______________________.

- - - - - - - - - - - - - - - - -

TPN solution or hyperosmolar dextrose/protein solution; fat emulsion solution.

CASE REVIEW

Mrs. Ryan, age 60, was admitted to the hospital for a possible intestinal obstruction. Diagnostic studies were ordered. Food, except for soup and tea, nauseated her. She had not had a bowel movement for a week and 2000 ml of 5% dextrose/0.33% saline, with 20 mEq/L of KCl in 1 liter (1000 ml) for 24 hours were ordered.

1. The objective of Mrs. Ryan's intravenous therapy was *________________________

   ____.

2. 5% dextrose/0.33% saline is what type of solution? *________________________

   ____.

3. What is the rationale for using 5% dextrose/0.33% saline? *________________________

   ____.

4. Was Mrs. Ryan's fluid and nutrient intake completely dependent on intravenous therapy? ______________. Explain. *________________________

   ____.

   After two days, Mrs. Ryan vomited when she took any oral fluid. Her IV fluid order for the day was 1 liter of 5% dextrose in lactated Ringer's, and 1 liter of 5% dextrose in water.

5. Name the types of solutions ordered:

   a. 5% dextrose in water *________________________

   b. 5% dextrose in lactated Ringer's *________________________

6. Explain the relationship of lactated Ringer's to body fluids. *________________________

   ____.

7. Is the potassium in lactated Ringer's sufficient to meet daily requirement? ______

   ______________Explain. *________________________

   ____.

8. During intravenous administration of fluid and electrolytes, the kidney function is extremely important. Why? *________________________

   ____.

   Mrs. Ryan's clinical condition did not improve. Her x-ray showed a complete intestinal obstruction. The following day, a major bowel resection was performed. Hyperalimentation was started with a hyperosmolar solution containing dextrose 25% per liter, protein hydrolysate, vitamins, and electrolytes.

9. One liter of dextrose 25% has 850 mOsm. The osmolality of this solution is (hypo-osmolar/hyperosmolar) (mOsm and osmolality are explained in Chapter 1).

   ____.

10. Why is hyperalimentation indicated following major bowel resection? *__________

____.

11. Mrs. Ryan should receive hyperalimentation therapy in what blood vessel? *_______

______________________________. Explain your reason. *__________________

____.

12. What four major complications can result from hyperalimentation therapy?

*_____________________, _____________________, ___________________,

and ____________________.

13. Name at least two nursing actions necessary when caring for Mrs. Ryan as she receives hyperalimentation.

a. *____.

b. *____.

14. How can an air embolus be prevented? *____________________________

____.

\- - - - - - - - - - - - - - - -

1. to provide mainenance requirements
2. hydrating solution
3. daily maintenance of body fluid when a minimal sodium and chloride are required
4. No. She was taking soup and tea
5. a. hydrating solution
   b. replacement solution
6. lactated Ringer's resembles the electrolyte struction of normal blood serum–plasma
7. No. The potassium is 4 mEq/L. Daily requirement is approximately 40–45 mEq/L
8. renal dysfunction can result in electrolyte retention
9. hyperosmolar
10. hyperalimentation increases the intestine's ability to absorb and to rest the intestines.
11. subclavian or internal jugular veins. High concentration of sugar is not as irritating to large veins; less chance of phlebitis.
12. air embolism; infection; hyperglycemia; hypoglycemia; also fluid overload or overhydration.
13. a. Observe for sepsis due to infection. Nutrient-rich hyperalimentation solution provides a good medium for growth of bacteria and yeast.
    b. Monitor blood glucose level with chemistrip bG. A rapid rate of infusion can cause hyperglycemia.

    Others: Observe for electrolyte imbalance, prevent air embolus, and use strict aseptic technique when dressing and tubing are changed.
14. Valsalva maneuver: taking a breath, holding it, and bearing down

NURSING DIAGNOSES

Potential alteration in fluid volume related to inadequate or excess fluid replacement.

Potential alteration in nutrition related to anorexia, nausea, vomiting, or NPO status.

Alteration in nutrition of less than body requirements related to possible bowel obstruction.

Potential impairment of skin integrity related to insertion site of catheter and protective taping.

Knowledge deficit related to signs and symptoms of complications and catheter care.

Potential for injury related to possible disconnection of IV tubing and air embolism resulting from air in venous line.

Disturbance in self-concept related to body image and dependence on continuous need for hyperalimentation.

Potential for ineffective family action related to long-term total parenteral nutrition therapy for patient, financial crisis, and undetermined needs.

NURSING ACTIONS

1. Recognize the need for intravenous therapy to provide maintenance requirements, to replace previous losses, and to meet concurrent losses.
2. Explain the purpose for hydrating solutions and give examples. Five percent dextrose in water, 2½% dextrose in ½ NSS (0.45% NaCl), and 5% dextrose in ½ NSS are examples; and they are used for daily maintenance of body fluid, rehydration, and in establishing effective renal output.
3. Give examples of replacement solutions and state their purposes. Lactated Ringer's, 5% or 10% dextrose in NSS (saline—0.9% NaCl), and Ionosol B are examples; and they are useful for replacing fluids, calories, and electrolyte deficits.
4. Monitor blood level of patients who are receiving hyperalimentation (TPN) every 4 to 6 hours with Chemistrip bG. The IV flow rate may need to be decreased or insulin, given.
5. Prevent complications of TPN. Use aseptic technique when changing dressing, tubing, and bottles to prevent infection. Have patient perform Valsalva maneuver when changing IV tubing to prevent air embolism.
6. Observe for signs and symptoms of overhydration and allergic reactions as the result of IV dextran solutions and hyperosmolar solutions. Symptoms of overhydration are a constant, irritated cough, dyspnea, neck and hand vein engorgement, and chest rales. Symptoms of allergic reactions are nausea and vomiting, urticaria, dyspnea, wheezing, and hypotension.
7. Observe for signs and symptoms of infection; i.e., elevated temperature, increased pulse rate, chills, sweating, redness, swelling, drainage at insertion site, pain in arm or shoulder, and positive urine test for bacteria or yeast.
8. Answer the questions of the patient and family members or refer them to the appropriate health professionals.

FLUID REPLACEMENTS

---

47 There are three groups of fluids that are used in restoring body fluids. They include

Crystalloids (lactated Ringer's, saline, and dextrose)
Blood (whole blood and red blood cells)
Colloids (albumin, plasma, plasmanate, and dextran)

Dextrose, saline, and lactated Ringer's are considered (crystalloids/colloids) ____________________.

\- - - - - - - - - - - - - - - - -

crystalloids

---

48 The first step in treatment of fluid and electrolyte disturbances should be the reconstitution of the extracellular fluid and blood volume. This can best be accomplished by iso-osmolar saline (normal saline), lactated Ringer's, or M/6 sodium lactate.

The iso-osmolar solutions that can be used in the reconstitution of the extracellular fluid and blood volume are:

1. *______________________________.
2. *______________________________.
3. *______________________________.

\- - - - - - - - - - - - - - - - -

1. normal saline
2. lactated Ringer's
3. M/6 lactated sodium

---

49 Do you think the volume of urine would increase or decrease following the administration of these iso-osmolar solutions? ____________________. Why?

*____.

\- - - - - - - - - - - - - - - - -

increase. There would be more fluid in the body to be excreted

---

WHOLE BLOOD AND BLOOD PRODUCTS

---

50 Hematocrit (volume of red blood cells in proportion to the intravascular fluid) reading would be one indication of gain or loss of the fluid volume. An increased hematocrit reading can indicate an intravascular fluid loss. Do you know the reason? ____________. Explain. *____________

____

______________________________.

- - - - - - - - - - - - - - - - - -

Yes? Good; no? Intravascular fluid loss increases the number of red blood cells in proportion to the fluid

---

51 Concentration of red blood cells is known as *hemoconcentration.*

A high hematocrit reading can be an indication of:

( ) a. dehydration

( ) b. overhydration

- - - - - - - - - - - - - - - - - -

a. X; b. – (due to hemoconcentration)

---

52 Giving whole blood or plasma will dilute the hemoconcentration, lower the hematocrit, raise the blood pressure, and establish renal flow.

How do you think whole blood would dilute the hemoconcentration?

*____

____.

- - - - - - - - - - - - - - - - - -

There would be a slight increase in the intravascular fluid, since 55% whole blood is plasma; however, plasma or crystalloids may be a better choice in dilution

---

53 For best results, whole blood should be given following initial dilution by an iso-osmolar solution.

Explain why you think an iso-osmolar solution should be given first for hemoconcentration before blood. *____________________

____.

- - - - - - - - - - - - - - - - - -

It will aid in immediate dilution and will also prevent clotting of blood

---

54 Various components of whole blood can be fractionated and administered separately. These components include red blood cells, plasma, platelets, white blood cells (leukocytes), albumin, and blood factors II, VII, VIII, IX, X.

Name three blood components that can be fractionated from whole blood.

1. ______________.

2. ______________.

3. ______________.

Red blood cells; plasma; platelets. Also WBC, albumin, and blood factors

---

55 Red blood cells are known as *packed cells.* A unit (200-250 ml) of red blood cells (packed cells) is composed of whole blood minus the plasma.

Could a unit of red blood cells be administered to dilute hemoconcentration? ______. When do you think a unit of red blood cells would be indicated? * ______________________________.

No. To restore red blood cells and not fluid.

---

56 The shelf life of whole blood, refrigerated, is 21 days. Red blood cells and plasma can be frozen, thus extending their shelf life to approximately 3 years. Platelets should be administered within 3 days after they have been extracted from whole blood.

As whole blood ages, potassium leaves the red blood cells (increases the K level in the blood), and platelets and red blood cells are destroyed. After 3 weeks of shelf life, serum potassium (in whole blood) could be increased to 20-25 mEq/L.

The shelf life of whole blood is ______ days; frozen red blood cells and plasma, ______ years; and platelets, ______ days.

21; 3; 3

57 Serum potassium level in 2–3-week-old whole blood (increases/decreases).

____________.

What could happen if a critically ill patient with poor renal function is given 3 or more units (pints) of "aging" whole blood? ________________________

________________________.

If a patient has a serum potassium of 6.0 mEq/L, should the patient receive a transfusion of 3-week-old whole blood? ____________. Why?

* ________________________________.

\- - - - - - - - - - - - - - - - -

increases. Cardiac arrest or extreme hyperkalemia.
No. Serum potassium level of whole blood could be 20–25 mEq/L (five times normal value), thus increasing patient's serum level, or a similar response

---

58 As blood volume is being restored, attention is directed to the osmolal changes, correcting the osmolality. 5% dextrose in water is effective in correcting water deficit.

Do you think 5% dextrose would be administered to a patient with hyponatremia? ____________. Why or why not * ________________________

________________________________.

\- - - - - - - - - - - - - - - - -

No. Body salt or sodium would be superdiluted

COLLOIDS

---

59 Name the colloids used in restoring body fluids. (Refer to Frame 46.)

a. ____________________ c. ____________________

b. ____________________ d. ____________________

\- - - - - - - - - - - - - - - - - -

a. albumin; b. plasma; c. plasmanate; d. dextran

60 Albumin concentrate is helpful in restoring body protein. It is considered to be a plasma volume expander. Too much albumin, or albumin administered too rapidly, can hold fluid in the lungs.

Albumin is used to restore * ______________________ and is considered a * ______________________.

What should the nurse assess when the patient is receiving albumin? * ______________________.

- - - - - - - - - - - - - - - - -

body protein; plasma volume expander;
lungs for fluid congestion (rales). Good!

---

61 Plasmanate is a commercially prepared protein product that is used in place of plasma and albumin.

When there is an insufficient supply of plasma, name the closely resembled (commercially prepared) solution used. ______________________.

- - - - - - - - - - - - - - - - -

Plasmanate

---

62 Dextran, in saline or dextrose, is another plasma volume expander. Dextran comes in two concentrations, Dextran 40 and 70. The stronger concentration is a colloid hyperosmolar solution. If large quantities are administered too rapidly, fluid will leave the cells and intestine, thus causing dehydration.

Dextran can affect clotting by coating the platelets, which reduces their ability to form clots. Also, dextran interferes with blood typing and cross-matching.

As a nurse, if your patient is to receive dextran and his blood is to be typed and cross-matched, what is the nursing action? * ______________________.

- - - - - - - - - - - - - - - - -

Draw blood for type and cross-match before administering dextran

---

63 Another test the physician uses to determine the need for intravenous therapy is the BUN. Blood urea nitrogen (BUN) is a serum test to determine the amount of urea, a by-product of protein metabolism, remaining in the blood that is normally excreted by the kidneys. Normal range for BUN is 10–25 mg/dl.

An elevated BUN frequently indicates poor renal function; however, a slightly elevated BUN may indicate a decrease in body fluid volume (dehydration).

After the patient has been rehydrated, if an elevated BUN does not return to normal range, would it be an indication of (poor renal function/decrease in body fluids)? *______________________.

- - - - - - - - - - - - - - - - -

poor renal function

---

64 The third method for determining the need for intravenous therapy is through checking the serum electrolytes.

To determine whether or not intravenous therapy is necessary, a physician will check:

1. ______________________.
2. ______________________.
3. ______________________.

- - - - - - - - - - - - - - - - -

1. hematocrit
2. BUN
3. electrolytes

---

65 Fluids and electrolytes for maintenance therapy should be continuous for 24 hours and rescheduled on a day-to-day basis.

Normally, infusion of fluids results in prompt excretion of any excess water and electrolyte. This response defends the body against *______________ in water and electrolyte balance.

How should fluids and electrolytes for maintenance therapy be scheduled?

*______________________________.

Why? *______________________________.

- - - - - - - - - - - - - - - - -

excess or significant alterations

CONTINUOUS on a day-by-day basis. This defends the body against significant alterations

---

66 If a patient receives his full 24-hour maintenance parenteral therapy in 8 hours, two-thirds of the water and electrolyte will be in (excess/deficit) ________ of current needs, and, normally, a large quantity of the excess maintenance fluid will be (excreted/retained) ________________.

— — — — — — — — — — — — — — — — —

excess; excreted

---

67 Tolerance for water and electrolytes is limited for the very ill patients, patients following surgery, aged patients, small children, and infants.

Rapid administration of maintenance fluids, which exceeds physiologic tolerance, can cause hyponatremia, pulmonary edema—accumulation of fluid in the chest—and other complications.

Maintenance parenteral therapy should be administered over a period of

*________________.

What are the two possible results from rapid administration of maintenance fluids? ____________ and *________________.

— — — — — — — — — — — — — — — — —

24 hours; hyponatremia; pulmonary edema

INTRAVENOUS FLOW RATE

68 Hyperosmolar solutions, those having osmolality greater than the body fluids, should be injected not faster than 2 to 4 ml per minute or as ordered by the physician. Two to 4 ml is equivalent to 30–60 drops (gtts), Abbott fluid set, or 30–40 gtts, Travenol fluid set.

List at least five hyperosmolar solutions to which this might apply. Refer back to Table 36 if needed.

1. *________________.
2. *________________.
3. *________________.
4. *________________.
5. *________________.

— — — — — — — — — — — — — — — — —

1. Dextrose 5% in normal saline
2. Dextrose 10% in water
3. Dextrose 10% in normal saline
4. Dextrose 5% in lactated Ringer's solution
5. Hyperosmolar saline
6. Multiple electrolyte solutions

Table 39 outlines the three types of therapy frequently employed with intravenous administration. The table includes the desired amount of solution and the rate of flow for each type of therapy. The symbol *ml* (milliliter) has the same equivalence as cc (cubic centimeter). The symbol for drops is *gtts.*

The asterisks refer to the number of drops equivalent to one milliliter, according to the type of intravenous sets used. The single asterisk is the Abbott set: 15 gtts = 1 ml, and the dagger refers to the Travenol set: 10 gtts = 1 ml.

Today many institutions use IV controllers to deliver most IV fluids, but the nurse still needs to know how to calculate and regulate them.

Take several minutes to study this table carefully. Memorize the equivalent drops per milliliter for both intravenous sets. Know the three types of therapy, the recommended dosage, and the rate at which the fluid should be administered. Refer to this table as needed.

TABLE 39. RATES OF INTRAVENOUS ADMINISTRATION

Type of Therapy	Amount of Solution Desired (ml)	Rate of Flow
Maintenance therapy	1500–2000	2 ml/minute (30 gtts*, 20 gtts†) if given over 24 hours; 3 ml/minute (45 gtts*, 30 gtts†) in two infusions.
Replacement with maintenance therapy	2000–3000	2 or 4 ml/minute (30–60 gtts*, 20–40 gtts†). (Depends on individual and his physician.)
Hydration therapy	1000–3000 (Iso-osmolar)	8 ml/minute (120 gtts*, 80 gtts†) for first 45 minutes. If urinary output is not established, then 2 ml/minute the next hour. After urinary output is established, continue as in maintenance therapy.

These guidelines may be adapted to individual circumstances. The physician sets the 24-hour requirements and the registered nurse computes 1-hour requirements from this. The amount of solution to be administered and the rate of flow can vary greatly with the very sick, the aged person, the small child, and the infant.

*Abbott set = 15 gtts per 1 ml.

†Travenol set = 10 gtts per 1 ml.

---

69 The names of the three types of intravenous therapy are:

1. * ______________________.
2. * ______________________.
3. * ______________________.

— — — — — — — — — — — — — — — —

1. maintenance therapy
2. replacement therapy
3. hydration therapy

---

70 The type of solution most frequently used in hydration therapy is:
( ) a. hyperosmolar
( ) b. iso-osmolar

Once urinary output is reestablished, then (maintenance/replacement) ______ ______ therapy is started.

- - - - - - - - - - - - - - - - -

a. –; b. X
maintenance

---

71 The amount of solution to be administered and the rate of flow can vary greatly for what patients? * ______, the * ______, the * ______, and the ______.

- - - - - - - - - - - - - - - - -

a very sick person; aged person; small child; infant

---

72 The physician sets the 24-hour requirements, and the registered nurse computes * ______.

- - - - - - - - - - - - - - - - -

1-hour requirements from this

---

73 The number of drops per ml of the commercial administration sets for intravenous fluids will vary according to the manufacturer of the set. Some sets as manufactured by a given company are calculated to give 10 gtts (drops) per 1 ml (cc), whereas sets of other manufacturers are calculated at 15 gtts per 1 ml.

The drops per ml can vary according to * ______ ______.

- - - - - - - - - - - - - - - - -

the manufacturer of the intravenous sets

74 The formulas for calculating the intravenous flow rate are the following:

Formula I:

a. Amount of fluid ÷ hours to be administered

b. $\dfrac{\text{Number ml per hour} \times \text{drops per ml}}{\text{60 minutes per hour}}$

Formula II:

$\dfrac{\text{Amount of fluid} \times \text{drops per ml}}{\text{Hours to be administered} \times \text{minutes per hour}}$

Example A:

Your patient is to receive 1000 ml of 5% dextrose in water in 8 hours. Calculate the number of drops per minute according to the Abbott set (15 gtts/ml).

Formula I: a. 1000 ÷ 8 = 125 ml per hour

b. $\dfrac{\text{125 ml per hour} \times \text{15 drops per ml}}{\text{60 minutes per hour}} = \dfrac{1875}{60} =$

Formula II: $\dfrac{1000 \times \text{15 drops per ml}}{8 \times \text{60 minutes per hr}} = \dfrac{15000}{480} =$

30–32 drops (gtts) per minute.

75 Example B:

Your patient is to receive 1000 ml of 5% dextrose in 0.45% saline in 10 hours. Calculate the number of drops per minute according to the Travenol macrodrop set.

If the physician orders the amount of fluid to be administered per hour or the nurse knows how much fluid is to be delivered per hour use formula I and *omit* step A. Step B gives the answer in drops per minute.

Formula I: a. 1000 ÷ 10 = 100 ml per hour

b. $\dfrac{\text{100 ml per hour} \times \text{10 drops per ml}}{\text{60 minutes per hour}} = \dfrac{1000}{60} =$

Formula II: $\dfrac{1000 \times 10}{10 \times 60} = \dfrac{10{,}000}{600} =$

16–17 drops (gtts) per minute.

76 Here is another problem in intravenous therapy. Your patient's intravenous order is 500 ml of 5% dextrose in normal saline to be administered in 4 hours, using the Travenol set. Calculate the number of drops per hour and per minute. Work space is provided.

Example C:

Formula I:

$500 \div 4 = 125$ ml per hour

$$\frac{125 \times 10}{60} = \frac{1250}{60} = 20\text{–}21 \text{ gtts per minute}$$

Formula II: $$\frac{500 \times \cancel{10}^{1}}{4 \times \cancel{60}_{6}} = \frac{500}{24} = 20\text{–}21 \text{ gtts}$$

---

77 Write the formula for calculating intravenous flow rate. Work space is provided.

a. * ____.

b. * ____.

---

a. Amount of fluid ÷ hours to be administered

b. $$\frac{\text{Number ml per hour} \times \text{drops per ml (set)}}{\text{60 minutes per hour}}$$

CASE REVIEW

Mr. Deale, age 84, was admitted to the hospital because of malnutrition. His hematocrit and BUN were elevated and his serum electrolytes were decreased. His blood gases were: pH of 7.3, P_{CO_2} of 40, and HCO_3 of 19. Mr. Deale's urine output was decreased. He was to receive 3 liters of fluid for 24 hours, 2000 ml of 5% dextrose in lactated Ringer's, and 1000 ml of 5% dextrose in 0.2% saline with 30 mEq/L of KCl.

1. The first step in treating Mr. Deale's nutritional, fluid, and electrolyte deficits should be the reconstitution of * ____________________________ and * __________________________________.

2. Along with the fluid replacement, name two other replacements that Mr. Deale needs.

   a. *________________________________.

   b. ________________________________.

3. Mr. Deale's elevated hematocrit and BUN may be indicative of *____________________________________.

4. After hydration, if Mr. Deale's BUN does not return to normal, what then would the elevated BUN indicate? *____________________________________.

5. Concerning Mr. Deale's blood gases, his low pH and HCO_3 may indicate *________________________________.

6. If the intravenous fluids were administered at a rapid rate, what could occur to Mr. Deale's physical state, taking into consideration his age and his state of dehydration? *____________________________________.

7. Mr. Deale's intravenous fluid orders were 3000 ml (3 liters) in 24 hours or 1000 ml (1 liter) in 8 hours. If you use the Travenol set, calculate the drops per minute.
8. The physician will determine the need for intravenous therapy by checking on the:

   a. ________________________.

   b. ________________________.

   c. *________________________.
9. If Mr. Deale was ordered to receive a plasma volume expander, name three solutions that could be used. __________________, *__________________, and *________________________.
10. Give two adverse effects with using dextran. *________________________________ and *________________________.

\- - - - - - - - - - - - - - - -

1. extracellular fluid; blood volume
2. a. nutrients such as glucose
   b. electrolytes
3. dehydration or a decrease in blood volume
4. renal impairment or failure

5. acidosis (metabolic)
6. overhydration or pulmonary edema
7. 1000 ÷ 125 ml per hour

$$\frac{125 \times 10 \text{ gtts per ml}}{60 \text{ minutes}} = \frac{1250}{60} = 20\text{–}21 \text{ gtts per minute}$$

8. a. hematocrit
   b. BUN
   c. serum electrolytes
9. albumin; plasma or plasmanate; dextran
10. reduces clotting ability; interferes with type and cross-matching

NURSING DIAGNOSES

Potential alteration in fluid volume related to fluid replacement (deficit or excess) or incorrect flow rate.

Potential knowledge deficit related to lack of understanding of intravenous fluid administration, types of fluid (crystalloids, blood, colloids), and laboratory findings.

Potential fluid volume excess related to excessive amounts of fluid infused in the critically ill, elderly, the very young, and the debilitated.

Potential alteration in tissue perfusion related to decreased blood circulation or inadeuqate fluid replacement.

NURSING ACTIONS

1. Check the patient's hematocrit and BUN; if these are elevated, it could be due to dehydration (hypovolemia). If the BUN > 60 mg/dl, kidney impairment is most likely the cause.
2. Administer intravenous fluids rapidly for the first 30 to 45 minutes to reestablish urinary flow (with physician's permission).
3. Assess patient's lungs for rales when administering fluid rapidly. Overhydration commonly occurs to the young, the older adult, the debilitated patient, and those persons with poor kidney function.
4. Monitor IV flow rate according to the type of IV therapy ordered. For maintenance therapy, IV fluids should run for 12 to 24 hours at 2 ml per minute (20 gtts per minute, Travenol; or 30 gtts per minute, Abbott).
5. Calculate the flow rate according to physician's orders.
6. Check the age of whole blood for transfusion before giving it to a critically ill patient or to a patient with poor kidney function. Blood that is 2 to 3 weeks old has a high serum potassium, and hyperkalemia could result.
7. Check the age of platelets before administering. Platelets should be given within 3 days after they have been extracted from whole blood.

CLINICAL APPLICATIONS

78 The nurse has many responsibilities in intravenous therapy, especially when there are serious fluid, electrolyte, and nutritional imbalances.

The nurse must keep careful records of what the patient is receiving and assume responsibility for accurate record of intake and output of the individual.

The nurse must know what intravenous fluids her patient is receiving and keep an accurate record of * ______________________________.

- - - - - - - - - - - - - - - - -

intake and output

NURSING ASSESSMENT

Table 40 discusses the nursing responsibility in the assessment of intravenous therapy. The physician orders the IV fluids, and the nurse must have knowledge and understanding of the various solutions, needles, catheters, tubing, injection sites, and complications in order to assess and monitor intravenous therapy. This tables gives the assessment, nursing actions, and rationale. Study the table carefully and refer to it as needed.

TABLE 40. NURSING ASSESSMENT OF INTRAVENOUS THERAPY

Assessment	Nursing Action	Rationale
Types of intravenous solutions	Note the types of intravenous fluid ordered, hypo-osmolar, iso-osmolar, or hyperosmolar solution.	Osmolality of solutions should be known, for an excessive use of hypo-osmolar solutions can cause water intoxication, and excess use of hyperosmolar solutions could cause dehydration.
		An iso-osmolar solution has 240–340 mOsm/L; less than 240 is hypo-osmolar, and greater than 340 is hyperosmolar (see Chapter 1, Frames 46–48).
	Report if the patient is receiving continuous IV's of dextrose in water.	Five percent dextrose in water administered continuously becomes a hypo-osmolar solution. Dextrose is metabolized rapidly and the water remaining will decrease the serum osmolality. Correction would be to alternate D/W with D/NSS (saline).
	Observe for signs and symptoms of dehydration, i.e., dry mucous membranes, poor skin turgor, increased	Continuous use of hyperosmolar solutions pulls fluid from the cells to the vessels, and the fluid is excreted by

TABLE 40. (Continued)

Assessment	Nursing Action	Rationale
	pulse and respiration rates when using hyperosmolar solutions.	the kidneys. If poor kidney function is present, fluid "back-up" in the circulation and overhydration occur.
Intravenous tubing and bag	Inspect IV bags for leaks by gently squeezing.	Microorganisms can enter intravenous bags at small leak sites, contaminating the fluid.
	Check drop size on the equipment box. Use IV tubing with macrodrip chamber (10 gtts per 1 ml or 15 gtts per 1 ml) for administering IV fluids in 8 hours or less.	Do not use a macrodrip chamber (IV tubing) for fluids that are ordered to run for 10 hours or longer. The flow rate would be too slow and inaccurate.
	Use microdrip chamber (60 gtts per 1 ml) for administering IV fluids in 10 hours or more.	An infusion pump is helpful in regulating correct number of drops per minute, especially for IV fluids to run for 12–24 hours.
	Change IV tubing every 24–48 hours with a new liter of fluids.	Studies have shown that IV tubes hanging for 48 hours are free of bacteria when proper aseptic technique is used.
		An IV bag should not be used for longer than 24 hours. If the order is for KVO (keep vein open) a 250–500 ml bag with a microdrip chamber set is suggested.
Needles and IV catheters (cannulas)	Recognize the types of IV needles and catheters used for IV fluids: Straight needles Scalp vein needles (butterfly needles) Heparin lock Over-the-needle catheter Through-the-needle catheter	Needles (straight and scalp vein) are used for short-term IV therapy and for patients with autoimmune problems. Occurrence of phlebitis is less with the use of needles than catheters; however, this is still debatable. Catheters made of silicone and Teflon are less irritating than polyvinylchloride and polyethylene catheters.
	Change injection site every 2 to 3 days according to agency policy.	Needles and catheters in longer than 72 hours could cause a phlebitis.
	Check the over-the-needle catheter for placement and function.	There are many types of over-the-needle catheter, i.e., Angiocath, A-Cath, Vicra Quik-Cath, etc. Catheter length can be 2.5 to 36 inches. Care should be taken to avoid severing the catheter with the needle tip.

TABLE 40. (Continued)

Assessment	Nursing Action	Rationale
	Check for fluid leak at the injection site after the insertion of a through-the-needle catheter.	Through-the-needle catheter is used for CVP monitoring, TPN (hyperalimentation), etc. It is frequently inserted in large veins, i.e., subclavian vein. Leaks result from needle punctures that are larger than the catheter.
Injection site	Insert needle or catheter in the hand or the distal veins of the arm. Use the antecubital fossa (elbow) site last.	The upper extremity is preferred for the injection site, since occurrence of phlebitis and thrombosis is not as prevalent as it is in the lower extremities.
	Avoid using the leg veins if possible.	Circulation in the leg veins is reduced, and thrombosis formation could occur.
	Avoid using affected limbs due to CVA or mastectomy for IV sites.	Circulation is usually decreased in affected extremities.
	Apply arm board and/or soft restraints to the extremity with the IV when the patient is restless or confused.	Prevention of extremity movement with an IV decreases the chance of dislodging the needle.
Flow rate and irrigation	Check types of solutions patients are receiving.	Knowledge of tonicity (osmolality) of fluids will aid in determining rate of flow. Rate of hyperosmolar solution should be slower than iso-osmolar solutions.
	Observe drip chamber and regulate accordingly.	Regulation of intravenous fluids is important to prevent overhydration, i.e., cough, dyspnea, neck vein engorgment, and chest rales. Do not play "catch-up" with IV fluids.
	Regulate KVO rate to run 10 drops a minute (microdrip, 60 gtts = 1 ml)	KVO (keep vein open) should run approximately 10 ml per hour or 240 ml per day. The microdrip chamber set should be used, calculated as 10 drop s(gtts) per minute.
	Label IV bag for ml to be received per hour. Check rate of flow every 30 minutes to 1 hour with hyperosmolar and toxic solutions, and every hour with iso-osmolar solutions.	Hyperosmolar solutions administered rapidly could cause cellular dehydration and, if the kidneys are properly functioning, vascular dehydration. Hyperosmolar fluids act as an osmotic diuretic and can cause diuresis. When toxic solutions are administered rapidly,

TABLE 40. (Continued)

Assessment	Nursing Action	Rationale
		speed shock could occur; and if infiltration occurs, necrotic tissue could result.
	Restore IV flow if stopped by opening flow clamp, milking the tubing, raising the height of IV bag, or reposition extremity.	If IV flow has stopped and will not start by opening clamp, milking tubing, raising the bag, or repositioning extremity, then the IV catheter should be removed.
		Irrigating IV catheters are prohibited in some institutions. Forceful irrigation could dislodge clot(s) and cause an embolus to the lungs.
Position of intravenous line	Position and tape IV tubing to prevent kinking.	Kinking of the tubing may cause the IV to be discontinued and to be restarted at a different site.
	Hang IV bag 2½ to 3 feet above patient's infusion site.	The higher the IV bag, the faster the flow rate. If the IV bag is too low, IV fluids may stop or the patient may receive insufficient amount.
Infusion problems and complications 1. Infiltration	Observe injection site for infiltration, i.e., swelling, coolness, and soreness.	Infiltration is fluids accumulating in the subcutaneous tissue. When infiltration occurs, IV should be discontinued and restarted at a different site.
2. Phlebitis	Observe injection site for phlebitis, i.e., red, swollen, hard, pain, and warm to touch. Apply moist heat to area as ordered.	Phlebitis is an inflammation of the vein that could be due to irritating substances. Drugs that are not accurately diluted and hyperosmolar solutions may cause phlebitis. Application of moist heat decreases inflammation.
3. Systemic infection	Observe for pyrogenic reactions (septicemia), i.e., chills, fever, headache, fast pulse rate. Check vital signs q4h for shock-like symptoms. Utilize aseptic technique when inserting IV catheters, changing IV tubing and IV bag.	Aseptic technique should be used at all times with intravenous therapy. Prevention of systemic infections is of utmost importance.
4. Speed shock	Observe for signs and symptoms of speed shock, i.e., tachycardia, syncope, decreased blood pressure.	Speed shock occurs when solutions with drugs are given rapidly. High drug concentration accumulates rapidly in the body and can cause shocklike symptoms.
5. Air embolism	Remove air from tubing to prevent air embolism	Air can be removed from tubing by: (1) inserting a

TABLE 40. (Continued)

Assessment	Nursing Action	Rationale
		needle with syringe into side arm of tubing set and withdrawing the air; (2) Placing pen or pencil on tubing, distal to the air, and rolling tubing until air is displaced in drip chamber.
	Observe for signs and symptoms of air embolism. These include pallor, dyspnea, cough, cyncope, tachycardia, decreased blood pressure.	Air embolism occurs when air inadvertently enters the vascular system. Injection of more than 5 ml of air could be fatal. It occurs more frequently in the central veins, i.e., subclavian veins, and symptoms usually appear within 20 minutes.
	Place patient immediately on his or her left side in Trendelenburg's position.	Air is trapped in the right atrium, which prevents it from going to the lungs.
6. Pulmonary embolism	Report signs and symptoms of pulmonary embolism, i.e., restlessness, chest pain, cough, dyspnea, tachycardia.	Thrombus, originating in the peripheral vein, becomes an embolus and lodge in a large pulmonary vessel.
	Administer oxygen, analgesics, anticoagulants, and fluids as ordered.	Preventive measures should be taken, such as *never* forcefully irrigating an IV catheter to reestablish flow and avoiding the use of veins in the lower extremities.
7. Pulmonary edema	Check breathe sounds for chest rales. Check neck vein engorgement. Decrease IV flow rate.	Intravenous fluids administered too rapidly or in large amounts can cause overhydration Fluids "back-up" into lung tissue.
8. "Runaway" IV fluids	Monitor IV fluid every hour if not on an IV controller. Check IV controllers: flow rate, alarm-set.	Flow valve on IV tubing is opened. Alarm was not set on IV controller to note incorrect flow rate—rapid fluid infusion.
9. Hematoma	Observe for hematoma with unsuccessful attempts to start IV therapy.	Hematoma (blood tumor) are raised ecchymosed areas.
	Apply cold compresses followed with heat.	Cold causes vasoconstriction and decreases bleeding.
Additives to intravenous fluids	Recognize the untoward reactions of drugs in IV fluids: potassium, Levophed, low pH drugs, vitamins, antibiotics, antineoplastic drugs.	Potassium, antineoplastic drugs, and Levophed irritate the blood vessels and body tissue. Phlebitis is common with these drugs; and if infiltration occurs, sloughing of tissues results. Vitamins and antibiotics should not be mixed. They are incompatible.

TABLE 40. (Continued)

Assessment	Nursing Action	Rationale
	Stay with the patient 10 to 15 minutes when the patient is receiving drugs intravenously.	Allergic reactions can occur when drugs are administered IV.
	Inject drugs into IV bag and invert bag several times before administering.	Drugs distributed throughout the solutions insure proper dilution. *Do not* inject drugs, i.e., potassium, into the IV bag while it is being administered unless the IV is temporarily stopped and the bag is inverted several times.
Intake and output	Check urine output every 4 to 8 hours. If a critically ill patient is receiving potassium, urine output should be checked every hour.	If urine output is poor, overhydration can occur when excessive or continuous IV fluids are given. Potassium is excreted by the kidneys, and with decreased urine, hyperkalemia could result.

---

79 Refer back to Table 40 as needed.

When the patient receives IV therapy the nursing responsibility begins with assessment of the intravenous solutions ordered.

Intravenous solutions with < 240 mOsm/L (less than) are considered (hypo-osmolar/hyperosmolar) ____________ solutions. Continuous use of this type of solution could cause *____________.

Give an example of a solution that could become hypo-osmolar. *____________________.

— — — — — — — — — — — — — — — — — —

hypo-osmolar; water intoxication; 5% dextrose in water

---

80 Continuous use of hyperosmolar solutions could cause:

( ) a. Overhydration

( ) b. Dehydration

( ) c. Water intoxication

It is usually recommended that IV solutions with different osmolality be alternated. Give example of alternating solutions. *________________________.

— — — — — — — — — — — — — — — — — —

a. –; b. X; c. –;

5% D/W and 5% D/NSS or 5% D/½ NSS (0.45% NaCL)

---

81 Intravenous bags should be inspected for ____________________

If IV fluids are to run for 12 hours, the IV tubing with (macrodrip/microdrip) ____________ chamber should be used.

The macrodrip chamber delivers *__________ gtts (drops) per 1 ml. The microdrip chamber delivers __________ gtts per ml.

– – – – – – – – – – – – – – – – – –

leaks; microdrip; 10 or 15; 60

---

82 IV tubing should be changed at least every *__________ hours.

KVO means *____________________.

An IV bag should not be used longer than __________ hours.

– – – – – – – – – – – – – – – – – –

48 (24–48 hours); keep vein open; 24

---

83 Needles or IV catheters should be changed every *__________ days.

Which of the following needles/catheters are irritating to the veins and could cause phlebitis?

( ) a. Scalp vein needles
( ) b. Straight needles
( ) c. Polyvinyl-chloride catheters
( ) d. Polyethylene catheters
( ) e. Teflon catheters
( ) f. Silicone catheters

– – – – – – – – – – – – – – – – – –

3 days;
a. –; b. –; c. X; d. X; e. –; f. –

---

84 A problem with over-the-needle catheter is *____________________.

What can happen at the skin site with through-the-needle (in-the-needle) catheter? *____________________.

– – – – – – – – – – – – – – – – – –

severing the catheter with the needle tip. A leak can occur at the injection site

---

---

85 Which of the following body areas are preferred for insertion of needle or catheter?

( ) a. Hand veins
( ) b. Distal arm veins
( ) c. Leg veins

What two sites (body areas) should be avoided?

1. *____________________.

2. *____________________.

Explain why. *____________________________________.

\- - - - - - - - - - - - - - - - - -

a. X; b. X; c. –; 1. leg veins 2. affected limbs resulting from a CVA or mastectomy; poor circulation

---

86 IV fluids running too fast could cause (dehydration/overhydration) __________

____________________________________.

Give two symptoms of hypervolemia (overhydration).

1. ____________________.

2. ____________________.

\- - - - - - - - - - - - - - - - - -

overhydration
cough; dyspnea. Also neck vein engorgement and chest rales

---

87 KVO should run approximately ______________ ml per hour or __________ gtts per minute with microdrip chamber.

Name two types of solutions whose flow rate should be checked every 30 minutes to 1 hour.

1. *____________________.

2. *____________________.

\- - - - - - - - - - - - - - - - - -

10; 10
1. hyperosmolar solutions 2. toxic solutions, i.e., solution with potassium or Levophed

88 Give two ways for restoring IV fluids that are not running.

1. * ______________________________.

2. * ______________________________.

Explain the danger of forceful IV catheter irrigation when the intravenous fluid has stopped. * ______________________________.

- - - - - - - - - - - - - - - - -

1. opening flow clamp 2. milking the tubing. Also raise bag, reposition extremity.
Irrigation could dislodge clot(s) and cause an embolus (emboli) to the lungs

---

89 How high should an IV bag hang above the infusion site?
( ) a. 2½ to 3 inches
( ) b. 2½ to 3 feet
( ) c. 5 to 6 feet

- - - - - - - - - - - - - - - - -

a. –; b. X; c. –

---

90 Indicate which of the following problems/complications can result from intravenous therapy.
( ) a. Infiltration
( ) b. Phlebitis
( ) c. Infections (septicemia)
( ) d. Bradycardia
( ) e. Speed shock
( ) f. Air embolus
( ) g. Pulmonary embolus
( ) h. Hematoma
( ) i. Pulmonary edema

- - - - - - - - - - - - - - - - -

a. X; b. X; c. X; d. –; e. X; f. X; g. X; h. X; i. X

---

91 If IV fluids infiltrate:

( ) a. the flow rate should be decreased

( ) b. the IV fluids should be discontinued and restarted

Phlebitis (inflammation of the vein) could result from an IV needle or catheter that has been in the vein too long; solutions with irritating drugs (potassium, Levophed); or hyperosmolar solutions (25% dextrose) for TPN-hyperalimentation. Give three symptoms of phlebitis.

______________, ______________, and * ______________

______________.

– – – – – – – – – – – – – – – – – –

a. –; b. X

redness; edema (swelling); skin warm to touch. Also, pain

---

92 How can pyrogenic reactions (septicemia) be prevented? ______________.

Give two symptoms of systemic infections.

1. ______________.

2. ______________.

– – – – – – – – – – – – – – – – – –

Aseptic technique

1. chills 2. fever. Others: headache, tachycardia

---

93 What is speed shock? * ______________.

What is a hematoma? * ______________.

– – – – – – – – – – – – – – – – – –

Drugs given rapidly in solution. This increases the drug concentration in the body.

Blood tumor *or* raised ecchymosed area

---

94 Air embolus could be fatal if more than ______________ ml of air is injected in the vein.

If an air embolus is expected, what should be done? * ______________

______________.

– – – – – – – – – – – – – – – – – –

Place the patient immediately on his or her left side

---

95 Pulmonary embolus results when a thrombus in the peripheral veins becomes an embolus. It travels to what organ? ______________________.

Give two ways in which a pulmonary embolus could occur.

1. * ______________________.
2. * ______________________.

- - - - - - - - - - - - - - - - -

lungs
1. forcefully irrigating a clotted IV needle or catheter; 2. IV fluids given in the lower extremities

---

96 Match the symptoms for air embolus and pulmonary embolus. Refer to Table 40 as needed.

1. Air embolus
2. Pulmonary embolus

( ) a. Restlessness
( ) b. Chest pain
( ) c. Pallor
( ) ( ) d. Cough
( ) ( ) e. Dyspnea
( ) ( ) f. Tachycardia

- - - - - - - - - - - - - - - - -

a. 2; b. 2; c. 1; d. 1 & 2; e. 1 & 2; f. 1 & 2

---

97 Which of the following drug additives in IV solutions could cause phlebitis or, if infiltrated, could cause sloughing of the tissue.

a. Potassium
b. Decadron (cortisone)
c. Antineoplastic agents
d. Low pH drugs (tetracycline)
e. Levophed

- - - - - - - - - - - - - - - - -

a. X; b. –; c. X; d. X; e. X

98 In critically ill patients who are receiving potassium in IV solutions, the urine output is monitored hourly. Why? * ____

____.

Do you know why potassium should not be injected in an IV bag while it is being administered? * ____.

To determine kidney function and also 80 to 90% of body potassium is excreted by the kidney and, with kidney impairment or shutdown, hyperkalemia could occur.

Potassium would not be properly diluted in the IV bag and its highest concentration would be in the lower part of the bag. High concentration of potassium is toxic to the myocardium (heart)

CASE REVIEW

Ms. McCann, age 74, has orders to receive 2 liters of 5% dextrose in saline (0.9% NaCl) daily. After several days, Ms. McCann's skin turgor was poor and her mucous membranes were dry. Her pulse rate had increased 26 beats per minute.

1. Give the type of solution tonicity (osmolality) of 5% dextrose in saline. ________

   ____________________.

2. What type of fluid imbalance did Ms. McCann have according to her symptoms?

   ____________________ Give three symptoms that were indicative of the fluid imbalance. * ____________________, * ____________________, and * ____________________.

The nurse has many responsibilities with assessment and monitoring intravenous therapy. Describe the nursing action in regard to:

3. Needle or catheter. * ____

   ____.

4. IV tubing. * ____

   ____.

5. Flow rate. * ____

   ____.

6. Injection site. * ____

   ____.

7. Prevention of infection. *____

____.

8. Prevention of air and pulmonary emboli.

   a. *______________________________.

   b. *______________________________.

9. Drug additives. *____

____.

10. Urine output. *____

____.

- - - - - - - - - - - - - - - -

1. hyperosmolar
2. dehydration; poor skin turgor, dry mucous membrane, and increased heart rate
3. Needles and catheters should be changed every 3 days with exceptions. Check for leaks with through-the-needle catheter and severing the catheter with over-the-needle catheter
4. Change IV tubing every 2 days
5. Regulate drip chamber to deliver specified drops per minute with macrodrip and microdrip chambers
6. Use hands or distal veins in the arm. Avoid using leg veins, antecubital fossa (elbow), or affected limb
7. Use aseptic technique when administering and caring for IV therapy
8. a. Remove air from the IV tubing
   b. Avoid irrigating a clotted catheter (not running IV)
9. Check for infiltration and phlebitis
10. Monitor urine output q4 to 8 hours and qh when potassium is in the solution

NURSING DIAGNOSES

Potential alteration in fluid volume (deficit or excess) related to excessive fluid infusion, incorrect flow-rate regulation, insufficient fluid replacement, and continuous use of hypo-hyperosmolar solutions.

Potential for injury (body tissues and vessels) related to infiltration, phlebitis of concentrated,solutions, or additives in solutions; e.g., potassium chloride.

Potential knowledge deficit related to a lack of understanding of intravenous fluid administration and/or infusion problems and complications.

Potential effective breathing patterns related to chest pain, dyspnea secondary to pulmonary embolus, pulmonary edema, and air embolus.

Potential for infection related to septic technique with intravenous preparation and administration.

NURSING ACTIONS

1. Assess the osmolality of the intravenous solutions. Recognize hypo-osmolar solutions < 240 mEq/L used continuously could cause water intoxication. Five percent dextrose in water if used continuously becomes a hypo-osmolar solution. Hyperosmolar solution can cause dehydration.
2. Observe for signs and symptoms of water intoxication, i.e., headaches, behavioral changes (confusion, restlessness, delirium); and of dehydration, i.e., dry mucous membrane, poor turgor, shocklike symptoms.
3. Check expiration date of IV fluids and blood. Return outdated solutions.
4. Use IV tubing with macrodrip chamber (10 or 15 gtts per 1 ml) for IV fluids to run 8 hours or less, and microdrip chamber (60 gtts per 1 ml) for IV fluids to run 10 hours or longer. KVO (keep vein open) should run no less than 10 ml per hour (10 gtts per minute with microdrip).
5. Invert intravenous bag with drugs several times before administering. It promotes adequate drug dilution.
6. Hang IV bag 2½ to 3 feet above the infusion site. Raise IV bag if there is backflow of blood in the tubing (IV bag could have been too low).
7. Label bag for ml to be given per hour.
8. Regulate intravenous flow rate according to physician's order and patient's age. Rapidly administered IV fluids can cause overhydration, i.e., irritated cough, dyspnea, neck vein engorgement, and chest rales.
9. Check the needle tip of over-the-needle catheter after insertion. The needle tip could sever the catheter, break it off. Check for leaks at the injection site when using through(in)-the-needle catheter. The needle is larger than the catheter, causing leaks.
10. Change intravenous tubing every 24–48 hours and intravenous bag every 24 hours.
11. Change IV site every 3 days. Recognize that needles and Teflon and silicone IV catheters cause less phlebitis than polyvinyl-chloride and polyethylene catheters unless they are not removed after 4 or more days.
12. Avoid the use of leg veins, elbow (antecubital) site, and affected limb for intravenous therapy.
13. Use aseptic technique when inserting IV catheter, changing IV tubing and bag. Recognize symptoms of septicemia, i.e., chills, fever, tachycardia.
14. Assess the injection site hourly for infiltration and phlebitis when patient is receiving drugs that can cause vein and tissue irritation, i.e., potassium, Levophed, antineoplastic agents, antibiotics. Most of these drugs, if infiltrated, can cause sloughing of the tissue.
15. Assess for signs and symptoms of phlebitis, i.e., red, warm to touch, edematous, pain.
16. Observe the injection site for infiltration, i.e., swelling, coolness at site, pain. Discontinue IV and restart at another site.
17. Do not forcefully irrigate IV catheters. Clots can be dislodged at catheter site

and become an embolus. To restore IV flow, open clamp, milk the tubing, raise height of IV bag, or reposition extremity.

18. Avoid administering drugs rapidly in intravenous solutions. Speed shock could occur due to drug concentration and accumulation.
19. Remove air from IV tubing to prevent an air embolism. Have patient with a subclavian catheter perform valsalva procedure when changing tubing. Injection of 5 ml of air could be fatal.
20. Place patient immediately on his left side with feet higher than head when air embolism is suspected.
21. Recognize symptoms of pulmonary embolism, i.e., restlessness, chest pain, cough, dyspnea, and tachycardia.
22. Apply cold compresses followed later with heat to hematoma site, according to the physician's orders.
23. Monitor urine output hourly when giving potassium in IV solution to a critically ill patient.

REFERENCES

Abbott Laboratories: *Fluid and Electrolytes.* North Chicago, 1970, pp 25-30.

Anderson MA, Aker, SN, Hickman, RO: The double-lumen Hickman catheter. *Am J Nurs* 82(2): 272–273, February 1982.

Birdsall C: When is TPN safe? *Am J Nurs* 83(1):73, January 1983.

Blood Volume. *Pitoclinic.* Ames Co, 1962, pp 5–10.

Borgen L: Total parenteral nutrition. *Am J Nurs* 78(2):224–228, February 1978.

Brunner LS, Suddarth DS: *Textbook of Medical-Surgical Nursing,* ed 5. Philadelphia, JB Lippincott Co, 1984, pp 152–161, 780–786.

Burgess RE: Fluid and electrolytes. *Am J Nurs* 65 (10): 93-94, 1965.

Coco CD: *Intravenous Therapy.* St. Louis, CV Mosby Co, 1980, pp 36-52, 112-119, 135-145.

Colley R, Phillip K: Helping with hyperalimentation. *Nursing '73* 3 (7): 6-17, 1973.

Fox B, Stegall B: Take precautions now. *Nursing, 85* 15(5):48–49, May 1985.

Geolot DH, McKinney NP: Administering parenteral drugs. *Am J Nurs* 75 (5): 788-793, 1975.

Josephson A, Kliman A, Shively J: Transfusions. *Patient Care* 1:118-139, August 15, 1971.

——: IV therapy, a special feature: Fundamentals of IV maintenance. *Am J News* 79(7):1274+, July 1979.

Kurdi WJ: Refining your IV therapy techniques. *Nursing '75* 5 (11):40-47, 1975.

Levenstein BP: Intravenous therapy: A nursing specialty. *Nurs Clin North Am* 1:259-265, June 1966.

Luckmann J, Sorenesen KC: *Medical-Surgical Nursing,* ed 2. Philadelphia, WB Saunders Co, 1980, pp 216-225.

Maxwell H, Kliman CR (ed): *Clinical Disorders of Fluid and Electrolyte Metabolism.* New York, McGraw-Hill Book Co, 1972, p 593.

McGaw Laboratories: *Guide to Parenteral Fluid Therapy.* Glendale, CA, 1963, pp 28-48; 52-55.

Metheney NM, Snively WD: *Nurses' Handbook of Fluid Balance,* ed 4. Philadelphia, JB Lippincott Co, 1983, pp 135–184.

Montag M, Swenson R: *Fundamentals in Nursing Care,* ed 3. Philadelphia, WB Saunders Co, 1959, pp 297-314.

Sager DP, Bomar SK: *Intravenous Medications.* Philadelphia, JB Lippincott Co, 1980, pp 29-134.

Schakenback LH, Dennis M: And now, a quad-lumen IV catheter. *Nursing, 85* 15(11):50–51, November 1985.

Snider MA: Helpful hints on IV's. *Am J Nurs* 74: 1978-1981, 1974.
Statland H: *Fluid and Electrolytes in Practice*, ed 3. Philadelphia, JB Lippincott Co, 1963, pp 155-164.
Transfusion: What blood component does your patient really need? *Nursing Update* 4 (3):1-11, 1973.
Travenol Laboratories, Inc: *Guide to Fluid Therapy*, Deerfield, IL, 1970, pp 91-107.
Ungvarski PJ: Parenteral therapy. *Am J Nurs* 76: 1974-1977, 1976.
Wilhelm L: Helping your patient settle in with TPN. *Nursing, 85* 15(4):60, April 1985.
Wilson JA: Infection control in intravenous therapy. *Heart and Lung* 5(3):430–436, 1976.
Wool, NL et al: Hickman catheter placement simplified. *Am J Surg* 145:283–284, February 1983.

CHAPTER FIVE

Clinical Conditions of Fluid and Electrolyte Imbalance

BEHAVIORAL OBJECTIVES

Upon completion of this chapter, the student will be able to:

Explain the physiologic factors leading to dehydration (extracellular fluid volume deficit), edema (extracellular fluid volume excess), water intoxication (intracellular fluid volume excess), and shock.
Explain the extracellular fluid volume shifts in relation to dehydration and overhydration.
State various clinical causes associated with dehydration, edema, water intoxication, and shock.
Explain the clinical management needed to alleviate the four clinical conditions.
Explain and observe clinically the signs and symptoms of these four conditions. (You may need some assistance in recognizing all these symptoms.)
Differentiate between the four types of shock and be able to identify the type of shock in a clinical situation.
Explain how clinical management will differ according to the type of shock.

INTRODUCTION

Many of the disease entities are inclined to have some sort of fluid and electrolyte imbalance. Much of the imbalance is the result of certain clinical conditions, i.e., dehydration, edema, water intoxication, and shock. Each of these clinical conditions is discussed in this chapter, including the physiologic factors involved, causes, clinical applications with clinical symptoms, and clinical management. Several of the clinical conditions have examples that add to the clarification and understanding of the fluid and electrolyte imbalance present and clinical management needed.

Again, it must be emphasized that the physician computes and orders fluid replacements. However, the nurse should understand these clinical conditions and reasons for their occurrence and should assess physical changes that occur before and during clinical management.

There are four situational reviews and four sets of nursing diagnoses and nursing actions for dehydration, edema, water intoxication, and shock.

An asterisk (*) on an answer line indicates a multiple-word answer. The meanings for the following symbols are: ↑ increased, ↓ decreased, > greater than, < less than.

DEHYDRATION–EXTRACELLULAR FLUID VOLUME DEFICIT

---

1 Dehydration means the loss of water from the body. The water is lost mainly from the extracellular fluid compartment. Severe water loss from the body can cause a condition known as ______________.

- - - - - - - - - - - - - - - - -

dehydration

---

2 Another name for dehydration is *______________

______________.

- - - - - - - - - - - - - - - - -

extracellular fluid volume deficit

---

3 The fluid loss leads to a drop in blood volume and an increase in serum sodium. This is known as *hyperosmolar dehydration.*

With the fluid loss, there is actually a decrease in the total amount of sodium remaining in the body, but the serum sodium level will be elevated since the water has been lost in excess of sodium.

Hyperosmolar dehydration occurs when the serum sodium level is (elevated/decreased) ______________ with the fluid loss.

- - - - - - - - - - - - - - - - -

elevated

---

4 When water has been lost in excess of sodium, the serum sodium level would be

________________, even though the total amount of sodium remaining in

the body would be ________________.

\- - - - - - - - - - - - - - - - -

elevated; decreased

---

5 The plasma/serum osmolality increases with the retained sodium, causing water to be drawn from the cells. With the elevation of serum sodium, the extracellular

fluid becomes (hyperosmolar/hypo-osmolar), ________________

resulting in a(an) (increase/decrease) ________________ in plasma/serum osmo-

lality, which will cause a withdrawal of fluid from the *________________

________________.

\- - - - - - - - - - - - - - - - -

hyperosmolar; increase; cells or intracellular compartment

---

6 The hyperosmolar extracellular fluid will cause:

( ) a. intracellular dehydration
( ) b. intracellular hydration

Explain. *________________

________________.

\- - - - - - - - - - - - - - - - -

a. X; b. –
The hyperosmolar extracellular fluid will pull intracellular fluid from the cells by osmosis

---

7 Potassium, magnesium, phosphates, and some protein are lost with the intracellular water.

Increased serum osmolality results in intracellular fluid (loss/excess) ________.

With intracellular fluid loss, the following electrolytes are also lost:

( ) a. potassium
( ) b. sodium
( ) c. magnesium
( ) d. chloride
( ) e. phosphate

Why? *____.

---

loss
a. X; b. –; c. X; d. –; e. X
These electrolytes are mainly found in the cells

Table 41 describes the degrees of dehydration in relation to the percentage of body weight loss, symptoms, and body water deficit by liter for a man weighing 150 pounds. Study this table carefully; be able to name the degrees of dehydration, their symptoms, the percentage of body weight loss, and an estimation of body fluid loss in liters. Hopefully, you will be able to recognize and identify degrees of dehydration that can occur to your patients during your clinical experience. Refer back to this table as you find it necessary.

TABLE 41. DEGREES OF DEHYDRATION

Degrees of Dehydration	Percentage of Body Weight Loss (%)	Symptoms	Body Water Deficit by Liter
Mild dehydration	2	1. Thirst	1–2
Marked dehydration	5	1. Marked thirst 2. Dry mucous membranes 3. Dryness and wrinkling of skin—poor skin turgor 4. Acid-base equilibrium toward greater acidity 5. Temperature—low grade elevation, e.g., 99° F (37.2° C) 6. Tachycardia (pulse greater than 100) as blood volume drops 7. Respiration 28 and ↑ 8. Systolic BP 10–15 mm Hg ↓ in standing position	3–5

TABLE 41. (Continued)

Degrees of Dehydration	Percentage of Body Weight Loss (%)	Symptoms	Body Water Deficit by Liter
		9. Hand veins: slow filling with hand lowered 10. Urine volume: <25 ml/hr and highly concentrated 11. Specific gravity: >1.030 12. Body weight loss 13. Hct ↑, Hgb ↑, BUN ↑	
Severe dehydration	8	1. Same symptoms as marked dehydration, plus: 2. Skin becomes flushed 3. Systolic BP 60 or ↓ 4. Behavioral changes, e.g., restlessness, irritability, disorientation, and delirium	5-10
Fatal dehydration	22-30 total body water loss can prove fatal	1. Anuria 2. Coma leading to death	

---

8 If water loss reaches 8% of body weight, then severe dehydration results. Life will not continue with a water loss of *________ % of body weight.

The four degrees of dehydration are:

1. ________________.
2. ________________.
3. ________________.
4. ________________.

- - - - - - - - - - - - - - - - -

22-30; 1. mild; 2. marked; 3. severe; 4. fatal

---

9 The percentage of body weight loss is a guide for replacement fluid therapy.

What percentage loss is associated with mild dehydration? ____________.

The symptom for mild dehydration is ________________.

- - - - - - - - - - - - - - - - -

2%; thirst

---

10 Symptoms for marked dehydration would be:

( ) a. Disorientation
( ) b. Dry mucous membranes
( ) c. Dryness and wrinkling of skin
( ) d. Irritability
( ) e. Marked thirst
( ) f. Delirium
( ) g. Tachycardia as blood volume drops
( ) h. Body temperature 97°F
( ) i. Urine volume decreased
( ) j. Specific gravity of urine of 1.010
( ) k. Urine highly concentrated
( ) l. Respiration ↑
( ) m. Systolic BP ↑
( ) n. Slow-filling hand veins

What percentage loss is associated with marked dehydration? ________.

- - - - - - - - - - - - - - - - -

a. –; b. X; c. X; d. –; e. X; f. –; g. X; h. –; i. X; j. –; k. X; l. X; m. –; n. X; 5%

---

11 With marked and severe dehydration, the hematocrit, hemoglobin, and blood urea nitrogen (BUN) would be (increased/decreased) ________.

- - - - - - - - - - - - - - - - -

increased

---

12 Symptoms for severe dehydration would be:

( ) a. Bradycardia
( ) b. Tachycardia as blood volume drops
( ) c. Temperature 99.6°F
( ) d. Urine volume is increased
( ) e. Specific gravity of urine of 1.035 and higher
( ) f. Skin flushed
( ) g. Irritability
( ) h. Restlessness and disorientation
( ) i. Specific gravity of urine lower than 1.020
( ) j. Marked thirst

What percentage loss is associated with severe dehydration? ________.

- - - - - - - - - - - - - - - - -

a. –; b. X; c. X; d. –; e. X; f. X; g. X; h. X; i. –; j. X
8%

---

13 The percentage of body weight loss is a guide for *________________ therapy.

- - - - - - - - - - - - - - - - - -

replacement fluid

---

14 For the average body weight of 150 pounds, the amount of body water loss for mild dehydration is *____________, for marked dehydration is *____________, and for severe dehydration is *____________________.

- - - - - - - - - - - - - - - - - -

1-2 liters; 3-5 liters; 5-10 liters

CLINICAL APPLICATIONS

---

15 In early dehydration, the fluid is lost in equal quantities from both the extracellular and intracellular fluid spaces.

As dehydration continues, fluid is lost in greater quantities from the extracellular than from the intracellular fluid space and so there would be an ECF (excess/deficit) ____________________.

- - - - - - - - - - - - - - - - - -

deficit

---

16 When dehydration is severe, cellular dehydration from the intracellular fluid space occurs.

ECF deficit leads to ICF ____________________.

- - - - - - - - - - - - - - - - - -

deficit

---

17 As rule, hypernatremia results from water depletion. The amount of water loss is in excess of the amount of ________________ loss.

If the urinary excretion of potassium is decreased, then (hypokalemia/hyperkalemia) ____________________ will result.

- - - - - - - - - - - - - - - - - -

sodium, hyperkalemia

---

18 The hematocrit and the plasma proteins are elevated as a result of hemoconcentration (increased blood cells and a decrease of fluid).

Hemoconcentration occurs with dehydration because *________________

____.

The hematocrit and the plasma proteins are (elevated/decreased) __________

____________ in marked dehydration.

- - - - - - - - - - - - - - - - -

there is an increase of red blood cells and a decrease of fluid; elevated

---

19 In dehydration, the urine soon becomes concentrated with a specific gravity of 1.030 or higher. The urinary output is (increased/decreased) ______________.

Hypernatremia frequently results from *________________.

- - - - - - - - - - - - - - - - -

decreased; water depletion or water loss

Mild dehydration can result from fever, sweating, or insufficient water intake. Vomiting and diarrhea are common causes of marked dehydration.

Table 42 lists the causes of extracellular fluid volume (ECFV) deficit or dehydration. Study the table carefully and refer to it as needed.

TABLE 42. CONDITIONS CAUSING ECFV DEFICIT/DEHYDRATION

Conditions	Rationale
Insufficient water intake	Decreased water intake is a common cause of mild dehydration, especially in the aged.
Vomiting	Vomiting is a common cause of marked dehydration. Persistent vomiting could cause severe fluid loss, which often leads to shock. Potassium, sodium, chloride, plentiful in the GI tract, are also lost.
Diarrhea	Diarrhea is also a cause of marked dehydration. Continuous diarrhea can produce severe loss of fluids and electrolytes (K, Na, HCO_3)
Sweating, increased environmental temperature, fever, and muscular exercise	These conditions usually produce mild dehydration which could become marked if they persist. Sodium is also lost through the skin.
Fistulous drainage	Fluid loss can result from fistulas (abnormal passageways from the GI tract to the external space).
Gastric or gastrointestinal suction	Fluid loss occurs from GI suction. If IV fluid replacement is not given or is inadequate, moderate/marked dehydration sets in.

TABLE 42. (Continued)

Conditions	Rationale
Hypovolemic Causes:	
Hemorrhage	Excessive blood loss decreases intravascular fluid volume which causes marked or severe dehydration. If hemorrhage occurs rapidly, fluid shifts to compensate blood loss could be inadequate.
Burns	Fluid shifts from the vascular to the burn site/interstitial space (third space fluid) are responsible for a circulating fluid volume deficit.
Ascites	Fluid shifts to the peritoneal space form ascites (third space fluid) and a decrease in the circulating fluid volume.
Intestinal obstruction	Fluid accumulates at the intestinal obstruction site, thus decreasing the fluid volume.

---

20 Hypovolemia may be a new term to you. Hypo indicates ____________
Volemia comes from the Latin word volumen, meaning "volume."

Actually, hypovolemia is a diminished volume of circulating blood or intravascular fluid. It is frequently referred to as low blood volume.

\- - - - - - - - - - - - - - - - - -

loss, less, deficit, or diminished

---

21 Some causes for mild dehydration are __________, __________,
and * __________________.

Other causes of dehydration are:

( ) a. Diarrhea
( ) b. Persistent vomiting
( ) c. Excessive water intake
( ) d. Hemorrhage
( ) e. Congestive heart failure
( ) f. Metabolic acidosis
( ) g. Increased caloric intake
( ) h. Burns

\- - - - - - - - - - - - - - - - - -

fever; sweating; insufficient water intake
a. X; b. X; c. –; d. X; e. –; f. X; g. –; h. X

---

22 The nurse can make a quick assessment of dehydration or hypovolemia by checking the peripheral veins in the hand. Hold the hand above heart level and then lower the hand below heart level. The peripheral veins in the hand below heart level should be engorged within 5-10 seconds for normal blood volume and circulating blood flow.

If the peripheral veins do not engorge in 10 seconds, this may be an indication of *________________.

\- - - - - - - - - - - - - - - - -

dehydration or hypovolemia (low blood volume)

---

23 Body weight is an important tool for assessing fluid imbalance. Two and one-half pounds (2½) of body weight loss is equivalent to 1 liter of water loss.

Intake and output give the approximate amount of body fluid intake and output; however, *________________ gives the more accurate assessment of fluid balance.

\- - - - - - - - - - - - - - - - -

body weight

---

24 Vital signs comprise another tool for assessing dehydration or hypovolemia.

With dehydration the:

temperature would be *________________.

pulse *________________.

respirations ________________.

blood pressure *________________.

The urine volume would be *________________.

\- - - - - - - - - - - - - - - - -

temperature: low grade elevation
pulse: tachycardia (rate over 100)
respirations: increased
systolic blood pressure: ↓ 10-15 mm Hg (standing position)
urine volume: decreased or small amount and highly concentrated
(Kidney damage can occur if the systolic blood pressure is less than 60 for several hours.)

CLINICAL MANAGEMENT

In replacing body water loss, the total fluid deficit is estimated according to the percentage of body weight loss. The physician computes the fluid replacement for his patient. The following is only an example. Many physicians use this method for replacement of fluid loss.

Mr. Smith, who was admitted to the hospital, had a weight loss of 10 pounds due to dehydration. His weight had originally been 154 pounds, or 70 kg (kilograms). To determine the percentage of body weight loss, divide the weight loss by the original weight; therefore, 10 ÷ 154 = .06 or 6%. To determine the total fluid loss, multiply the percentage of body weight loss by kg of body weight; therefore, .06 X 70 kg = 4.2 liters.

---

25 Clinically, Mr. Smith would be considered to have:
( ) a. Mild dehydration
( ) b. Marked dehydration
( ) c. Severe dehydration

\- - - - - - - - - - - - - - - - -

a. –; b. X; c. –

---

26 To determine the percentage of body weight loss, one would *______________

____.

To determine the total fluid loss, one would *______________

____.

\- - - - - - - - - - - - - - - - -

divide the weight loss by the original weight
multiply the percentage of body weight loss by kilograms of body weight

---

27 One-third of body water deficit is from ECF (extracellular fluid), and two-thirds of body water deficit is from ICF (intracellular fluid) (Chapter 1). To determine replacement therapy for the first day, you would multiply:
(a) 1/3 X 4.2 L = 1.4 L (ECF replacement)
Replacement fluid needed for ECF would be ______ L or ______ ml.
(b) (2/3 X 4.2 L = 2.8 L (ICF replacement)
Replacement fluid needed for ICF would be ______ L or ______ ml.
(c) 2.5 L or 2500 ml are added to replace the current day's losses (constant daily amount)
The total fluid replacement for the first day would be ______ L or ______ ml (sum of ECF and ICF and current day's losses).

\- - - - - - - - - - - - - - - - -

1.4; 1400; 2.8; 2800; 6.7; 6700

---

28 One-third of the water deficit would be from the * ______________________,
and two-thirds of the water deficit from the * ______________________.

- - - - - - - - - - - - - - - - -

extracellular fluid; intracellular (cellular) fluid

---

29 The sodium deficit would be the amount contained in the extracellular fluid loss of 1.4 L. Can you explain the deficit? * ______________________
______________________.

The potassium deficit would be the amount contained in the intracellular fluid loss of 2.8 L. Can you explain the potassium deficit? * ______________________
____
______________________.

- - - - - - - - - - - - - - - - -

Sodium is the main cation of ECF, so with ECF loss Na would accompany it. However, the serum sodium level may be elevated if the fluid loss is greater than Na loss
Potassium is the main cation of ICF, so with ICF loss K would accompany it. However, the serum potassium level may be elevated because the K leaves the cells and accumulates in the ECF. If diuresis occurs, the serum K may be low

---

30 In severe dehydration, cellular breakdown usually occurs, and acid metabolites, such as lactic acid, are released from the cells; thus, metabolic acidosis results. The serum CO_2 and the arterial bicarbonate (HCO_3) levels would be decreased

Bicarbonate is usually added to a liter or two of IV fluids. Constant use of saline (NaCl) would not be indicated. Do you know why? * ______________________
____
______________________.

- - - - - - - - - - - - - - - - -

A low HCO_3 or serum CO_2 indicates a lack of HCO_3 ion, which is needed to neutralize the body's acidotic state. If Cl^- was administered, it would combine with H^+ and increase acidosis

---

31 The following suggested solution replacement would be needed in correcting dehydration:

1500 ml of lactated Ringer's to replace ECF losses.

500 ml of M/6 sodium lactate to make up sodium deficit and help correct metabolic acidosis. This solution will increase serum CO_2.

4700 ml of 5% dextrose in water to replace water deficit and increase urinary output. Potassium chloride, 40-90 mEq/L, is added to 3 L to replace potassium loss, according to the serum potassium level.

When the potassium is restored in the cells, would the intracellular fluid be increased or decreased? ______________. Why do you think this might happen? *____

____.

When potassium is being administered in the form of KCl (potassium chloride), explain your concern about the patient's urinary output. *______________

____

____.

- - - - - - - - - - - - - - - - - -

increased. Fluid would flow into the cells as potassium returns to the cells. Cellular function would be restored.

Poor urinary output leads to potassium excess, so urine output should be 250 ml/8 hours

---

32 In correcting dehydration, the steps would be to *______________________

______________________ and *______________________.

- - - - - - - - - - - - - - - - - -

replace fluid volume; reduce osmolality

33 Lactated Ringer's solution is helpful in treating dehydration because:
( ) a. It resembles the electrolytic structure of normal blood serum.
( ) b. It will replace the extracellular fluid volume.
( ) c. It will replace all of the electrolyte loss.
( ) d. It will replace potassium loss.
M/6 sodium lactate is helpful in treating dehydration because:
( ) a. It replaces the sodium loss.
( ) b. It will aid in decreasing the CO_2 combining power of the plasma.
( ) c. It will aid in increasing the CO_2 combining power of the plasma.
( ) d. It is helpful in the correction of metabolic acidosis.
( ) e. It is helpful in the correction of metabolic alkalosis.
Dextrose 5% in water is helpful in treating dehydration because:
( ) a. It will replace water deficit.
( ) b. It will aid in increasing urine output.
( ) c. It will aid in decreasing urine output.
( ) d. It will replace the sodium deficit.

- - - - - - - - - - - - - - - -

a. X; b. X; c. –; d. –
a. X; b. –; c. X; d. X; e. –
a. X; b. X; c. –; d. –

---

34 Mild dehydration is frequently treated with dextrose, water, and small amounts of electrolytes.

Dextrose 5% in water is frequently given first, followed by a solution of low electrolyte content with 5% dextrose. An example of this solution could be

* ______________________________. (These solutions could be given in reverse, according to patient's condition and physician's choice.)

When administering 5% dextrose in water, dextrose is metabolized quickly, leaving ______________ to replace fluid deficit.

- - - - - - - - - - - - - - - -

lactated Ringer's or Ringer's solution; water

CASE REVIEW

Mr. Cooper, age 55, has been vomiting persistently for 3 days. On admission, he weighed 153 pounds. His original weight was 165 pounds (75 kg). The nurse assessed his fluid state and noted that his mucous membranes and skin were dry. His temperature was 99.4°F (37.5°C), pulse 112, respirations 32, blood pressure 110/88, and urine output in 8 hours 125 ml with a specific gravity of 1.036. Electrolyte findings were serum K 3.5 mEq/L, Na 154 mEq/L, and Cl 102 mEq/L. His hematocrit and BUN were elevated.

1. What is the type of Mr. Cooper's dehydration? * ______________________ . Explain your answer. * ______________________ .
2. What is another name for dehydration? * ______________________ .
3. The nurse assesses Mr. Cooper's body fluid state. Name four of his symptoms and laboratory findings that are suggestive of fluid imbalance (dehydration).
   a. * ______________________ .
   b. * ______________________ .
   c. * ______________________ .
   d. * ______________________ .
4. Determine the percentage of Mr. Cooper's body weight loss. * ______________________ .
5. Clinically, Mr. Cooper would have:
   ( ) a. Mild dehydration
   ( ) b. Marked dehydration
   ( ) c. Severe dehydration
6. Mr. Cooper's total fluid loss is * ______________________ .
   (Work space is provided.)
7. a. Replacement fluid needed for ECF would be * ______________________ .
   b. Replacement fluid needed for ICF would be * ______________________ .
   c. Replacement for current day's losses would be * ______________________ .
   (constant daily amount)
8. What two laboratory results were indicative of dehydration other than the electrolytes? * ______________________ and * ______________________ .
9. Hypernatremia frequently results from * ______________________ .
10. Mr. Cooper's serum potassium level of 3.5 mEq/L is considered low average. Do you think his cellular potassium is (increased/decreased)? Explain your answer.
    * ______________________ .
11. If Mr. Cooper is hydrated without potassium added, the nurse would expect his serum potassium to be (increased/decreased). ______________________ .

Why? *____

____.

12. What intravenous solution resembles the electrolyte structure of plasma? *______

____.

13. Name three causes of mild dehydration.

    a. ____.

    b. ____.

    c. *____.

---

1. Hyperosmolar dehydration. Serum sodium is elevated with the fluid loss
2. extracellular fluid volume deficit, or hypovolemia
3. a. Dry mucous membrane and dry skin.
   b. Vital signs—temperature slightly elevated, tachycardia, respiration increased, systolic blood pressure ↓
   c. elevated sodium level
   d. Hct and BUN increased

   Others—weight loss, urinary output ↓
4. 12 ÷ 165 = 0.07 X 100 = 7%
5. b. marked dehydration
6. 75 kg X 0.07 = 5.25 L loss
7. a. 1/3 X 5.25 L = 1.75 L or 1750 ml

   b. 2/3 X 5.25 L = 3.5 L or 3500 ml

   c. 2.5 L or 2500 ml
8. elevated Hct; BUN
9. water depletion
10. decreased. With dehydration, K leaves cells
11. decreased. With hydration, K moves from ECF back into cells; thus, serum K would be lowered
12. lactated Ringer's
13. a. fever
    b. sweating
    c. insufficient water intake

NURSING DIAGNOSES

Potential alteration in fluid volume deficit related to vomiting, diarrhea, lack of fluid intake, and hemorrhage.

Potential alteration in fluid volume excess related to excessive amounts of fluid infusion while correcting ECFV deficit/dehydration.
Potential alteration in tissue perfusion related to decreased blood circulation from inadequate fluid volume.
Potential alteration in urinary elimination related to inadequate fluid volume resulting from dehydration.

NURSING ACTIONS

1. Recognize causes of dehydration, i.e., insufficient water intake, fever, diaphoresis, diarrhea, persistent vomiting, metabolic acidosis, severe diabetic acidosis, hemorrhage, and burns.
2. Assess for signs and symptoms of mild dehydration, i.e., thirst; marked dehydration, i.e., marked thirst, dry mucous membranes and secretions, poor skin turgor, decreased urine output, shocklike symptoms (↑ P, ↑ R, ↑ Tsl, ↓ BPsl) and ↑ BUN, ↑ Hct, ↑ Hgb; severe dehydration, i.e., same symptoms as marked dehydration, ↓ BP60, and behavioral changes.
3. Check laboratory results and report abnormal findings. Serum sodium may be elevated due to fluid loss in excess to sodium. The serum potassium level may be normal or elevated depending on the severity of the dehydration and the degree of cellular breakdown. Once the patient has been hydrated, serum potassium decreases (hypokalemia). Potassium is diluted, and, with correction of clinical cause, potassium moves back into the cell.
4. Calculate the degree of body fluid loss according to loss of weight.
5. Regulate intravenous fluid to run as ordered, or as calculated according to degree of dehydration (see Frame 27).
6. Assess for signs and symptoms of metabolic acidosis, such as deep, rapid breathing, flushed skin, restlessness, ↓ serum CO_2, ↓ arterial HCO_3. Acidosis can occur from cellular breakdown due to dehydration. Acid metabolites (lactic acid and others) are released from the cells.

EDEMA—EXTRACELLULAR FLUID VOLUME EXCESS

35 Extracellular fluid volume (ECFV) excess could be the result of fluid excess in the interstitial spaces (tissues) and/or intravascular space (vessels). It usually relates to the excess fluid in tissues of the extremities (peripheral edema) or lung tissues (pulmonary edema).

Hypervolemia and overhydration are interchangeable terms for ECFV excess/edema. Actually they can contribute to the cause of ECFV excess or edema.

Another name for edema is * ______________________.

- - - - - - - - - - - - - - - - - -

extracellular fluid volume excess

36 Edema is the abnormal retention of fluid in the interstitial spaces of the extracellular fluid compartment or in serous cavities. Frequently, edema results from sodium retention in the body, causing a retention of water and an increase in extracellular fluid volume. Three names that can mean the same as extracellular fluid volume excess are ________________, ________________, and ________________.

- - - - - - - - - - - - - - - - -

hypervolemia; overhydration; edema

---

37 Edema is the abnormal retention of fluid in the *________________ ________________.

- - - - - - - - - - - - - - - - -

interstitial spaces of the ECF compartment or in serous cavities

---

Diagram 10 demonstrates the make-up of normal body fluid versus abnormal body fluid, such as with edema. As you recall from Chapter 1, 60% of the adult body weight is water; 40% of that is intracellular or cellular water, and 20% is extracellular water. Of the extracellular fluid, 15% is interstitial fluid and 5% is intravascular fluid or plasma. Note that with edema there is an increase of fluid in the interstitial space, which is between tissues and cells. The intracellular fluid may be decreased in extreme cases. Refer to this diagram as needed.

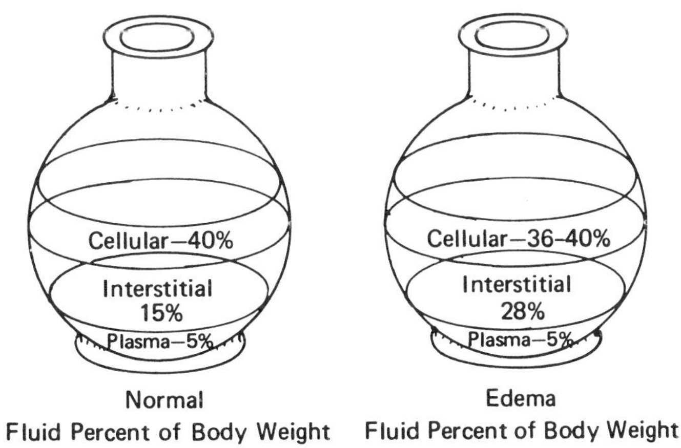

Diagram 10. Body fluid compartments and edema. (Adapted with permission, from H Statland: *Fluid and Electrolytes in Practice*, ed 3. Philadelphia, J.B. Lippincott Co., p 177.)

38 In edema, the greatest increase in volume occurs in which of the fluids? *______

______________.

---

interstitial fluid

---

39 The intracellular fluid may be (increased/decreased) ______________ in extreme cases.

---

decreased

PHYSIOLOGIC FACTORS

Table 43 gives five physiologic factors that can cause edema. With each physiologic factor, the rationale and examples of clinical situations in which edema will occur are given.

Study this table carefully; note whether there is an increase or decrease in the physiologic factors that can serve as an indication of edema. Be able to explain how edema occurs in the various clinical situations. Refer to this table as you find it necessary.

TABLE 43. PHYSIOLOGIC FACTORS LEADING TO EDEMA

Physiologic Factors		Rationale	Clinical Conditions
Plasma hydrostatic pressure in the capillaries	↑ Increased	Blood dammed in the venous system will cause "back" pressure in capillaries, thus raising capillary pressure. Increased capillary pressure will force more fluid into tissue areas, thus producing edema.	1. Congestive heart failure from increased venous pressure. 2. Kidney failure due to sodium and water retention. 3. Venous obstruction leading to varicose veins. 4. Pressure on veins because of swelling, constricting bandages, or casts.
Plasma colloid osmotic pressure	↓ Decreased	Decrease in plasma colloid osmotic pressure results from diminished plasma protein concentration. Decrease in protein content will cause water to flow from plasma into tissue spaces, thus causing edema.	1. Malnutrition due to lack of protein in diet. 2. Chronic diarrhea resulting in loss of protein. 3. Burns leading to loss of fluid containing protein through denuded skin. 4. Kidney disease, particularly nephrosis. 5. Cirrhosis of liver due to lack of protein. 6. Loss of plasma proteins through urine.

TABLE 43. (Continued)

Physiologic Factors	Rationale	Clinical Conditions
Capillary permeability	↑ Increased Increase in permeability of capillary membrane will allow plasma proteins to leak out of capillaries into interstitial space more rapidly than lymphatics can return them to circulation. Increased capillary permeability is predisposing factor to edema.	1. Bacterial inflammation causes increased porosity. 2. Allergic reactions. 3. Burns causing damage to capillaries. 4. Acute kidney disease, e.g., nephritis.
Sodium retention	↑ Increased Kidneys regulate level of sodium ions in extracellular fluid. Kidney function will depend on adequate blood flow. Inadequate blood flow, presence of excess aldosterone, or diseased kidneys, are predisposing factors to edema since they cause sodium retention.	1. Congestive heart failure (CHF) causing inadequate circulation of blood. 2. Renal failure—inadequate circulation of blood through kidneys. 3. Increased production of adrenal cortical hormones—aldosterone, cortisone, and hydrocortisone—will cause retention of sodium 4. Cirrhosis of liver. Diseased liver cannot destroy excess production of aldosterone. 5. Trauma resulting from fractures, burns, and surgery.
Lymphatic drainage	↓ Decreased Blockage of lymphatics will prevent return of proteins to circulation. Obstructed lymph flow is said to be high in protein content. With inadequate return of proteins to circulation, plasma colloid osmotic pressure will be decreased, thus causing edema.	1. Lymphatic obstruction e.g., cancer of lymphatic system. 2. Surgical removal of lymph nodes. 3. Elephantiasis, which is parasitic invasion of lymph channels, resulting in fibrous tissue growing in nodes, obstructing lymph flow. 4. Obesity because of inadequate supporting structures for lymphatics in lower extremities. Muscles are considered the supporting structures.

---

40 Place I for increased and D for decreased beside the physiologic factors as they occur in edema:

____ a. Capillary permeability

____ b. Sodium retention

____ c. Lymphatic drainage

____ d. Plasma colloid osmotic pressure

____ e. Plasma hydrostatic pressure in the capillaries

- - - - - - - - - - - - - - - - -

a. I; b. I; c. D; d. D; e. I

---

41 Blood backed up in the venous system will increase the capillary pressure, forcing more fluid into *____

______________________________________.

The clinical situations in which edema may occur as a result of an increase of plasma hydrostatic pressure are:

1. *____.
2. *____.
3. *____.
4. *____.

- - - - - - - - - - - - - - - - -

the tissue spaces (interstitial spaces)

1. CHF (congestive heart failure)
2. kidney failure
3. venous obstruction
4. pressure on the veins, e.g., from casts or bandages

---

42 A decrease in plasma protein will result in a decrease in the plasma colloid osmotic pressure. This will cause water to flow from the plasma into *______________________________.

Name at least four situations in which edema occurs as a result of a decrease in plasma colloid osmotic pressure:

1. *______________________________.
2. *______________________________.
3. *______________________________.
4. *______________________________.

— — — — — — — — — — — — — — — —

the tissue spaces (interstitial spaces)

1. malnutrition
2. chronic diarrhea
3. kidney diseases
4. cirrhosis of the liver
5. burns
6. urinary excretion of protein

---

43 An increase in the capillary membrane permeability will permit plasma proteins to escape from ______________, causing more water to flow into *______________________________.

Name at least three situations in which edema occurs as a result of increased capillary permeability:

1. *______________________________.
2. *______________________________.
3. *______________________________.

— — — — — — — — — — — — — — — —

capillaries; the tissue spaces (interstitial spaces)

1. bacterial inflammation
2. allergic reactions
3. acute kidney disease, e.g., nephritis
4. burns

---

44 The kidneys regulate the level of ________________ ions in the extracellular fluid. An inadequate blood flow, the presence of excess aldosterone, or diseased kidneys will result in sodium (excretion/retention) ________________________.

Name at least three clinical situations that can cause sodium retention:

1. *____.
2. *____.
3. *____.

– – – – – – – – – – – – – – – – –

sodium; retention

1. CHF (congestive heart failure)
2. renal failure
3. adrenal cortical hormones, e.g., cortisone
4. cirrhosis of the liver
5. trauma

---

45 Obstruction of the lymph flow will prevent the return of proteins to the circulation. The obstructed lymph flow is high in ________________ content.

A decrease in protein content in the plasma will cause the water to flow from *________________________________ into *________________________ ____.

Name at least three clinical situations that cause a decrease in lymphatic drainage:

1. *____.
2. *____.
3. *____.

– – – – – – – – – – – – – – – – –

protein; the plasma (intravascular space); the tissue spaces

1. cancer of the lymphatic system
2. removal of the lymph nodes
3. obesity
4. elephantiasis

CLINICAL APPLICATIONS

---

46 The influence of gravity has an effect on the distribution of edema fluid in the edematous person. In a lying-down position, there is a more equal distribution of edema, whereas in an upright position the edema would be more prevalent in the lower extremities. This is called *dependent edema.*

The eyelids of a person with generalized edema would be swollen in the morning, but by afternoon, with increased activity, the swelling would be (more/less) ________________ marked.

\- - - - - - - - - - - - - - - - -

less

---

47 In the edematous person, the distribution of edema fluid is influenced by

________________.

\- - - - - - - - - - - - - - - - -

gravity

---

48 With patients who are up and about, the edema fluid frequently will be found in the (ankles and feet/sacrum and buttocks) * ________________

________________. For those who are bedridden, edema fluid will most likely be found at the * ________________

________________.

\- - - - - - - - - - - - - - - - -

ankles and feet
eyes or sacrum and buttocks, or more equally distributed

---

49 The type of edema associated with gravity and the person's body position is called * ________________.

\- - - - - - - - - - - - - - - - -

dependent edema

---

50 The nurse should assess for edema in the ankles and feet early in the morning.

Why? * ____

____

____.

— — — — — — — — — — — — — — — — —

Dependent edema should not be present after the patient has been in a prone or supine position for the night. If edema is present in the morning, it is most likely due to cardiac, renal, or liver disease and can be called *nondependent edema* (to differentiate between edema due to gravity versus edema due to cardiac, renal, or liver dysfunction; it can also be called *refractory edema* when edema does not respond to diuretics)

---

51 Edema of the lungs, often called *pulmonary edema,* can occur in patients with limited cardiac or renal reserve. If the heart is not able to beat adequately and the kidneys cannot excrete a sufficient amount of urine, then the fluid will back up into the pulmonary circulatory system.

When the hydrostatic pressure of the blood in the pulmonary capillaries rises to equal or exceed the plasma colloid osmotic pressure, then the water will

flow from plasma into the * ______________________________, leading to pulmonary edema.

— — — — — — — — — — — — — — — — —

lung tissues

---

52 Giving excessive intravenous infusions to a person with pulmonary edema would cause the blood volume to increase. What do you think increased blood volume

is called? ______________________.

Intravenous infusions should be regulated so that the rate of flow is not in

excess of the urinary ______________.

— — — — — — — — — — — — — — — — —

hypervolemia; output

53 Giving excessive IV fluids to a person in severe congestive heart failure will cause an increase in pulmonary venous pressure.

This will predispose the person to * ______________________.

A name for increased blood volume is ______________________.

— — — — — — — — — — — — — — — —

pulmonary edema; hypervolemia

Table 44 gives the clinical signs and symptoms of extracellular fluid volume excess, which is also known as *hypervolemia* and *overhydration.*

Study the table. A constant and irritating cough is one of the early symptoms of hypervolemia. As a nurse, you will observe for symptoms of hypervolemia or overhydration frequently, so be sure you know the signs and symptoms and their rationale. Refer to the table as needed.

TABLE 44. CLINICAL SIGNS AND SYMPTOMS OF EXTRACELLULAR FLUID VOLUME EXCESS—HYPERVOLEMIA, OVERHYDRATION

Signs and Symptoms	Rationale
Constant and irritating cough	Cough is frequently first clinical symptom of hypervolemia. It is caused by fluid, "backed up" into lungs (lung congestion).
Dyspnea—difficulty in breathing	Breathing is labored and difficult due to fluid congestion in lungs.
Neck vein engorgement	Jugular vein remains engorged when patient is in semi-Fowler's or sitting position.
Sublingual vein engorgement	Engorged veins under tongue may indicate hypervolemia.
Hand vein engorgement	Peripheral veins in hand will remain engorged with hand above heart level for 10 seconds.
Bounding pulse	Full, bounding pulse is present with hypervolemia. Pulse rate can increase.
Moist rales in lung	Lungs are congested with fluid. Moist rales in lung can be heard with stethoscope.
Cyanosis	A late symptom of pulmonary edema from hypervolemia as result of inadequate heart and kidney function.
Puffy eyelids	Swollen eyelids from generalized edema.
Pitting in peripheral edema of lower extremities	Peripheral edema that is present in morning is nondependent edema, resulting from inadequate heart, liver, or kidney function (Frame 50).
Weight gain	Gain of 2½ pounds is equivalent to gain of 1 L of body water.
Central venous pressure (CVP)	An increase in CVP of more than 12–15 cm H_2O is indicative of hypervolemia.
Hemoglobin and hematocrit ↓	Decrease in Hgb and Hct can indicate hypervolemia due to decrease in RBCs and increase in fluid (hemodilution).

---

54 One of the first clinical symptoms of ECFV excess or hypervolemia or overhydration is *______________________________

\- - - - - - - - - - - - - - - - -

constant, irritating cough

---

55 The person with pulmonary edema (fluid throughout the lung tissue) has irritating cough, dyspnea, and moist rales. The alveoli (air sacs) are filled with fluid.

What effect do you think fluid throughout the lung tissue would have on ventilation? *______________________________

____.

Pulmonary edema is due to what type of fluid imbalance? *______________

____.

\- - - - - - - - - - - - - - - - -

poor or inadequate ventilation; extracellular fluid volume excess or hypervolemia or overhydration (all have the same meaning)

---

56 If the person is in a semi-Fowler's position (45° elevated) and the jugular vein remains engorged, what can this indicate? *______________________________

____.

\- - - - - - - - - - - - - - - - -

extracellular fluid volume excess or hypervolemia

---

57 The nurse can make a quick assessment of hypervolemia or overhydration by checking the peripheral veins in the hand. The nurse instructs the patient to hold the hand above the heart level. If the peripheral veins of the hand remain engorged after 10 seconds, this can be an indication of

______________________.

How does peripheral vein assessment for hypervolemia differ from peripheral vein assessment for hypovolemia? * ______________________________

____

____.

— — — — — — — — — — — — — — — —

hypervolemia
Hypervolemia is assessed with the hand above the heart level for vein engorgement, and hypovolemia is assessed with the hand below heart level for flat veins or no engorgement

---

58 The nurse can assess the lungs for evidence of hypervolemia by listening for

* ______________________ with a stethoscope.

— — — — — — — — — — — — — — — —

moist rales

---

59 Cyanosis is a(n) (early/late) ______________ symptom of pulmonary edema due to hypervolemia.

— — — — — — — — — — — — — — — —

late

---

60 Another tool for assessing edema and hypervolemia is body weight.

If the patient has edema and has gained 2½ pounds, the weight gain would be equivalent to ______________ liter(s) of water.

— — — — — — — — — — — — — — — —

one

---

61 If hemoglobin and hematocrit had been in normal range and then decreased with fluid imbalance, it would be indicative of ______________________.
If the hemoglobin and hematocrit increased, the fluid imbalance might be
* ____.

- - - - - - - - - - - - - - - - - -

hypervolemia
hypovolemia or dehydration

---

62 Many edematous persons are malnourished due to loss of proteins or electrolytes.
Unless contraindicated, the nurse should encourage the edematous patient to eat ________________ Why? * ________________________
____.

- - - - - - - - - - - - - - - - - -

protein. It would increase plasma colloid osmotic pressure, and thus pull fluid out of the tissues.

---

63 Edematous persons may suffer from dehydration. Can you explain why?
* ____
____.

- - - - - - - - - - - - - - - - - -

The edema fluid is trapped in the interstitial space (tissue) and is not circulating; e.g. ascites and peripheral edema

---

64 The tissues of an edematous person are said to be more vulnerable to injury, resulting in tissue breakdown.
A bedfast person with edema of the sacrum and buttocks is apt to develop
________________________ due to * ________________________
____.

- - - - - - - - - - - - - - - - - -

decubiti (bedsores); tissue breakdown or constant pressure on tissues

---

---

65 With the edematous person, what is a nursing intervention to prevent decubiti?

*______________________________.

\- - - - - - - - - - - - - - - - - -

frequent change of body position

---

66 Generalized edema is called *anasarca.* The following terms are given in describing edema fluid in the various body cavities:

Peritoneal cavity, *ascites*
Pleural cavity, *hydrothorax*
Pericardial sac, *hydropericardium*

Anasarca means *______________________.
Give the names of edema fluid in the following body cavities:

Peritoneal cavity. ______________________.

Pleural cavity. ______________________.

Pericardial sac. ______________________.

\- - - - - - - - - - - - - - - - - -

generalized edema; ascites; hydrothorax; hydropericardium

CLINICAL MANAGEMENT

---

67 With nephrotic edema, cirrhosis of the liver with ascites, or any other marked hypoproteinemic state the administration of albumin, which is (hypo-osmolar/hyperosmolar) ________________, will (raise/decrease) ________________ the plasma colloid osmotic pressure; this will cause the fluid to flow from the *________________ into the ________________.

\- - - - - - - - - - - - - - - - - -

hyperosmolar; raise; tissue space; plasma

---

68 When edema is present, salt and water intake will probably increase the edema.

Do you think water intake alone probably (will/will not) ______________

increase the edema? Why? * ______________________________

____.

- - - - - - - - - - - - - - - - -

will not. Salt (sodium) has a water-retaining effect and without the sodium the water would not increase the edema. However, caution should be taken with giving excess amounts of water

CASE REVIEW

Mrs. Shea, age 72, was admitted to the hospital with complaints of shortness of breath, coughing, and swollen ankles and feet. Her blood pressure was 190/110, pulse 96, and respirations 28 and labored. Her hemoglobin and hematocrit were slightly low. She has a history of a "heart condition" and hypertension.

1. The nurse assesses Mrs. Shea's physical state. Her shortness of breath, coughing, and swollen ankles and feet may be indicative of * ______________.
   Other names for extracellular fluid volume excess include ______________, ______________, and ______________.
2. An early symptom of extracellular fluid volume excess or hypervolemia is
   * ______________________________.
3. The five main physiologic factors leading to edema are:
   a. * ______________________________.
   b. * ______________________________.
   c. * ______________________________.
   d. * ______________________________.
   e. * ______________________________.
4. The two physiologic factors that can cause Mrs. Shea's edema are
   * ______________________________ and
   * ______________________________.
5. The nurse assesses Mrs. Shea's ankles and feet in the morning to define the type of edema. If Mrs. Shea's ankles and feet remain swollen before she arises in the

morning, this would be *__________________ edema. Explain what can cause the edema. *____________________________________.

6. The nurse assesses Mrs. Shea's peripheral veins. Her veins are still engorged after holding her hand above the heart level for 30 seconds. This can be indicative of

   *____.

7. Mrs. Shea's shortness of breath or dyspnea and coughing may be due to

   __________________ edema. Give two ways in which the nurse can determine if this type of edema is present.

   a. *____.

   b. *____.

8. What can cause pulmonary edema? *__________________________

   __________________________________.

9. Mrs. Shea gained 5 pounds in 2 days. Her fluids and salt were not restricted.

   This weight gain would be equivalent to __________ liter(s) of body water

   gain. This would be equal to __________ ml.

10. Mrs. Shea's hemoglobin and hematocrit have been normal. At present, they are

    decreased, which may indicate _____________________________.

    Why? *____

    ____.

11. If Mrs. Shea developed generalized edema and was bedfast, what skin complication

    might result? ____.

    What precautions can be taken to prevent this complication? *____________

    ____________________________________.

12. Give the name for generalized edema. ____________________________.

\- - - - - - - - - - - - - - - - -

1. extracellular fluid volume excess or edema. An irritating cough is an early symptom of extracellular fluid volume excess; edema; hypervolemia; overhydration
2. constant, irritating cough
3. a. increased hydrostatic pressure
   b. decreased colloid osmotic pressure
   c. increased capillary permeability
   d. increased sodium retention
   e. decreased lymphatic drainage

4. Increased hydrostatic pressure; increased sodium retention
5. nondependent. It can result from inadequate heart, kidney, or liver function (see Frame 50 for further clarification)
6. extracellular fluid volume excess or hypervolemia
7. pulmonary
   a. observe the jugular veins for engorgement when Mrs. Shea is in semi-Fowler's position
   b. assess chest sounds for moist rales
8. inadequate heart and kidney function
9. two (2); 2000
10. hypervolemia. Dilution of RBCs with a decrease in RBCs and an increase in water
11. decubiti. Changing body position
12. anascara

NURSING DIAGNOSES

Potential alteration in fluid volume excess related to excessive amount of fluid infusion or fluid trapped in the interstitial spaces; i.e., peripheral edema, pulmonary edema.

Potential ineffective breathing patterns related to fluid overload in the lung tissues.

Potential impaired physical mobility related to excessive fluid in the tissues in the extremities or lungs.

Potential impairment of urinary elimination related to fluid retention and edema.

Knowledge deficit related to a lack of understanding of the disease process or medical treatment regime.

Potential for noncompliance to treatment and dietary regime related to inadequate explanation.

Anticipatory anxiety related to dependency on family and other people or on drug therapy and dietary restrictions for an extended time.

NURSING ACTIONS

1. Recognize physiologic factors that can cause edema: increased hydrostatic pressure in the capillaries; decreased serum protein, which decreases the serum colloid osmotic pressure; increased capillary permeability; sodium retention; and decreased lymphatic drainage.
2. Assess for signs and symptoms of overhydration (hypervolemia), i.e., constant, irritated cough, dyspnea, neck and hand vein engorgement, bounding pulse, chest rales, increased CVP, and decreased hemoglobin and hematocrit (with no history of anemia or blood loss).
3. Check peripheral edema in the morning to determine if nondependent edema or refractory edema due to cardiac, renal, or liver dysfunction is present.
4. Make a quick assessment of overhydration by checking the peripheral veins in the hand. First lower the hand and then raise it above heart level. Overhydration is present if the peripheral veins remain engorged after 10 seconds.

5. Turn an edematous patient frequently to prevent decubiti. Edematous persons are more prone to tissue breakdown.
6. Weigh an edematous patient daily to determine fluid gain or loss.
7. Teach patient to avoid using excess salt on foods. Salt (sodium) holds water and will increase the edematous condition.
8. Encourage the patient to eat foods rich in protein. Protein increases plasma/serum colloid osmotic pressure and, thus, will pull fluids from the tissue spaces and decrease edema.
9. Monitor intake and output. Fluid and salt restrictions may be necessary if edematous condition remains the same or is increased.
10. Monitor intravenous flow rate. Excessive fluid infusion could increase the hypervolemic state.

EXTRACELLULAR FLUID VOLUME SHIFTS

In the extracellular fluid compartment fluid volume with electrolytes and protein can shift from the intravascular to the interstitial fluid space; this fluid is referred to as *third-space fluid*. Later it shifts back from the interstitial spaces to the intravascular space. It is considered nonfunctional and a physiologically useless fluid.

---

69 The extracellular fluid is found in the ________________ space and ________________ space.

- - - - - - - - - - - - - - - - - -

intravascular; interstitial

---

70 Extracellular fluid will shift between the intravascular and interstitial spaces.

Water, electrolyte, and protein shift from the intravascular (blood vessels) to the interstitial (tissue) spaces is referred to as *shift to the third space.* Excess fluid in the tissue spaces or third space is physiologically useless fluid (the body can not use the fluid). Clinical causes could be simple, such as a blister or sprain; or serious, due to massive injuries, burns, perforated peptic ulcer, and intestinal obstruction.

When the fluid leaves the blood vessels and goes into the tissue spaces, this fluid is considered to be physiologically ________________.

What type of fluid imbalance occurs when the fluid shifts out of the intravascular spaces? ________________.

- - - - - - - - - - - - - - - - - -

useless; dehydration or hypovolemia

---

71 Clinical signs and symptoms of fluid shift from blood vessels to tissue space are similar to those of dehydration or shocklike symptoms.

The signs and symptoms include:

Pallor and cold skin
Pulse rate ↑
Blood pressure ↓ (later)
Disorientation

Do you think the hemoglobin, hematocrit, and BUN would be (increased/decreased) ______________________.

- - - - - - - - - - - - - - - - - -

increased

---

72 Later water and electrolyte shift from interstitial (tissue space) to intravascular space (blood vessels) is caused by clinical conditions such as rapid infusion of hyperosmolar solutions to a debilitated person and during recovery phase of injury, e.g., 2–3 postburn days when fluid shifts from the traumatized, edematous area back into the intravascular space.

Why do you think a rapid infusion of hyperosmolar solution(s) can cause fluid shift to intravascular space? *____

____.

In the intravascular space could (hypovolemia/hypervolemia) ______________ ______________________ occur from the fluid shifting back.

- - - - - - - - - - - - - - - - - -

Because of osmosis, fluid shifts from the lesser concentration, or tissue spaces and cells, to the greater concentration, or vascular space due to hyperosmolar solution.
hypervolemia

73 Clinical signs and symptoms of fluid shift from the tissues to the blood vessels are similar to those of hypervolemia or overhydrationlike symptoms.
The signs and symptoms are:

Constant cough
Dyspnea
Vein engorgement
Full bounding pulse

Can you name other overhydrationlike symptoms? *______________

and ______________.

- - - - - - - - - - - - - - - - -

moist rales; cyanosis

WATER INTOXICATION—INTRACELLULAR FLUID VOLUME EXCESS

74 Water intoxication occurs as a result of an excess water intake and a lack of sodium intake. Also associated with this condition is a rapid decrease in serum osmolality.
With water intoxication, would the sodium intake be greater or less than the water intake? ______________.
As the result of this imbalance, explain what would happen to the serum osmolality? *______________.

- - - - - - - - - - - - - - - - -

lesser. It would decrease, or hypo-osmolality

75 What is another name for water intoxication? *______________

______________.

- - - - - - - - - - - - - - - - -

intracellular fluid volume excess

76 Edema may result from excess of ________________, whereas water intoxication results from excess of ________________.

With edema, there is excessive fluid in the *________________ compartment, whereas with water intoxication there is excess fluid in the *________________ compartment.

- - - - - - - - - - - - - - - - - -

sodium; water; extracellular fluid; intracellular fluid

---

77 With dehydration, the water loss can be greater than the sodium loss; so with with dehydration, what would be the serum osmolality? *________________ ________________.

- - - - - - - - - - - - - - - - - -

It would be increased, or hyperosmolality

---

78 Water intoxication is not the same as edema. Edema is the accumulation of fluid in the interstitial spaces. With water intoxication, the excess fluid first enters the extracellular space, lowering the osmotic pressure. Then, due to osmosis, water moves from the extracellular fluid into the cells, causing the cells to swell.

In water intoxication, the water first enters the *________________ ________________.

Why would the water then leave the extracellular space and move into the cells? *________________ ________________.

- - - - - - - - - - - - - - - - - -

extracellular spaces. Due to osmosis, the cells have the greatest concentration, drawing water into the cells

---

79 Water intoxication (is/is not) ________________ the same as edema. Edema is an extracellular accumulation of fluid, whereas with water intoxication the end result is an (extracellular/intracellular) ________________ ________________ accumulation of fluid. This causes the cells to ________________.

— — — — — — — — — — — — — — — —

is not; intracellular; swell

CLINICAL APPLICATIONS

---

80 It is difficult for a person to drink himself into water intoxication unless the renal mechanisms for elimination fail.

If excessive water has been given and the kidneys are not functioning properly, what would likely occur? *________________ ________________.

— — — — — — — — — — — — — — — —

water retention or water intoxication

---

81 The commonest occurrence of water intoxication is seen is postoperative patients when fluids orally and intravenously have been forced without compensatory amounts of salt. The amount of water exceeds that which the kidneys can excrete.

If your patient postoperatively was receiving several liters of 5% dextrose in water and ordered crushed ice and sips of water PO (by mouth), what type of fluid imbalance may occur? *________________ ________________.

— — — — — — — — — — — — — — — —

water intoxication. Intake of a large amount of hypo-osmolar fluids

---

82 Although 5% dextrose in water is an iso-osmolar solution, it can become a hypo-osmolar solution. Explain how. * ______________________________

____

______________________________.

— — — — — — — — — — — — — — — —

Dextrose is metabolized by the body, leaving water without any solutes to increase osmolality. The body fluids become hypo-osmolar

---

83 Also, after surgery, an overproduction of the antidiuretic hormone (ADH), known as the syndrome of inappropriate ADH secretions (SIADH), usually occurs due to trauma, anesthesia, pain, and narcotics. Because of the overproduction of ADH, the water excretion will (increase/decrease) ______________, causing the urine volume to (rise/drop) ______________ and the vascular (intravascular) fluid volume to (rise/drop) ______________.

— — — — — — — — — — — — — — — —

decrease; drop; rise

---

84 When does water intoxication commonly occur? * ______________________________

____

______________________________.

— — — — — — — — — — — — — — — —

After surgery or trauma. Also after several liters of 5% dextrose in water. After surgery, more water is reabsorbed from the kidney due to SIADH. Several liters of 5% D/W without any other solutes will dilute the vascular fluid.

---

85 If excess ADH (SIADH) is the cause, the specific gravity of urine is high and the urinary output in 24 hours may be as low as 500–700 ml.

If drinking excess water (as in psychogenic polydipsia) or forced fluids postoperatively is the cause, the specific gravity may be low. Name two factors that might cause an overproduction of ADH (antidiuretic hormone) after surgery?

______________ and ______________.

In case of excess ADH, will specific gravity of urine be increased or decreased?

______________.

Will urine volume be increased or decreased? ______________.

— — — — — — — — — — — — — — — —

trauma, anesthesia also pain, narcotics; increased; decreased

---

86 If the circulation through the kidneys is impaired and there is excessive amount of plain water intake, the fluid imbalance would most likely be

* ______________________________.

Impairment of renal circulation can be due to arteriosclerosis. If the kidneys do not receive sufficient blood circulation kidney dysfunction can result.

— — — — — — — — — — — — — — — —

water intoxication from water taken without solutes.

Table 45 summarizes the causes of intracellular fluid excess/water intoxication. Study the table carefully and refer back to it as needed.

TABLE 45. CONDITIONS CAUSING INTRACELLULAR FLUID VOLUME EXCESS/WATER INTOXICATION

Conditions	Rationale
Continuous administration of 5% D/W	Dextrose is metabolized rapidly by the body, thus leaving water without solutes.
Overproduction of ADH (SIADH)	Stress, surgery, trauma, pain, and narcotics cause syndrome of inappropriate antidiuretic hormone secretions, or SIADH. Excess ADH causes water retention hemodilution.
Psychogenic polydipsia	Compulsive drinking of plain water almost continuously could result in water intoxication.
Kidney dysfunction	Renal impairment can decrease water excretion.
Protein/albumin deficit	Decreased serum protein and albumin levels promote sodium and water loss to interstitial spaces. Excess water may remain in the vascular space.

---

87 Name the five causes of water intoxication.

1. * ______________________.
2. * ______________________.
3. * ______________________.
4. * ______________________.
5. * ______________________.

— — — — — — — — — — — — — — — —

1. Continuous administration of 5% D/W
2. Overproduction of ADH or SIADH
3. Psychogenic polydipsia
4. Kidney dysfunction
5. Serum protein or albumin deficit

The clinical signs and symptoms and rationale of water intoxication or intracellular fluid volume excess are explained in Table 46. Study the table and be cognizant of the types of symptoms. Refer to the table as necessary.

TABLE 46. CLINICAL SIGNS AND SYMPTOMS OF INTRACELLULAR FLUID VOLUME EXCESS–WATER INTOXICATION

Type of Symptoms	Signs and Symptoms	Rationale
Early	Headache Nausea and vomiting Excessive perspiration Acute weight gain	Cerebral cells absorb hypo-osmolar fluid more quickly than other cells.
Progressive		
Central nervous system (CNS)	Behavioral changes: Progessive apprehension Irritability Disorientation Confusion Drowsiness Incoordination Blurred vision Elevated intracranial pressure (ICP)	Hypo-osmolar body fluids pass into cerebral cells first. Swollen cerebral cells will cause behavioral changes and elevate ICP.
Vital signs (VS)	Blood pressure ↑ Bradycardia (slow pulse rate) Respiration ↑	Vital signs are the opposite of shock. VS are similar to those in increased ICP.
Later (CNS)	Neuroexcitability (muscle twitching) Projectile vomiting Papilledema Delirium Convulsions, then coma	Severe CNS changes occur when water intoxication is not corrected.
Skin	Warm, moist, and flushed	

88 What are the early symptoms of water intoxication?

1. ____________.
2. *____________.
3. *____________.
4. *____________.

- - - - - - - - - - - - - - - -

1. headache
2. nausea and vomiting
3. excessive perspiration
4. weight gain

---

89 Central nervous systems are (least/most) ______________ prominent with water intoxication. Explain why? * ______________________________

____.

- - - - - - - - - - - - - - - - - -

most. Hypo-osmolar body fluids pass into cerebral cells; swollen cerebral cells cause behavioral changes

---

90 Name three of the behavioral changes that occur with progressive symptoms of water intoxication.

a. ______________________.

b. ______________________.

c. ______________________.

- - - - - - - - - - - - - - - - - -

a. apprehension
b. irritability
c. disorientation
d. confusion

---

91 Is the intracranial pressure (increased/decreased) ______________________ with water intoxication?

- - - - - - - - - - - - - - - - - -

increased

---

92 With progressive intracellular fluid volume excess, the vital signs are:

a. blood pressure ______________________.

b. pulse rate * ______________________.

c. respiration ______________________.

- - - - - - - - - - - - - - - - - -

a. increased
b. decreased, or bradycardia
c. increased

---

93 Name five later symptoms of water intoxication or intracellular fluid volume excess.

a. * ______________________________.

b. * ______________________________.

c. ______________________________.

d. ______________________________.

e. ______________________________.

— — — — — — — — — — — — — — — — —

a. muscle twitching
b. projectile vomiting
c. papilledema
d. delirium
e. convulsions

---

94 The skin in later stages of water intoxication is * ______________________________

______________________.

— — — — — — — — — — — — — — — — —

warm, moist, flushed

CLINICAL MANAGEMENT

Overall Objective: To reduce excess water in the body.

There are two ways to reduce this water:

1. Reduce water intake
2. Promote water excretion

---

95 In *less* severe cases of water intoxication, water restriction may be sufficient, or an extracellular replacement solution such as lactated Ringer's may be given to increase the osmolality of the extracellular fluid.

What is the overall objective in the clinical management of water intoxication?

* ____.

Name two ways in which this objective is accomplished.

1. * ______________________________.

2. * ______________________________.

— — — — — — — — — — — — — — — — —

To reduce excess water in the body

1. reduce water intake
2. promote water excretion

---

96 Concentrated saline may be given in severe cases of water intoxication to raise extracellular electrolyte concentration in hope of drawing water out of the

(intracellular space/interstitial space) * ______________________

and (increasing/decreasing) ______________ urinary output.

– – – – – – – – – – – – – – – – –

intracellular space; increasing

---

97 However, administration of additional salt to a person who already has too much water can result in expansion of the interstitial fluid and blood volume,

and the development of (water intoxication/edema) ______________.

An osmotic diuretic, e.g., mannitol, includes diuresis and a loss of retained fluid, especially from the cerebral cells.

– – – – – – – – – – – – – – – – –

edema

---

98 From Frames 95, 96, and 97, can you give three methods for promoting water excretion?

1. * ______________________.

2. * ______________________.

3. * ______________________.

– – – – – – – – – – – – – – – – –

1. water restriction
2. extracellular replacement solution, e.g., lactated Ringer's
3. concentrated saline solution
4. osmotic diuretics, e.g., mannitol

---

99 For less severe cases of water intoxication, what would the clinical management be? * ______ and/or * ______.

For more severe cases of water intoxication, what would the clinical management be? * ______ and/or * ______.

- - - - - - - - - - - - - - - - - -

water restriction; extracellular replacement solution;
concentrated saline solution; osmotic diuretics

---

100 An osmotic diuretic induces * ______.

- - - - - - - - - - - - - - - - - -

diuresis and a loss of retained fluid, especially from cerebral cells

---

CASE REVIEW

Ms. Cline, age 19, returned from having an appendectomy performed. She received 1 liter of 5% dextrose in water during the procedure and another liter postoperatively. She was allowed to have crushed ice and sips of water. That evening she became nauseated, and the third liter of 5% dextrose in water was added. The following day she received 2 more liters of 5% dextrose in water. Ms. Cline took several glasses of crushed ice. Her first day postoperatively she complained of headache. Later she was drowsy, disoriented, and confused. Her blood pressure had a slight increase, and there was a drop in her pulse rate.

1. The nurse assessed Ms. Cline's fluid state. From the history and her symptoms, what type of fluid imbalance was present? * ______.
2. Excessive amounts of 5% dextrose in water along with glasses of crushed ice without any other solute intake can cause * ______. Explain why. * ______.
3. Name Ms. Cline's early symptoms of intracellular fluid volume excess. * ______.

4. As fluid imbalance progressed, name the symptoms that indicated water intoxication or intracellular fluid volume excess. ______________________, ______________________, ______________________, and ______________________.

5. Her vital signs were similar to those of *______________________ ______________________.

6. Can an overproduction of ADH increase her water intoxication? __________. Explain how. *______________________ ______________________.

7. What type of intravenous solution can be administered to correct water intoxication? *______________________.

8. In what way could this fluid imbalance have been prevented? *______________________ ______________________.

- - - - - - - - - - - - - - - -

1. water intoxication or intracellular fluid volume excess
2. water intoxication. With 5% dextrose in water, the dextrose is metabolized by the body, leaving water. The intravenous solution and crushed ice cause the plasma to be hypo-osmolar.
3. headache. If your answer was nausea—possibly; however, early nausea was most likely the result of the surgery and anesthesia.
4. drowsiness; disorientation; confusion
5. increased intracranial pressure
6. yes. After surgery, there could be an increased secretion of ADH, due to trauma, anesthesia pain, and narcotics. This increases water retention and, with the hypo-osmolar fluids she received, could increase the state of water intoxication.
7. concentrated saline solution, e.g., 3% saline, hyperosmolar solution to "pull" water out of the cells
8. Ms. Cline should have received intravenous fluids containing saline (solute) together with dextrose

NURSING DIAGNOSES

Potential for fluid volume excess related to infusion or ingestion of hypo-osmolar solutions that causes intracellular fluid volume excess/water intoxication.

Potential alteration in urinary elimination related to increased secretions of ADH secondary to surgery, trauma, stress, and pain.

Potential alteration in tissue perfusion (cerebral) related to intracellular fluid volume excess that causes cerebral cellular swelling.

NURSING ACTIONS

1. Describe the physiological factor associated with water intoxication. Hypo-osmolar fluid shifts into the cells and causes the cells to swell. This fluid shift increases intracellular (cellular) fluid. Causes of water intoxication are excess IV 5% dextrose in water, overproduction of ADH (SIADH), psychogenic polydipsia, and kidney dysfunction.
2. Monitor intravenous therapy and report if patient is receiving only 5% dextrose in water continuously. Dextrose is metabolized rapidly by the body leaving water, a hypo-osmolar solution. Due to the process of osmosis, fluid diffuses from the lesser concentration (blood vessels) to the greater concentration (cells).
3. Monitor the urine output after surgery or trauma. An excess ADH, known as the syndrome of inappropriate antidiuretic hormone (SIADH), is secreted following surgery, which causes more water to be reabsorbed from the kidney, and the vascular fluid to be diluted. Urine output would be decreased due to water reabsorption.
4. Check specific gravity. If urine output is decreased due to SIADH, specific gravity would be elevated, >1.030. Drinking excess water could cause the specific gravity to be low.
5. Observe for signs and symptoms of early water intoxication, i.e., headache, nausea and vomiting, excessive perspiration, and weight gain.
6. Observe for progressive signs and symptoms of water intoxication, i.e., headache, behavioral changes (irritability, drowsiness, disorientation, confusion, delirium), elevated intracranial pressure, changes in vital signs (↑Bp, ↓P, ↑R), and warm, moist, flushed skin.
7. Teach patient to reduce water intake. Offer fluids containing solutes, i.e., broth, juice.
8. Assess for signs and symptoms of overhydration (see section on edema) when concentrated saline solution is given to correct water intoxication. Concentrated saline pulls fluid from the cells (by osmosis) to the vessels, and the fluid could overload the vascular system.

SHOCK

---

101 The state of circulatory collapse, known as *shock,* occurs when the hemostatic circulatory mechanism, which regulates circulation, fails to maintain adequate circulation. With shock, the cardiac output is insufficient to provide vital organs and tissues with blood.

Shock is a state of *______________________________.

Shock occurs when the hemostatic circulatory mechanism fails to

*____

____.

— — — — — — — — — — — — — — — —

circulatory collapse; maintain adequate circulation or provide adequate blood to vital organs or tissues

---

---

102 A common feature of shock, regardless of the cause, is a low circulating blood volume in relation to the vascular capacity. There is a loss of blood, not necessarily from hemorrhaging, but from "pooling" in body areas so that the blood does not circulate. This causes inadequate tissue perfusion.

A low blood volume is known as ______________.

A disproportion between the volume of blood and the capacity (size) of the vascular chamber is the essential feature of *______________ ______________.

- - - - - - - - - - - - - - - - -

hypovolemia; shock or circulatory collapse

---

103 A common feature of shock is *______________.

With shock, is hypovolemia always due to hemorrhaging? ______________.

Explain. *______________.

- - - - - - - - - - - - - - - - -

low blood volume or loss of blood
NO! Can be due to pooling of blood in body areas

---

104 When the blood volume becomes too small for the vascular capacity, venous blood return to the heart is reduced, and there is a drop in cardiac output and systemic arterial blood pressure. This will lead to an inadequate return of blood to the right side of the heart.

The cardiac output is dependent on:

( ) a. arterial return
( ) b. venous return

With a decrease in venous return, the amount of circulating blood would be ______________.

- - - - - - - - - - - - - - - - -

a. –; b. X
decreased

Table 47 outlines the physiologic factors resulting from shock. It includes whether an increase or decrease of the physiologic factor would lead to shock. The rationale gives an explanation of the reasons, causes, and results of shock according to the physiologic factors.

Study this table carefully, noting if it is an increase or a decrease in the physiologic factor that can lead to shock. Be able to give one explanation for each from the rationale column. Refer to this table as needed.

TABLE 47. PHYSIOLOGIC FACTORS RESULTING FROM SHOCK

Physiologic Factors		Rationale
Arterial blood pressure	Decreased ↓	Reduced venous return to heart will decrease cardiac output and arterial blood pressure (BP). Decrease in BP is sensed by pressoreceptors in carotid sinus and aortic arch which leads to immediate reflex increase in systemic vasomotor activity. (This center is found in medulla.) Result will be in cardiac acceleration and vasoconstriction in order to maintain homeostasis with respect to blood pressure. This may be sufficient for early or impending shock.
Vasoconstriction of blood vessels	Increased ↑	Increased sympathetic nerve activity will cause vasoconstriction. Vasoconstriction will tend to maintain blood pressure and reduce discrepancy between blood volume and vascular capacity (size). Vasoconstriction is greatest in skin, kidneys, and skeletal muscles and not as significant in cerebral vessels. Coronary arteries actually dilate with decrease in blood volume. This will provide sufficient blood to heart muscle (myocardium) for heart function.
Heart rate	↑ Increased	Heart rate is increased to overcome poor cardiac output and to increase circulation. Rapid, thready pulse will be one of first identifiable signs of shock.
Metabolic changes	Decreased ↓	Fall in plasma hydrostatic pressure reduces urinary filtration. Unopposed plasma colloidal osmotic pressure draws interstitial fluid into vascular bed. Blood loss results in loss of serum potassium, phosphate, and bicarbonate. Inadequate oxygenation of cells prevents their normal metabolism and leads to formation of acid metabolites, thus lowering serum pH values. With loss of K, HCO_3, and PO_4 and fall in serum pH, metabolic acidosis results. Rise in blood sugar will first be seen due to release of epinephrine; later, blood sugar will fall due to decline in liver glycogen.

TABLE 47. (Continued)

Physiologic Factors		Rationale
Kidney function	D↓ Decreased	Low blood pressure causes inadequate circulation of blood to the kidneys. Renal ischemia is the result of a lack of O_2 to the kidneys. Renal insufficiency follows prolonged hypotension. Systolic blood pressure must be 60 mmHg and above to maintain kidney function. One of the body's compensatory mechanisms in shock is to shunt blood around kidney to maintain intravascular fluid. Deficient blood supply makes tubule cells of kidneys more susceptible to injury. Urine output of less than 25 ml per hour may be indicative of shock and/or decrease in renal function.

---

105 Place I for increase and D for decrease beside the physiologic factors as they occur with shock.

____ a. Arterial blood pressure

____ b. Kidney function

____ c. Heart rate

____ d. Metabolic changes

____ e. Vasoconstriction

\- - - - - - - - - - - - - - - - -

a. D; b. D; c. I; d. D; e. I

---

106 When there is a low blood pressure, the pressoreceptors in the carotid sinus and aortic arch will cause an increase in the systemic vasomotor activity that will lead to what two activities in order to maintain homeostasis? ________________ and *________________.

Increased systemic vasomotor activity occurs in order to maintain ________________.

\- - - - - - - - - - - - - - - - -

vasoconstriction; cardiac acceleration; homeostasis

107 Increased sympathetic activity will result in (vasoconstriction/vasodilation)

______________________.

Vasoconstriction is greatest in what three parts of the body? ______________,

______________, and * ______________.

The coronary arteries will (dilate/constrict) ______________ with a decrease in blood volume.

- - - - - - - - - - - - - - - - -

vasoconstriction; skin; kidneys; skeletal muscles; dilate

---

108 Heart rate in shock will be (increased/decreased) ______________ to overcome poor cardiac output and to increase circulation.

The pulse rate would be ______________ and ______________.

A person with a pulse rate above 120 would have (bradycardia/tachycardia)

______________________.

- - - - - - - - - - - - - - - - -

increased; rapid; thready; tachycardia

---

109 The following metabolic changes would occur with shock:

Fluid would be drawn from the interstitial space into the vascular space due to what kind of pressure? * ______________.

Inadequate oxygenation of cells will lead to the formation of acid metabolites, causing the pH to (rise/fall) ______________.

A fall in pH and a loss of K, HCO_3, PO_4 will lead to (metabolic acidosis/metabolic alkalosis) * ______________.

In shock, there will be a release of epinephrine, which will cause the blood sugar to (rise/fall) ______________. Later, there will be a (rise/fall) ______________ due to a decline in liver glycogen.

- - - - - - - - - - - - - - - - -

colloidal osmotic pressure; fall; metabolic acidosis; rise; fall

---

110 In shock, the compensatory mechanisms will shunt the blood around the kidney in order to maintain the volume of *____________________________

______________ . This results in a lack of oxygen in the kidneys known as

*________________________, causing a decrease in kidney function.

The systolic blood pressure for kidney function must be at least

*________________________ .

An indication of shock related to kidney dysfunction would be a urine output of less than *________________________ .

\- - - - - - - - - - - - - - - - - -

intravascular fluid; renal ischemia; 60 mm Hg; 25 ml per hour

---

111 Extracellular fluid volume shifts occur during shock. In *early* shock, fluid is shifted from the interstitial space to the intravascular space to compensate for fluid deficit in the vascular system. More fluid in the vascular system will increase the venous return to the heart; thus it increases cardiac output.

As the interstitial fluid becomes depleted, tissue (dehydration/edema) ________________________ occurs.

\- - - - - - - - - - - - - - - - - -

dehydration

---

112 With *late* shock, fluid is forced from the intravascular space (blood vessels) back into the interstitial space (tissues).

In early shock, fluid is shifted from *________________________ to the *________________________ Why? *________________

____ .

\- - - - - - - - - - - - - - - - - -

interstitial space; intravascular space. This shift compensates for fluid deficit in vascular system

---

CLINICAL APPLICATIONS

---

113 Normal blood pressure is the usual level of blood pressure in a person and will vary to some extent from person to person.

Therefore, a systolic blood pressure of less than 90 mm Hg is significant of shock in most people.

A systolic pressure of 60 to 70 mm Hg is necessary to maintain the coronary circulation and renal function (urinary output). A person with a systolic pressure of 50–60 mm Hg is said to be in ________________.

\- - - - - - - - - - - - - - - - -

shock

---

114 A low pulse pressure, which is the difference between the systolic and diastolic pressures, is indicative of shock. The systolic blood pressure usually decreases before the diastolic.

To maintain coronary circulation and renal function the systolic pressure should be at least * ________________________________.

A pulse pressure of 20 mm Hg is indicative of ____________________.

\- - - - - - - - - - - - - - - - -

60–70 mm Hg; shock

---

115 Blood supply to the organs most susceptible to acute anoxia (absence or lack of oxygen), i.e., the brain and the heart, is maintained as long as possible at the expense of the less vital organs and tissues.

The two organs most susceptible to anoxia are ________________ and ________________.

The brain can survive 4 minutes in an anoxic state before cerebral damage occurs.

\- - - - - - - - - - - - - - - - -

heart; brain

116 Which of the following would indicate shock?
( ) a. Arterial blood pressure of less than 90
( ) b. Pulse pressure of 55 mm Hg
( ) c. Pulse pressure of 20 mm Hg

The two organs most susceptible to acute anoxia are:
( ) a. Heart
( ) b. Brain
( ) c. Intestines

The organ that cannot survive anoxia longer than 4 minutes without permanent damage is the ________.

- - - - - - - - - - - - - - - - -

a. X; b. –; c. X;
a. X; b. X; c. –
brain

Table 48 describes four types of shock: hypovolemic, also called hematogenic when hemorrhage is the cause; cardiogenic; septic, also known as endotoxic or vasogenic; and neurogenic. Clinical causes, rationale, and physiologic results of these shocks are listed in the table.

Study the table carefully. Refer to the glossary for unfamiliar terms and refer to this table as needed.

TABLE 48. TYPES OF SHOCK

Type of Shock	Clinical Causes	Rationale and Physiologic Results
Hypovolemic: Hematogenic (from hemorrhage)	Severe vomiting or diarrhea-acute dehydration Burns, intestinal obstruction, fluid shift to third space Hemorrhage that results from internal or external blood loss	Blood, plasma, and fluid loss from decreased circulating blood volume *Physiologic Results:* 1. Decreased circulation 2. Decreased venous return 3. Reduced cardiac output 4. Increased afterload 5. Decreased preload 6. Decreased tissue perfusion
Cardiogenic	Myocardial infarction Severe arrhythmias Congestive heart Cardiac tamponade Pulmonary embolism	Because of these clinical problems, the pumping action of the heart is inadequate to maintain circulation. (Pump failure of myocardium.) *Physiologic Results:* 1. Decreased circulation 2. Decreased stroke volume 3. Decreased cardiac output 4. Increased preload

TABLE 48. (Continued)

Type of Shock	Clinical Causes	Rationale and Physiologic Results
		5. Increased afterload 6. Increased venous pressure 7. Decreased venous return 8. Decreased tissue perfusion
Septic: Endotoxic Vasogenic	Severe systemic infections Septic abortion Peritonitis Debilitated conditions Immunosuppressant therapy	Septic shock is characterized by increased capillary permeability that permits blood, plasma, and fluid to pass into surrounding tissue. Usually caused by a gram negative organism. *Physiologic Results:* 1. Vasodilatation and peripheral pooling of blood 2. Decreased circulation 3. Decreased preload, early shock, and increased preload, late shock 4. Decreased afterload, early shock, and increased afterload, late shock 5. Decreased tissue perfusion
Neurogenic	Mild to Moderate Neurogenic Shock: Emotional stress Acute pain Drugs: narcotics, barbiturates, phenothiazines High spinal anesthesia Acute gastric dilatation Severe Neurogenic Shock: Spinal cord injury Trauma: Extensive operative procedure	Neurogenic shock is caused by loss vascular tone. *Physiologic Results:* 1. Decreased circulation 2. Vasodilatation and peripheral pooling of blood 3. Decreased cardiac output 4. Decreased venous return 5. Decreased tissue perfusion

---

117 The four types of shock are:

1. *_________________________.
2. *_________________________.
3. *_________________________.
4. *_________________________.

- - - - - - - - - - - - - - - - - -

1. hypovolemic shock
2. cardiogenic shock
3. septic shock
4. neurogenic shock

118 Match the types of shock with the clinical causes.

a. Hypovolemic shock
b. Cardiogenic shock
c. Septic shock
d. Neurogenic shock

____ 1. High spinal anesthesia, emotional factors, or trauma from an extensive operative procedure

____ 2. Hemorrhaging from surgery or injury, burns, or gastrointestinal bleeding

____ 3. Severe bacterial infection, immunosuppressant therapy

____ 4. Myocardial infarction, cardiac failure, and cardiac tamponade

\- - - - - - - - - - - - - - - - - -

1\. d; 2. a; 3. c; 4. b

---

119 Match the types of shock with the rationale.

a. Hypovolemic shock
b. Cardiogenic shock
c. Septic shock
d. Neurogenic shock

____ 1. Failure of the myocardium causes a decrease in the circulating blood volume

____ 2. Loss of vascular tone with vasodilation

____ 3. Decrease in blood volume due to loss of blood and plasma

____ 4. Increase in capillary permeability resulting from an infection

\- - - - - - - - - - - - - - - - - -

a. b; 2. d; 3. a; 4. c

120 Match the types of shock with the physiologic results. One response may be used more than once.

a. Hypovolemic shock
b. Cardiogenic shock
c. Septic shock
d. Neurogenic shock

____, ____, ____, ____ 1. Decreased circulation

____, ____, ____ 2. Decreased cardiac output

____, ____ 3. Vasodilatation

____, ____, ____ 4. Decreased venous return

____, ____, ____, ____ 5. Decreased tissue perfusion

- - - - - - - - - - - - - - - - - -

1. a, b, c, d; 2. a, b, d; 3. c, d; 4. a, b, d; 5. a, b, c, d.

Table 49 lists the signs and symptoms of shock, types of shock, and rationale. Study the table carefully and be able to explain the signs and symptoms that frequently occur in shock. Refer to the table as needed.

TABLE 49. SIGNS AND SYMPTOMS OF SHOCK

Signs and Symptoms	Types of Shock	Rationale
Skin: Pallid and/or cold and moist	Hypovolemic Cardiogenic Neurogenic Septic (LATE)	Pale, cold, and/or moist skin results from increased sympathetic action. Peripheral vasoconstriction occurs and blood is shunted to vital organs. Skin is warm and flushed in early septic shock.
Tachycardia (pulse fast and thready)	Hypovolemic Cardiogenic Septic	Increased pulse rate is frequently one of the early signs, except in neurogenic shock, in which the pulse is often slower than normal. Norepinephrine and epinephrine, released by the adrenal medulla, increase the cardiac rate and myocardial contractility. Tachycardia, pulse >100, occurs before arterial blood pressure falls.
Apprehension, restlessness	Hypovolemic Cardiogenic Septic	Apprehension and restlessness, early signs of shock, result from cerebral hypoxia. As the state of shock progresses disorientation and confusion occurs.
Muscle weakness, fatigue	Hypovolemic Cardiogenic Septic Neurogenic	Muscle weakness and fatigue, which occur early in shock, are the result of a buildup of acid metabolites.

TABLE 49. (Continued)

Signs and Symptoms	Types of Shock	Rationale
Arterial Blood Pressure: Early, a rise in or normal BP; late, a fall in BP	Hypovolemic Cardiogenic Septic Neurogenic	In early shock blood pressure rises or is normal as a result of increased heart rate. As shock progresses blood pressure falls because of a lack of cardiac and peripheral vasoconstriction compensation.
Pulse Pressure: Narrowed <20 mmHg		Narrowing of pulse rate occurs because the systolic BP falls more rapidly than the diastolic BP.
Pressures: CVP, PAP, PCWP: Decreased in hypovolemic, septic, neurogenic. Increased in cardiogenic	Hypovolemic Cardiogenic Septic Neurogenic	Normal Values: 1. Central venous pressure/CVP: 5–12 cm/H_2O. With decreased blood volume the CVP <5 cm H_2O. 2. Pulmonary artery pressure/PAP: 20–30mmHg systolic 10–15 mm Hg diastolic. With blood volume depletion or pooling of blood the PAP in hypovolemic <10 mm Hg, septic <10 mm Hg, neurogenic <10 mm Hg. In cardiogenic shock the PAP >30 mm Hg. 3. Pulmonary capillary wedge pressure/PCWP: 4–12 mmHg. With blood volume depletion or peripheral pooling the PCWP in hypovolemic, septic, and neurogenic is <10 mm Hg and in cardiogenic, >20 mm Hg.
Respiration: Increased rate and depth (tachypnea)	Hypovolemic Cardiogenic Septic Neurogenic	Increased hydrogen ion concentration in the body stimulates the respiratory centers in the medulla, thus increasing the respiratory rate. Acid metabolites, e.g., lactic acid from metabolic catabolism, increases the rate and depth of respiration. Rapid respiration acts as a compensatory mechanism to decrease metabolic acidosis.
Temperature: Subnormal	Hypovolemic Cardiogenic Neurogenic	Body temperation is subnormal in shock because of decreased circulation and decreased cellular function. In septic shock the temperature is elevated.
Urinary Output: Decreased	Hypovolemic Cardiogenic Septic Neurogenic	Oliguria (decreased urine output) occurs in shock because of decreased renal blood flow caused by renal vasoconstriction. Blood is shunted to the heart and brain. Urine output should be >25 ml/hr.

121 Early mental changes occurring in shock are ______________________ and ______________________.

They generally result from * ______________________.

— — — — — — — — — — — — — — — — —

apprehension and restlessness; cerebral hypoxia

122 The CVP, PAP, PCWP are decreased in which types of shock?
( ) a. Hypovolemic
( ) b. Cardiogenic
( ) c. Septic
( ) d. Neurogenic

— — — — — — — — — — — — — — — — —

a. X; b. –; c. X; d. X

123 Increased rate and depth of respirations are present in shock. Why? * ______________________.

— — — — — — — — — — — — — — — — —

Increased hydrogen ion concentration stimulates the respiratory center in the medulla; OR acid metabolites from cellular catabolism increase respiratory rate and depth.
Note: The purpose of increased rate and depth of respiration is to decrease the acidotic state.

124 Frequently the urinary output is decreased in all types of shock. Why? * ______________________.

— — — — — — — — — — — — — — — — —

Decreased urinary output is the result of decreased renal blood flow caused by renal vasoconstriction.

125 In shock, tachycardia is frequently seen before the arterial blood pressure begins to fall.

The heart beats faster to (increase/decrease) ________ the circulating blood volume. This is an early compensatory mechanism to overcome shock.

With shock, what happens to the arterial blood pressure? *________

________.

— — — — — — — — — — — — — — — — —

increase. It will first rise and then fall

---

126 Below are some statements about signs and symptoms of shock. Check the true statements. Correct the false ones.

( ) a. Arterial blood pressure is low
( ) b. Bradycardia
( ) c. Respiration—slow and deep
( ) d. Apprehension, restlessness
( ) e. Temperature low in all types of shock
( ) f. Increased urine output
( ) g. Central venous pressure is high in cardiac shock and in hypovolemic shock
( ) h. Skin is pallid and hot

— — — — — — — — — — — — — — — — —

a. Arterial blood pressure rises, then falls
b. Tachycardia
c. Fast and deep
d. X
e. Not in early septic shock
f. Decreased output
g. It is low in hematogenic shock
h. Pallid or cold and moist, or both

CLINICAL MANAGEMENT

Table 50 outlines the calculation of blood loss and fluid replacement for three states of hypovolemic shock from hemorrhage.

In planning parenteral replacement therapy for patients in shock, factors responsible for the conditions are first considered. Factors responsible are frequently referred to as causes or causative agents.

Study this table very carefully, noting the states of hypovolemic shock, the systolic blood pressure ranges, the estimated blood volume loss according to a person weighing 70 kg, and the fluid replacement needed. Refer to this table whenever it is necessary.

TABLE 50. CALCULATION OF BLOOD LOSS AND FLUID REPLACEMENT FOR THREE STATES OF HYPOVOLEMIC SHOCK FROM HEMORRHAGE

State of Hypovolemic Shock	Systolic Blood Pressure (mm Hg)	Estimated Blood Volume Loss (%)	Replacement Needed for Blood Volume Loss in Person Weighing 70 kg (cc [ml])
Mild	90-95	15-20	750-1100
Moderate	75-90	20-30	1100-1700
Severe	Below 75	30-50	2000-3000 and up

Note: When blood is not available, large amounts of balanced salt solution, 100-150 ml/kg of body weight, are given.

There should be 50 ml replacement for each percentage of blood volume loss in 70-kg person or as determined by physician.

---

127 If a patient is said to be in *mild* hypovolemic shock, her blood volume loss is * ________%, and her systolic blood pressure is probably ________mm Hg. The replacement needed for this blood volume loss is * ________ml.

If a patient is said to be in *moderate* hypovolemic shock, her blood volume loss is * ________%, and her systolic blood pressure is probably * ________mm Hg. The replacement needed for this blood volume loss is * ________ml.

Or if a patient is said to be in *severe* hypovolemic shock, her blood volume loss is * ________%, and her systolic blood pressure is probably * ________mm Hg. The replacement needed for this blood volume loss is * ________ ml.

— — — — — — — — — — — — — — — — — —

15-20; 90-95; 750-1100; 20-30; 75-90; 1100-1700; 30-50; below 75; 2000-3000

---

128 There should be ________ml replacement for each percentage blood volume loss in a man of average height, 150 pounds or ________ kg.

— — — — — — — — — — — — — — — — — —

50; 70

129 Place the word Mild, Moderate, or Severe in the space provided as it relates to:

_____________ Systolic blood pressure below 75 mm Hg

_____________ Systolic blood pressure of 86 mm Hg

_____________ Systolic blood pressure of 92 mm Hg

_____________ 50% blood volume loss

_____________ 15-20% blood volume loss

_____________ 20-30% blood volume loss

_____________ 1100-1700 ml replacement needed

_____________ 2000-3000 ml replacement needed

_____________ 750-1100 ml replacement needed

- - - - - - - - - - - - - - - - -

Severe; Moderate; Mild; Severe; Mild; Moderate; Moderate; Severe; Mild

---

Table 51 outlines the clinical management for alleviating four types of clinical shock: hypovolemic, cardiogenic, septic, and neurogenic. Years ago the first and foremost treatment of shock was to administer a vasopressor drug. The drug would constrict the dilated blood vessels, which occur with shock, and raise the blood pressure. Vasopressors will act as only a temporary treatment for shock, and shock will continue to increase if the cause is not alleviated or removed. Today vasopressors are used for severe shock and types of shock nonresponsive to treatment. Note that vasopressors are *not* effective in the treatment of hypovolemic shock, for constricting blood vessels will not aid in the circulation of blood when the cause is most obvious—a lack of blood, causing hypovolemia. Replacing blood volume lost should correct this type of shock. Remember, removal of the cause is first and foremost in alleviating various types of shock.

Study this table carefully and be able to explain the treatments for each type of shock. Refer to this table as needed.

TABLE 51. CLINICAL MANAGEMENT FOR ALLEVIATING VARIOUS TYPES OF CLINICAL SHOCK

Hypovolemic Shock	Cardiogenic Shock	Septic Shock	Neurogenic Shock
1. Parenteral fluids, such as a. Lactated Ringer's b. Normal saline c. Whole blood d. Plasma or Plasmanate e. Dextran	1. Parenteral therapy is limited when edema is present and venous pressure is elevated. Close monitoring CVP and PCWP	1. Blood culture first and then parenteral fluids (salt solutions, plasma, dextran)	1. Parenteral therapy for severe shock
2. No vasopressors	*2. Vasopressors, if severe	*2. Vasopressors for nonresponsiveness	*2. Vasopressors, if severe
3. Electrolyte replacement	3. Lidocaine (to abort ventricular arrhythmias) Sodium nitroprusside/Nipride/Nitropress; Nitroglycerin/NTG; and prazosin/Minipress (decrease preload and decrease afterload). Digitalis Sedatives Diuretics	3. Antibiotics via IV fluids Steroids, e.g., hydrocortisone	3. Sedation for emotional shock
4. ABGs	4. O_2		
5. O_2			

*Examples of vasopressors are (1) levarterenol bitartrate/Levophed (2) Metaraminol bitartrate/Aramine, (3) dopamine hydrochloride/Intropin and (4) Dobutamine/Dobutrix (primary for cardiogenic shock.

---

130 What is shock? *______________________________.

What is a common feature of shock? *______________________________

______________________________.

- - - - - - - - - - - - - - - - -

State of circulatory collapse; low circulating blood volume or hypovolemia. This can be due to loss of blood from the body or "pooling" of blood.

131 The clinical management for *hypovolemic shock* may consist of:
( ) a. Dextran IV solution
( ) b. Whole blood
( ) c. Digitalization
( ) d. Vasopressors
( ) e. Oxygen
( ) f. Lactated Ringer's solution
( ) g. Normal saline
( ) h. Electrolyte replacement
( ) i. Lidocaine
( ) j. Lidocaine

- - - - - - - - - - - - - - - - -

a. X; b. X; c. –; d. –; e. –; f. X; g. X; h. X; i. –

---

132 When administering crystalloids, such as normal saline or lactated Ringer's solution, for hypovolemic shock, these intravenous solutions may be given rapidly at first to decrease shock symptoms and prevent fluid shift into the interstitial space at injured site. Later, the flow rate should be slowed.

What type of fluid imbalance could occur if massive quantities of crystalloids are rapidly administered intravenously? ______________________.

- - - - - - - - - - - - - - - - -

overhydration (hypervolemia). This could occur to the older adult, child, or debilitated person.

---

133 Clinical management for *cardiogenic shock* may consist of:
( ) a. Limited parenteral therapy
( ) b. No vasopressors
( ) c. Antibiotics
( ) d. Oxygen
( ) e. Digitalis product
( ) f. Sedation
( ) g. Sodium nitroprusside/Nipride

- - - - - - - - - - - - - - - - -

a. X; b. –; c. –; d. X; e. X; f. X; g. X

134 Clinical management for *septic shock* may consist of:

( ) a. Blood culture
( ) b. Antibiotics in intravenous fluids
( ) c. Hydrocortisone
( ) d. Vasopressors
( ) e. Digitalization
( ) f. Massive parenteral therapy with whole blood

- - - - - - - - - - - - - - - - -

a. X; b. X; c. X; d. X; e. –; f. –

---

135 Antibiotics and pain medications should be given intravenously in shock. Because circulation is poor, medications given intramuscularly (IM) are not fully absorbed. If given IM, after circulation is restored, the accumulated drug in the tissue spaces could be toxic.

Do you think the same dosage prescribed for IM should be given intravenously?

* ____.

- - - - - - - - - - - - - - - - -

Not always. Please check with the physician. Frequently, large doses are given diluted in 50–100 ml of IV solution. IV morphine is given slowly (approximately 5 minutes). In some cases, one-half of the IM dose is given IV.

---

136 Clinical management for *neurogenic shock* may consist of:

( ) a. Blood culture
( ) b. Vasopressors
( ) c. Intravenous therapy as needed
( ) d. Massive intravenous therapy
( ) e. Sedation

- - - - - - - - - - - - - - - - -

a. –; b. X; c. X; d. –; e. X

---

137 Explain the action of vasopressors (see introduction to Table 51, if necessary).

* ____

____.

The four vasopressors listed at the bottom of Table 51 are

______________________, ______________________________, and

______________________________.

— — — — — — — — — — — — — — — — —

Vasopressors constrict blood vessels in hopes of improving circulation;

levarterenol bitartrate/Levophed; metaraminol bitartrate/Aramine; dopamine HCl/Intropin; and dobutamine/Dobutrex.

---

138 Vasopressors should be used with care; however, they are helpful at the right time and with the right clinical problem.

When vasopressors are used, it is best to keep the systolic pressure no higher than 90 mm Hg to prevent cardiac arrhythmias.

Levophed (levarterenol bitartrate), a strong vasopressor, is norepinephrine, which increases blood pressure and cardiac output by constricting blood vessels.

Maintaining the blood pressure higher than 90 mm Hg with Levophed can cause * ______________________________________.

— — — — — — — — — — — — — — — — —

cardiac arrhythmia

---

139 Aramine (metaraminol bitartrate) triggers the body to release norepinephrine from storage sites to the blood vessels and heart. When discontinuing Aramine, the drug should not be completely stopped, but tapered gradually to prevent relapse into shock. The use of Aramine for a long period of time may deplete the body's norepinephrine, especially after abrupt stop.

What vasopressor do you think the physician might switch to for replacing the hormone norepinephrine when stopping Aramine? ____________________.

— — — — — — — — — — — — — — — — —

Levophed

---

140 Aramine does not tend to cause cardiac arrhythmia.

Levophed is a (strong/weak) ____________ vasopressor and can cause cardiac arrhythmia if systolic blood pressure is maintained above

____________ mm Hg.

Vasopressors are titrated according to the blood pressure and should be checked every 2–5 minutes.

— — — — — — — — — — — — — — — —

strong; 90

---

141 Dopamine HCl (Intropin) is a catecholamine precursor of norepinephrine. It increases blood pressure and cardiac output. It also dilates renal vessels, thus increasing renal blood flow and glomerular filtration rate.

Levophed and Aramine cause vasoconstriction, which affects the renal arteries and can decrease kidney function. The vasopressor that increases blood pressure, cardiac output, and urinary output is * ____________.

— — — — — — — — — — — — — — — —

dopamine HCl

---

142 Dopamine is helpful in cardiogenic shock, but not with severe hypotension.

What two vasopressors can be used in severe shock? ____________ and ____________.

— — — — — — — — — — — — — — — —

Aramine; Levophed

---

143 Dobutamine/Dobutrex is an adrenergic drug that increases blood pressure moderately and raises heart rate and cardiac output. Dobutamine is effective in increasing myocardial contractility without arrhythmias. It is frequently used with sodium nitroprusside.

Would dobutamine be used to treat severe hypotension? ____________.

Why? * ____________.

— — — — — — — — — — — — — — — —

No. Dobutamine has only a moderate effect on increasing blood pressure.

---

144 Identify the treatments listed below that may be used in various types of shock by placing:

H for hypovolemic shock
C for cardiogenic shock
S for septic shock
N for neurogenic shock

Some treatments may be used for more than one type of shock.

________ (a) IV therapy: lactated Ringer's, normal saline, dextran

________ (b) Digitalis products

________ (c) Electrolyte replacement

____, ____ (d) Sedatives

___, ___, ___ (e) Vasopressors

________ (f) Oxygen

____, ____ (g) Limited intravenous therapy

________ (h) Antibiotics in intravenous fluids

________ (i) Hydrocortisone

________ (j) Lidocaine

- - - - - - - - - - - - - - - -

a. H
b. C
c. H
d. C, N
e. C, S, N
f. C
g. C, N
h. S
i. S
j. C

CASE REVIEW

Mr. Martz, age 58, had diverticulitis. The diverticulum had ruptured, causing peritonitis and systemic septicemia. His vital signs were temperature 104°F (40°C), pulse 126 rapid and thready, respirations 32, and blood pressure 64/45. His urinary output was 25 ml per hour. His skin was warm and dry. He was markedly apprehensive and restless. He was diagnosed as being in septic shock.

1. Shock occurs when the hemostatic circulatory mechanism fails to maintain adequate *__________________________________.

2. Septic shock is characterized by *______________________________
______________________________.

3. Give Mr. Martz's clinical signs and symptoms of shock.
   a. *______________________________.
   b. *______________________________.
   c. *______________________________.
   d. *______________________________.

4. Mr. Martz's temperature was elevated due to *______________________________
______________________________.

5. His pulse pressure was 20 mm Hg (65 minus 45), which is indicative of
______________________________.

6. His urine output was __________ ml for 24 hours, which is in the low "normal" range. If his urine output goes below 25 ml per hour, what can occur? *______________________________
______________________________.

7. If Mr. Martz's systolic blood pressure drops below 60 mm Hg, what can occur to his renal function? *______________________________
______________________________.

8. In shock, which frequently occurs first, blood pressure decrease or pulse rate increase? *______________________________
______________________________. Why? *______________________________.

9. From Mr. Martz's blood pressure, the state of shock is (mild/moderate/severe)
______________________________.

10. Name the four methods for managing septic shock.
    a. *______________________________.
    b. *______________________________.
    c. *______________________________.
    d. *______________________________.

11. If vasopressors were used for Mr. Martz, which two would be indicated?
______________________________ or ______________________________.

12. What advantage does dopamine hydrochloride have on kidney function that other vasopressors do not have? *______________________________
______________________________.

1. circulation or blood volume
2. capillary permeability, permitting blood and plasma to pass into the surrounding tissues
3. a. pulse 126
   b. respiration 32
   c. blood pressure 65/45 and low pulse pressure
   d. apprehension and restlessness
4. septicemia (bacterial infection)
5. shock
6. 600; kidney dysfunction and (possibly) kidney failure
7. decreased renal function and output. It can lead to renal failure if prolonged.
8. pulse rate increase. Heart beats faster to maintain circulating blood volume. Increased pulse rate is an early compensatory mechanism to overcome shock
9. severe
10. a. blood culture
    b. antibiotics in intravenous fluids
    c. vasopressors as needed
    d. steroids, e.g., hydrocortisone
11. Aramine; Levophed
12. It dilates the renal arteries and increases blood flow and urine output

NURSING DIAGNOSES

Potential fluid volume deficit related to blood and fluid loss from trauma, injury, and surgery.

Potential alteration in tissue perfusion related to decreased blood circulation or inadequate blood volume.

Anxiety related to the uncertainty of outcome or perceived fear of death.

Potential alteration in comfort related to chest pain, awkward positioning, or life-support systems.

Potential for ineffective coping related to physiological changes or activity intolerance secondary to cardiac insufficiency.

Potential for impairment of urinary elimination (retention) related to inadequate blood volume and/or circulation.

Potential for self-care deficit related to an inability to take part in the daily routine of living because of life-support systems or prohibited physical activity.

Potential alteration in the family process related to an inability to deal with a crisis situation effectively.

NURSING ACTIONS

1. Recognize the physiologic factors resulting from shock: increased heart rate (rapid, thready pulse); increased vasoconstriction of the skin, skeletal muscles, kidneys; coronary artery dilation; decreased arterial blood pressure; decreased renal perfusion; loss of cellular potassium, phosphate, and bicarbonate; metabolic

acidosis from excess acid metabolites released from cells; and increased blood sugar.

2. Assess for changes in vital signs. The heart rate increases first, and as shock progresses the pulse rate becomes rapid and thready. The rate and depth of respiration are increased. Later the blood pressure drops and pulse pressure narrows <20 mm Hg. Severe shock ensues if systolic BP is <75 mm Hg.
3. Assess the skin and note changes. Pallor, gray, cold, and clammy are symptoms of shock.
4. Report behavioral changes such as apprehension, restlessness, confusion. Apprehension and restlessness are early symptoms of shock.
5. Report muscle weakness and fatigue, which are early signs of shock.
6. Monitor urine output. Hourly urine should be measured and, if less than 25 ml per hour, IV fluid rate should be increased. Don't forget to check for overhydration when pushing fluids—IV or orally. Renal artery vasoconstriction occurs in shock, which causes a decrease in kidney perfusion.
7. Report systolic blood pressure of 60 mm Hg or less immediately. Kidney damage can occur if systolic blood pressure is below 60 for several hours.
8. Monitor PAWP is necessary for adjusting fluid balance. Keeping the PAWP between 15–18 mm Hg in shock conditions provides the filling pressure required for adequate stroke volume and cardiac output. If the PAWP drops below 10 mm Hg fluid administration is usually needed. If PAWP is greater than 20 mm Hg fluid restriction may be necessary.
9. Administer certain drugs, i.e., morphine, antibiotics, steroids, intravenously as ordered during shock. Certain drugs, e.g., antibiotics, should be diluted in 50–100 ml of solution, others given slowly, and still others given as one-half the IM dose. Medications given IM during shock are poorly absorbed due to poor circulation.
10. Monitor blood pressure when administering vasopressors in IV fluids. The flow rate should be adjusted to keep the systolic pressure between 90 to 110. If the vasopressor is Levophed, the systolic pressure should be approximately 90 mm Hg to avoid cardiac arrhythmias. If urine output is poor, the choice of a vasopressor may be Dopamine.
11. Obtain a blood culture before administering antibiotics if the temperature is highly elevated and septic shock is suspected.
12. Check serum potassium. The cells lose potassium during shock, and normokalemia or hypokalemia with normal or excessive urine output could result. If oliguria or anuria occurs, hyperkalemia could result and IV potassium should not be given.
13. Stay with patient; give emotional support; explain care being given and answer patient's questions.

REFERENCES

Barrows JJ: Shock demands drugs. *Nursing, 82* 12(2):34–41, February 1982.

Bobb J: What happens when your patient goes into shock. *RN* 47(3):26–29, March 1984.

Borg N et al.: *Core Curriculum for Critical Care Nursing*. Philadelphia, WB Saunders Co, 1981.

Brunner LS' Suddarth DS: *Textbook of Medical-Surgical Nursing*, ed. 5. Philadelphia, JB Lippincott Co, 1984.

Burgess R: Fluids and electrolytes. *Am J Nurs* 65:94–95, 1965.
Guyton AC: *Textbook of Medical Physiology*, ed 5. Philadelphia, WB Saunders Co, 1981.
Jacob SW, Francone CA: *Structure and Function in Man*, ed 3. Philadelphia, WB Saunders Co, 1974
Kleinhenz TJ: Preload and afterload. *Nursing '85* 15(5):50–55, May 1985.
Luckmann J, Sorensen DC: *Medical-Surgical Nursing*, ed 2. Philadelphia, WB Saunders Co, 1980, pp 180–182, 231–243.
Meador B: Cardiogenic shock. *RN* 45(4):38–42, April 1982.
Metheney NM, Snively WD: *Nurses' Handbook of Fluid Balance*, ed 4. Philadelphia, JB Lippincott Co, 1983, pp 87–113.
Morris DG: The patient in cardiogenic shock. *Cardiovasc Nurs* 5: 15-17, 1969.
Moyer J, Mills L: Vasopressor agents in shock. *Am J Nurs* 75: 620-625, 1975.
Niemczura J: Eight Rules to remember when caring for the patient with swan-ganz catheter. *Nursing '85* 15(3): 38–45, March 1985.
Nursing care of patients in shock, Part 2. *Am J Nurs* 82(9): 1401–1403, September, 1982.
O'Donnell TF, Belkin SC: The pathophysiology, monitoring, and treatment of shock. *Orthopedic Clinics of North America* 9 (3): 589-610, July 1978.
Perry A, Potter P: *Shock*. St Louis, CV Mosby Co, 1983.
Rice V: Shock, a clinical syndrome. *Critical Care Nurse* 1(5): 34–43, September/October 1981.
Rice V: Shock management Part II. Pharmacologic interventions, *Critical Care Nurse* 5(1): 42–56, 1985.
Rose M: Shock: Fluids restore circulation. *Monitoring Fluid and Electrolytes Precisely: Nursing Skillbook.* Horsham, PA: Intermed Communications, Inc, 1983, pp 149-154.
Shock. *Hospital Focus.* Knoll Pharmaceutical Co, pp 16+, October 1, 1962.
Simeone FA: Shock: Its nature and treatment. *Am J Nurs* 66: 1286-1294, 1966.
Statland H: *Fluid and Electrolytes in Practice*, ed 3. Philadelphia, JB Lippincott Co, 1963, pp 76-83, 176-189.
Strickland WM: Replacement therapy in traumatic shock. *Semin Report* 6: 2-7, Spring 1961 (Merck, Sharp and Dohme).
Stude C: Cardiogenic shock. *Am J Nurs* 74: 1636-1640, 1974.
Thompson MA: *Shock Syndrome*. Reading MA: Addison-Wesley Publishing Co, 1978.
Twombly M: Shift to third space. *Monitoring Fluid and Electrolytes Precisely.* Horsham, PA: Intermed Communications, Inc, 1983, pp 49–54.
Wiley L: Shock—different kinds and different problems. *Nursing '74* 4: 43-53, 1974.

CHAPTER SIX

Clinical Situations

BEHAVIORAL OBJECTIVES

Upon completion of this chapter, the student will be able to:

Explain the physiologic factors that lead to fluid and electrolyte changes, including those of the aged, of children, and of gastrointestinal surgery, trauma, renal failure with dialysis, burns, cancer, COPD, CHF, cirrhosis, and diabetic acidosis.

Assess fluid and electrolyte changes in given clinical examples and case reviews, and in actual clinical situations in a hospital or a community situation.

Apply clinical management in given clinical situations.

Plan nursing actions/interventions to meet patient's needs in a given clinical situation.

INTRODUCTION

In a clinical setting the nurse provides care for persons, having fluid and electrolyte imbalance resulting from selected clinical situations. Eleven situations are discussed that includes the aged, the child, gastrointestinal surgery, trauma–acute injury, renal failure with dialysis, burns, cancer, congestive heart failure, cirrhosis, and diabetic acidosis. To assess the patients' needs and to provide the care needed for persons with disease entities, the nurse must have the knowledge and understanding of fluid and electrolyte balance and imbalance. Through his or her knowledge and understanding, the nurse can then assess the changes occurring with patients and plan nursing interventions to meet these changes.

In this chapter the participant will become acquainted with people who have fluid and electrolyte imbalances. Most of these people were presented as part of the clinical situations, and some were presented the first time in the case reviews. The participant in this program will gain an understanding of the physiologic factors involved in each clinical situation. Clinical applications, clinical management, case reviews, nursing diagnoses, and nursing actions are included with each. From the clinical problems presented, the nurse should be more cognizant of the fluid and electrolyte changes occurring in patients and should then be able to apply the knowledge and understanding gained to other clinical situations.

An asterisk (*) on an answer line indicates a multiple-word answer. The meanings for the following symbols are: ↑ increased, ↓ decreased, > greater than, < less than.

PROBLEMS OF THE AGED

1 Changes in the aged are in body structure, gradual decrease in body functions, and a decrease in the ability to recuperate from injury and stress.

The aged person has difficulty in maintaining homeostasis of fluid and electrolyte balance, especially with diminished pulmonary, renal, cardiac, and gastrointestinal functions.

The changes occurring in the aged are:

a. *____________________.

b. *____________________.

c. *____________________.

With diminished pulmonary, renal, cardiac, and gastrointestinal functions, the aged person is prone to *____________________

____________________.

\- - - - - - - - - - - - - - - -

a. body structure
b. decrease in body function
c. decreased ability to recuperate from injury and stress

fluid and electrolyte imbalance

PHYSIOLOGIC FACTORS

In the aged, there is a decrease in body function, causing fluid and electrolyte imbalance. Table 52 gives the four major physiologic changes in body function occurring with the aged, namely, pulmonary, renal, cardiac, and gastrointestinal. Rationale and nursing interventions are given for each physiologic change. Study the table and be able to state specific rationale and nursing interventions. Refer to the table as needed.

TABLE 52. MAJOR PHYSIOLOGIC CHANGES IN THE AGED AND NURSING INTERVENTION

Physiologic Factors		Rationale	Nursing Intervention
Pulmonary function	Decreased ↓	Maximal breathing capacity greatly reduced due to: 1. loss of elasticity of parenchymal lung tissue. 2. increased rigidity of chest wall. Poor diffusion of respiratory gases due to: 1. defective alveolar ventilation. 2. accumulation of bronchial secretions. Reduced ventilation: 1. causes increased CO_2 retention. 2. causes respiratory acidosis. 3. results from emphysema, asthma, chronic bronchitis, and bronchiectasis.	Increase breathing capacity with elimination of CO_2 through: 1. breathing exercise with prolonged expiration. 2. coughing after a few deep breaths. 3. changing positions. 4. chest clapping. 5. intermittent positive pressure breathing (IPPB). 6. preventing respiratory infections when possible.
Renal function	Decreased ↓	Persistent renal vasoconstriction causing: 1. reduced glomerular filtration rate. 2. decrease in functioning nephrons. Renal blood flow begins to decrease by age 40. Kidneys lose ability to retain or to excrete water and solutes. 1. decrease in H^+ excretion; thus, metabolic acidosis can occur. 2. inability to concentrate urine. 3. accumulation of waste products in body. 4. decreased renal ability to excrete drugs.	Determine and maintain renal function through: 1. checking fluid intake and output. 2. encouraging fluid intake. 3. checking acid-base balance according to serum CO_2 or HCO_3. 4. testing specific gravity to determine kidneys' ability to concentrate urine. 5. noting drugs that may be toxic to renal function. 6. observing for side effects from drug accumulation.
Circulation and cardiac function	Decreased ↓	Increased rigidity of arterial walls (arteriosclerosis) causing: 1. increased blood pressure. 2. decrease in elasticity of blood vessels, thus stasis of blood. Stasis blood in venous system causes back pressure on capillaries, raising capillary pressure:	Determine and maintain circulation and cardiac function through: 1. checking blood pressure for elevations resulting from arteriosclerotic changes. 2. determining blood flow in lower extremities by checking pulses.

TABLE 52. (Continued)

Physiologic Factors		Rationale	Nursing Intervention
		1. capillary pressure forces fluid into tissue areas, causing edema. Diminished strength of heart contraction: 1. decrease in cardiac output. 2. decreased blood flow. Decreased cardiac reserve (capacity of heart to respond to increased burden): 1. heart rate same as young adult except under stress; takes longer to return to normal. 2. congestive heart failure can result.	3. checking for edema from ↑ capillary pressure. 4. checking pulse rates (apical and radial) to determine heart contraction and cardiac output. 5. noting changes in heart rate following activity. 6. assessing chest sounds for moist rales.
Gastro-intestinal	Decreased ↓	Atrophy to gastric mucosa causing: 1. decrease in gastric secretions, especially loss of HCl. 2. metabolic alkalosis can result from decrease in HCl. Atrophic gastritis due to: 1. diffused inflammation of gastric mucosa. Muscular atrophy in small and large intestines: 1. reduces motility of gastrointestinal tract, causing constipation. 2. supportive structures in intestinal walls are weakened, which can cause diverticuli. Perception of bowel elimination is decreased: 1. constipation can occur.	Determine and maintain gastrointestinal function through: 1. discussing with patient his preference in foods. 2. suggesting diet to meet nutritional needs according to physiologic changes and individual preferences. 3. encouraging fluid intake to decrease constipation. 4. checking frequency of bowel elimination. 5. noting color and consistency of stool. 6. assessing bowel sounds to determine presence of peristalsis.
Others:			
Liver function	Decreased ↓	Liver cell degeneration. Decreased hepatic capacity to detoxify drugs.	Noting those drugs that patients are taking that may be toxic to liver.
Endocrine gland function	Decreased	Hormone output dwindles. Decreased overall metabolic capacity. Decreased endocrine gland function to react to adverse drug action.	

2 With the aging process, the pulmonary function is (increased/decreased).

____________________.

The maximal breathing capacity is reduced due to:

a. *____.

b. *____.

\- - - - - - - - - - - - - - - - - -

decreased
a. loss of elasticity of the parenchymal lung tissue
b. increased rigidity of the chest wall

---

3 Poor diffusion of respiratory gases are due to:

a. *____.

b. *____.

With reduced ventilation, there is:

( ) a. CO_2 retention
( ) b. CO_2 excretion
( ) c. respiratory alkalosis
( ) d. respiratory acidosis

Clinical diseases that cause a decrease in breathing capacity, poor diffusion, and reduced ventilation are ________________, ________________, ________________, and ________________.

\- - - - - - - - - - - - - - - - - -

a. defective alveolar ventilation
b. accumulation of bronchial secretions

a. X; b. –; c. –; d. X
emphysema; asthma; chronic bronchitis; bronchiectasis

4 Name five nursing interventions that will increase breathing capacity and help with the elimination of CO_2.

a. * ____.

b. * ________________________________.

c. * ____________________________.

d. * ____________________________.

e. * ______________________.

— — — — — — — — — — — — — — — — — —

a. breathing exercise with prolonged expiration
b. coughing and deep breathing
c. changing positions
d. chest clapping
e. using IPPB

---

5 Would the renal function in the aged be (increased/decreased)?

______________________.

There is a persistent renal vasoconstruction, mostly due to arteriosclerotic changes, which causes * ____ and * ____.

— — — — — — — — — — — — — — — — — —

decreased
reduced glomerular filtration rate; a decrease in functioning nephrons

---

6 The kidneys lose the ability to ________________ or ________________ water and solutes.

What happens to the kidneys' ability to excrete hydrogen? * ________________ ______________________. What type of acid-base imbalance results?

* ________________________________.

— — — — — — — — — — — — — — — — — —

retain; excrete;
decreased H^+ excretion;
metabolic acidosis

7 Can the kidneys concentrate urine? ________ Explain why. * ____________

____

____.

No
With aging, kidneys have a decreasing ability to concentrate urine; thus, waste products in the body accumulate

8 With decreased renal function, what may be the result of drug excretion?

* ____

____.

---

decrease in drug excretion and drug accumulation in the body

9 Name for nursing interventions for determining and maintaining renal functions.

a. * ______________________.

b. * ______________________.

c. * ______________________.

d. * ______________________.

---

a. checking fluid intake and output
b. encouraging fluid intake
c. testing specific gravity
d. noting drugs that may be toxic to renal function

Also, checking acid-base balance according to the serum CO_2 or HCO_3

---

10 With the aging process, circulation and cardiac function are (increased/decreased) ______________.
There is an increased rigidity of the arterial walls due to arteriosclerotic changes, which cause:

a. * ______________.

b. * ______________

______________.

- - - - - - - - - - - - - - - - -

decreased
a. increased blood pressure
b. decrease in elasticity of the blood vessels, causing stasis of the blood

---

11 Stasis of blood in the veins can cause back pressure on the capillaries, increasing

* ______________.

The result of capillary pressure is ______________.

- - - - - - - - - - - - - - - - -

capillary pressure
edema

---

12 Diminished strength of heart contraction causes a decrease in * ______________

______________ and * ______________.

- - - - - - - - - - - - - - - - -

cardiac output; blood flow

---

13 The aged person has a decrease in cardiac reserve. What is cardiac reserve?

* ____

____.

Under stress, the heart rate will increase both in the young adult and in the elderly. After stress, what happens to the heart rate in the young adult? *

* ____

and in the elderly? * ____

______________________________.

- - - - - - - - - - - - - - - - -

Capacity of the heart to respond to increased burden;
young adult: heart rate quickly returns to normal;
elderly: heart rate takes longer to return to normal

---

14 Name five nursing interventions for determining and maintaining circulation and cardiac function.

a. * ____.

b. * ____.

c. * ____.

d. * ____.

e. * ____.

- - - - - - - - - - - - - - - - -

a. checking blood pressure for elevation or hypertension
b. determining blood flow in lower extremities by checking pulses
c. checking for edema
d. checking apical and radial pulse rates
e. noting changes in heart rate following activities

Also, assessing chest sounds for moist rales

---

15 Gastrointestinal functions are (increased/decreased) ________________ in the aged.

Atrophy of the gastric mucosa occurs with aging. What important gastric juice is lost? ________________.

What type of acid-base imbalance may occur? * ________________.

- - - - - - - - - - - - - - - - - -

decreased
HCl
metabolic alkalosis

---

16 Muscular atrophy in the small and large intestines occurs in aging. Is the gastrointestinal motility (peristalsis) increased or decreased? ________________.

Would diarrhea or constipation occur? ________________.

- - - - - - - - - - - - - - - - - -

decreased
constipation

---

17 The aged frequently lack perception of a bowel movement. What type of bowel problem can result? ________________.

- - - - - - - - - - - - - - - - - -

constipation

---

18 The supportive structures in the intestinal wall are (strengthened/weakened) ________________ that can cause diverticuli. Do you know what a diverticulum is? * ____.

- - - - - - - - - - - - - - - - - -

weakened
Outpouch in the intestinal wall due to a weakened structural area. Diverticuli are many such outpouches.

---

19 Name at least five nursing interventions for determining and maintaining gastro-intestinal function.

a. *______________________________.

b. *______________________________.

c. *__________________.

d. *______________________________.

e. *______________________________.

- - - - - - - - - - - - - - - - - -

a. discussing with patients their food preference
b. suggesting diet to meet nutritional needs
c. encouraging fluid intake
d. noting color and consistency of bowel movements
e. checking frequency of bowel elimination
Also, assessing bowel sounds

CLINICAL APPLICATIONS

20 The total body water in the adult is ______ %. With the aged, the total body water decreases to approximately 54%. There is a slight increase in extracellular fluid and decrease in intracellular fluid.

- - - - - - - - - - - - - - - - - -

60

21 Hypokalemia is common with the aged since potassium is not conserved well. Many aged people receive diuretics (potassium-wasting) and steroids. What effect would these drugs have on serum potassium? *______________________________

______________________________.

(Review Chapter 2, potassium with drug relationship, if needed.)

- - - - - - - - - - - - - - - - - -

It would decrease serum potassium level, causing hypokalemia

22 The aged person's ECF is * ______________________, and the ICF is ______________.

The total body water in the aged is approximately ________%.

— — — — — — — — — — — — — — — — —

slightly increased; decreased; 54

There are six body fluid problems, including dehydration, edema, water intoxication, constipation, diarrhea, and diaphoresis, that are common to the aged.

Table 53 gives the six fluid problems with the aged, their causes, and nursing interventions. Refer to the table as needed.

TABLE 53. BODY FLUID PROBLEMS IN THE AGED

Fluid Problems	Causes	Nursing Intervention
Dehydration	1. Insufficient water intake 2. Increased urinary output 3. Decreased thirst mechanism 4. Diminished response to ADH (antidiuretic hormone) 5. Difficulty in concentrating urine	1. Assess fluid intake and output 2. Encourage adequate oral fluid intake. 3. Assess intravenous fluid according to osmolality. 4. Adjust rate of intravenous fluid according to age and physiologic state. 5. Assess for clinical signs and symptoms of hypovolemia, namely, dehydration. 6. Monitor other type of fluid therapy, e.g., clysis and tube feeding.
Edema	1. Slightly elevated ECF 2. Overhydration from intravenous therapy 3. Increased capillary pressure 4. Cardiac insufficiency	1. Assess fluid intake and output. 2. Adjust intravenous flow rate to prevent overhydration. 3. Assess for peripheral edema in morning. 4. Assess chest sounds for moist rales. 5. Observe signs and symptoms of hypervolemia—overhydration.
Water intoxication	1. Hypo-osmolar solutions with copious amounts of drinking water	Assess types of intravenous fluids, e.g., 5% dextrose in water. Observe for signs and symptoms of water intoxication.

TABLE 53. (Continued)

Fluid Problems	Causes	Nursing Intervention
Constipation	1. Decrease in water intake 2. Muscular atrophy of small and large intestines with decrease in GI motility 3. Perception loss for bowel elimination	1. Encourage fluid intake. 2. Assess bowel sounds for peristalsis. 3. Administer mild laxative or stool softeners. 4. Have patient eat at regular times. 5. Offer bedside commode.
Diarrhea	1. Tube feedings with too much carbohydrate 2. Constipation—with small amount of liquid stools 3. Partially digested nutrients 4. Viral or bacterial infection	1. Assess problem causing diarrhea. 2. Administer drug(s) to decrease motility, e.g., Lomotil, Kaopectate.
Diaphoresis	1. Excessive perspiration: a. fever b. high environmental temperature	1. Assess cause of problem.

---

23 The thirst mechanism is frequently (decreased/increased) ________ so that the older person drinks less water.

What two factors are mostly responsible for dehydration in the aged? *________ and ________.

- - - - - - - - - - - - - - - - -

decreased
insufficient water; increased urinary output

---

24 Give at least four nursing interventions for correcting dehydration in the older person.

a. *________.

b. *________.

c. *________.

d. *________.

- - - - - - - - - - - - - - - - -

a. assess fluid intake and output
b. encourage taking fluids orally
c. assess signs and symptoms of dehydration
d. assess intravenous fluid—type and rate

---

25 Overloading the vascular system with fluids can result in congestive heart failure. The type of edema occurring is * ______________________________

______________________________.

— — — — — — — — — — — — — — — — — —

pulmonary edema from overhydration

---

26 Peripheral edema can result from dependent or refractory edema. When the feet and ankles are edematous in the morning, what type of edema may be present?

* ______________________________

______________________________.

— — — — — — — — — — — — — — — — — —

refractory edema. Frequently the result of cardiac-renal impairment or insufficiency and with little to no diuretic effect. (May be called nondependent edema, since edema is present in the morning and is not necessarily due to gravity.)

---

27 Mark the correct nursing interventions for correcting edema in the aged. Correct the incorrect answers.

( ) a. Assess fluid intake and output
( ) b. Adjust intravenous flow rate
( ) c. Assess for peripheral edema in the evening for refractory or nondependent edema
( ) d. Assess bowel sounds
( ) e. Observe signs and symptoms for hypovolemia

— — — — — — — — — — — — — — — — — —

a. X; b. X; c. –in the morning; d. –chest sounds; e.–for hypervolemia (overhydration)

---

28 Why is 5% dextrose in water considered a hypo-osmolar solution?* ______________________________

____

______________________________.

— — — — — — — — — — — — — — — — — —

Five-percent dextrose in water is an iso-osmolar solution; however, if it is given without other solutes, the dextrose is utilized by the body, leaving water

29 Nursing interventions for correcting constipation include (correct the incorrect answers):
( ) a. encouraging fluid intake
( ) b. assessing bowel sounds for peristalsis
( ) c. administering harsh cathartics
( ) d. offering bedside commode
( ) e. having meals at regular times

- - - - - - - - - - - - - - - - -

a. X; b. X; c.—mild laxatives or stool softeners; d. X; e. X

---

30 Name four causes of diarrhea.

a. *________________________________.

b. *________________________________.

c. *________________________________.

d. *________________________________.

What is the most important nursing intervention for correcting diarrhea?

*____.

- - - - - - - - - - - - - - - - -

a. tube feedings with too much carbohydrate
b. constipation with small liquid stools
c. partially digested nutrients
d. viral or bacterial infections
Assessment of the problem causing diarrhea

---

31 The sixth fluid problem occurring with the aged is ________________, excessive perspiration.

What are the two causes of excessive perspiration?

a. ____________________________.

b. *____.

- - - - - - - - - - - - - - - - -

diaphoresis
a. fever
b. high environmental temperature

CASE REVIEW

Mrs. Palmer, age 89, is a guest in a nursing home. She has been hypoventilating and says that she has some difficulty with breathing. Her blood pressure was elevated, 168/100, and her pulse was 104. The nurse noted edema in the extremities. Her urine output was 500 ml per day. Mrs. Palmer's last bowel movement was 4 days ago.

1. As an aged person, Mrs. Palmer is prone to four major physiologic changes in body function, which are ____________, ____________, ____________, and ____________.
2. Mrs. Palmer's reduced breathing capacity may be due to:
   a. *____________.
   b. *____________.
3. Hypoventilation can result in *____________ retention. What type of acid-base imbalance can occur from hypoventilation? *____________.
4. Give at least three nursing interventions that can increase Mrs. Palmer's pulmonary function.
   a. *____________.
   b. *____________.
   c. *____________.
5. Mrs. Palmer's blood pressure was elevated due to *____________.
6. The physiologic reason for edema in her lower extremities is *____________.
7. Give three nursing interventions regarding Mrs. Palmer's blood pressure and edema.
   a. *____________.
   b. *____________.
   c. *____________.
8. Urine output should be ____________ ml per hour or ____________ ml per 24 hours to maintain adequate renal function. Mrs. Palmer's urine output was 500 ml in 24 hours. Is her urinary output adequate? ____________. Name two possible possible reasons for her poor urinary output.
   a. *____________.
   b. *____________.

9. Give two nursing interventions in regard to Mrs. Palmer's renal function:

   a. *____.

   b. *____.

10. What two physiologic factors can cause Mrs. Palmer's constipation?

   a. *____.

   b. *____.

11. The nursing interventions for alleviating constipation include

   *____ and

   *____.

Mrs. Palmer refused to eat. The nurse encouraged her to take liquids; however, she would spit the liquids out. Her temperature was 99° F (37.2° C), pulse 104, and respirations 28. Her laboratory findings showed an elevation in hemoglobin, hematocrit, and BUN. The serum K was 3.4 mEq/L, serum Na was 147 mEq/L and Cl was 105 mEq/L. (Review Chapter 2 for normal electrolyte ranges.) Her skin and mucous membranes were very dry. Her feet and ankles were still swollen. The problem of constipation was also present.

12. From this last history, what three fluid problems did Mrs. Palmer have?

   a. ____.

   b. ____.

   c. ____.

13. What was the clinical cause for Mrs. Palmer's dehydration? *________________

   ____

   ____.

14. What are Mrs. Palmer's clinical signs and symptoms of dehydration?

   a. *____.

   b. *____.

   c. *____.

   d. *____.

15. Her second fluid problem was ____________________. Is it possible to have dehydration and edema at the same time? ____________________.

   Can you explain why? *____

   ____.

Mrs. Palmer was given intravenous fluids for several days and then later given tube feedings daily.

16. While Mrs. Palmer was receiving intravenous fluids, two nursing interventions would be to *________________________________ and to *________________________________.

17. If Mrs. Palmer received intravenously 5% dextrose in water, what type of fluid problem might result? *________________________________.

18. If the intravenous fluids were administered too rapidly to Mrs. Palmer, the fluid imbalance that could develop would be *________________________________.

    Name three symptoms of this imbalance.

    a. *________________________________.

    b. *________________________________.

    c. ________________________________.

19. Tube feeding high in carbohydrate can cause what fluid problem? ________________________________.

20. Potassium level in the aged is *________________________________.

- - - - - - - - - - - - - - - - - -

1. pulmonary; renal; cardiac; gastrointestinal
2. a. loss of elasticity of the parenchymal lung tissue
   b. increased rigidity of the chest wall
3. CO_2 retention; respiratory acidosis
4. a. breathing exercise with prolonged expiration
   b. coughing after a few deep breaths
   c. changing positions. Also, chest clapping
5. increased rigidity of the arterial walls due to arteriosclerotic changes
6. increased capillary pressure forcing fluid into the tissues
7. a. check blood pressure
   b. determine blood flow in lower extremities by checking pulses
   c. check for edema in the morning to determine if it is dependent or nondependent edema
8. 25 ml per hour; 600 ml per 24 hours. No
   a. inadequate fluid intake
   b. kidneys unable to excrete water and solute. You could have answered: reduced glomerular filtration rate and decrease in functioning nephrons
9. a. check intake and output
   b. encourage fluid intake. Also, the answers could include: check acid-base balance and test specific gravity

10. a. reduce motility of the gastrointestinal tract
    b. loss of perception for bowel elimination
11. suggest diet (foods) to meet nutritional needs and maintain bowel function; encourage fluid intake. The answers could also include: check frequency of bowel elimination and determine the presence of peristalsis (bowel sounds)
12. a. dehydration
    b. edema
    c. constipation
13. insufficient water intake. You might have answered that a decreased thirst mechanism was present
14. a. Vital signs: temperature slightly elevated, pulse and respirations elevated
    b. Hgb, Hct and BUN elevated
    c. serum sodium elevated
    d. skin and mucous membranes very dry
15. edema; yes. Frequently a person can have edema and be dehydrated due to hypovolemia in the vascular system with increased fluids in the interstitial space
16. assess the intravenous fluid according to the type order and its osmolality; adjust rate of intravenous fluids according to her age and physiologic state
    The nurse should also assess fluid intake and output
17. water intoxication
18. hypervolemia or overhydration (pulmonary edema)
    a. constant, irritating cough
    b. engorged veins (neck and hand)
    c. dyspnea
19. diarrhea
20. poorly conserved

NURSING DIAGNOSES

Potential for impaired gas exchange related to the loss of elasticity of lung tissue and rigidity of chest wall due to the aging process.

Potential for ineffective breathing patterns related to the loss of elasticity of lung tissue, accumulation of mucous secretions, and/or CO_2 retention.

Potential alteration in the patterns of urinary elimination related to reduced glomerular filtration rate and a decrease in functioning nephrons.

Potential alteration in cardiac output related to diminishing strength of heart contractions and decreased cardiac reserve.

Potential alteration in tissue perfusion related to stasis blood in the venous system and arteriosclerotic changes.

Potential alteration in bowel elimination (constipation) related to a reduced motility of the gastrointestinal tract and a lack of perception of bowel elimination.

Potential alteration in nutrition related to insufficient intake of nutrients and lack of income.

Potential alteration in fluid volume (deficit) related to inadequate fluid intake, lack of thirst drive, and a decrease in the body's cellular fluid.

Alteration in the family process related to an inability to cope with the problems of an older family member or to participate in the care-giving role.
Potential for impaired physical mobility related to stiff joints, poor eyesight, or muscle weakness.
Potential for ineffective coping related to chronic disorder and dependence on others.

NURSING ACTIONS

1. Recognize the major physiologic changes in the aged (older adult), i.e., (1) decreased pulmonary function due to increased rigidity of chest wall, decreased alveolar ventilation, and accumulation of bronchial secretions; (2) decreased renal function due to reduced glomerular filtration rate, and decreased renal arterial flow; (3) decreased circulatory and cardiac function due to increased rigidity of arterial walls, increased capillary pressure, decreased cardiac output and blood flow; (4) decreased gastrointestinal function due to decreased gastric secretions, decreased GI tract motility and weakened intestinal wall.
2. Encourage the older adult to take deep breaths several times a day and, if bedridden, change positions every 2 hours. Chest clapping may be indicated to loosen mucous in the alveoli.
3. Teach older adults to avoid people who have colds and to seek medical care when they have a respiratory infection.
4. Check intake and output. If urine output is decreased, encourage fluid intake unless otherwise indicated (history of repeated heart failure); check specific gravity (↓ specific gravity with ↓ urine output could mean kidney dysfunction) and check BUN and creatinine levels. The physician should be notified if urine output is less than 400–600 ml per day.
5. Assess for circulatory changes by checking for blood pressure elevation, diminished pulses and edema in the lower extremities, and radial and apical pulse changes. Heart rate should be checked immediately after activity and at specified intervals. Pulse rate should increase with activity but return to the patient's normal range in approximately 30 minutes.
6. Check for changes in bowel movement regime, and note frequency, color, and consistency of stool.
7. Suggest foods to meet nutritional needs and to maintain normal bowel movements.
8. Monitor rate of intravenous fluids to run approximately 2 ml per minute (20 gtts per minute) or as ordered for the older adult. Fast-running IV fluids can overload the vascular area and cause overhydration and pulmonary edema. Signs and symptoms of overhydration are constant, irritated cough; dyspnea; neck and hand vein engorgement; and chest rales.
9. Check for peripheral edema in the morning. Morning peripheral edema is indicative of heart failure (right-sided). In the evening, peripheral edema could be due to venous stasis.
10. Assess for signs and symptoms of dehydration (cellular and vascular), such as thirst, dry mucous membrane, poor skin turgor, increased pulse rate, slightly ele-

vated temperature, decreased urine output. In the older adult, body water represents 54% of the body weight, whereas in young and middle-age adults, it is 60% of the body weight. Older adults may normally have poor skin turgor because of their decreased body water.

11. Offer fluids to the older adult to prevent dehydration. The thirst mechanism in the aged is diminished. They are not thirsty and do not realize they have ↓ ECFV (extracellular fluid volume deficit–dehydration).
12. Monitor other types of fluid therapy, i.e., tube feeding. Water should be given with tube feeding to prevent dehydration, since tube feeding alone does not hydrate a person but can pull fluid from the cells.

CHILDREN WITH FLUID IMBALANCE

32 The body is composed mostly of water. Body water in the early human embryo represents ________ % of body weight, in the newborn infant ________ % and in the adult ________ %. (Refer to Chapter 1, Frame 1.)

The low birth weight infant's (premature infant's) body water represents 80-90% of body weight.

— — — — — — — — — — — — — — — — —

97%; 77%; 60%

---

33 Complete the percentage of body weight that is representative of body water in the following:

Early human embryo ________ %

Low birth weight infant * ________ %

Newborn infant ________ %

Adult ________ %

— — — — — — — — — — — — — — — — —

97%; 80-90%; 77%; 60%

34 The infant needs proportionately more water due to: (1) the large body surface area, and (2) the infant's immature kidneys, which cannot concentrate urine effectively—urine volume is increased. More water is lost through the infant's skin because of the increased body surface area. Since the infant cannot concentrate urine, water is needed to maintain fluid volume because of the increased urine output.

Give two reasons why the infant needs a higher percentage of total body water:

a. *______________________________________.

b. *______________________________________.

- - - - - - - - - - - - - - - - - -

a. a large body surface
b. inability to concentrate urine with increased urine output

---

35 Water distribution in an infant is not the same as in an adult. The extracellular fluid (ECF) in the infant is 40% of body weight. Do you recall what percentage of body weight ECF is in the adult? ____________. (Refer to Chapter 1, Frame 11.)

The intracellular fluid (ICF) in the infant is 34% of body weight; whereas, in the adult it is __________%. (Refer to Chapter 1, Frame 11.)

- - - - - - - - - - - - - - - - - -

20%; 40%

---

36 At 1 year, the child's total body water is close in amount to the adult's (60%) but is not in the same extracellular fluid (ECF) and intracellular fluid (ICF) proportion as the adult. It is not until the child is between 3 to 5 years old that the proportions of ECF and ICF are similar to the adult's.

The extracellular fluid is composed of ________________________ and ________________________ fluid. Another name for intracellular fluid is (cellular/vascular) ________________ fluid. (Refer to Chapter 1, Frame 9.)

- - - - - - - - - - - - - - - - - -

interstitial and intravascular; cellular

---

37 Increased body surface area in the infant causes excess water loss through the

________________. The smaller the infant, the greater the body surface area in proportion to body weight. The infant's kidneys are (mature/immature)

________________, thus, the urinary volume is (increased/decreased)

________________.

- - - - - - - - - - - - - - - - -

skin; immature; increased

---

38 It may take two years before the child's kidneys are mature.

With infant's immature kidneys, the glomerular filtration rate (GFR) is decreased, thus, the ability to concentrate urine is (decreased/increased) ________

________________.

Giving too much water could cause (dehydration/overhydration). ________

________________.

- - - - - - - - - - - - - - - - -

decreased; overhydration

---

39 As the child grows, there is muscle growth and cellular growth. More water shifts from the extracellular to the intracellular fluid compartment.

When the child's ICF and ECF proportions become similar to the adult's, what do you think could be a contributing factor? *________________

________________________________.

- - - - - - - - - - - - - - - - -

Increased cellular growth causes water to shift from the ECF space to the ICF

---

40 The infant has less reserve of body fluid than the adult and is more likely to develop fluid volume deficit.

The infant loses one-half of his extracellular fluid daily; whereas the adult loses only one-sixth of his extracellular fluid.

Name two reasons why an infant loses one-half of his extracellular fluid daily.

1. *____________________.
2. *____________________.

— — — — — — — — — — — — — — — — — —

1. large body surface area causing water to be lost through the skin (insensible perspiration)
2. increased urinary output because immature kidneys cannot concentrate urine

---

41 Keeping an infant covered in a stable cool environment will (reduce/increase) ____________ insensible fluid loss through the body surface.

— — — — — — — — — — — — — — — — — —

reduce

---

42 Serum electrolytes do not vary greatly between infants and adults. Serum sodium level in a newborn fluctuates. It may be low the first 3 to 6 hours after birth and then rise slightly (2 to 6 mEq/L increase) during the first two days of life.

The infant's serum sodium level is 134–150 mEq/L and the child's level is 134–146 mEq/L. What is the normal adult serum sodium level? ____________

— — — — — — — — — — — — — — — — — —

135–146 mEq/L

---

43 If a 5-month-old infant consumes cow's milk and commercially prepared baby food, he will ingest five times more sodium than a breast-fed infant.

The name for an elevated serum sodium level is ____________.

— — — — — — — — — — — — — — — — — —

hypernatremia (serum sodium excess)

---

44 Low birth weight infants tend to develop hypernatremia with a normal to low sodium intake. Their body surface area is (greater/lesser) ________ than an average weight newborn's and their insensible water loss would be (increased/decreased) ________.

Also, low birth weight infant's kidneys are immature longer than average weight infants'; therefore, more diluted water is excreted. The loss of water is in excess of the loss of solutes.

- - - - - - - - - - - - - - - - - -

greater; increased

---

45 Hyponatremia can occur in infants and children. Another name for hyponatremia is serum sodium (deficit/excess) ________.

Causes of hyponatremia are:

1. Overhydration—water overloading
2. Continuous administration of oral or parenteral electrolyte-free solutions
3. Syndrome of inappropriate antidiuretic hormone secretion (SIADHS). This results in excess secretion of ADH causing excess water reabsorption from the distal tubules. Factors attributing to SIADHS are: CNS injuries or illness (head injuries, meningitis), pneumonia, neoplasma, stress, surgery, and drugs (narcotics, barbiturates).

- - - - - - - - - - - - - - - - - -

deficit

---

46 Name three causes of hyponatremia.

1. ________.
2. * ________.
3. ________.

- - - - - - - - - - - - - - - - - -

1. overhydration
2. continuous administration of electrolyte-free solutions
3. SIADH

---

47 A rapid decrease in serum sodium, 120 mEq/L or below, could cause central nervous system changes such as headache, twitching, confusion, and convulsion.

The nurse should observe for central nervous system changes when hyponatremia occurs suddenly. Give three CNS symptoms:

1. ______________________________.

2. ______________________________.

3. ______________________________.

— — — — — — — — — — — — — — — —

1. headache; 2. twitching; 3. confusion. Also convulsion

---

48 The serum potassium level in the infant is 3.5–5.8 mEq/L. The top range is slightly higher than adult's and remains in the upper level for the first few months of the infant's life.

Do you recall where the greatest concentration of potassium is found in the body? ______________________________.

— — — — — — — — — — — — — — — —

cells or intracellular fluid. In various institutions, laboratory values will vary slightly.

---

49 Infants and children may develop hypokalemia (serum potassium deficit) when cellular breakdown occurs from injury, starvation, dehydration, diarrhea, vomiting, diabetic acidosis, and steroids for treating nephrosis. Children do not conserve potassium well and the kidneys will continue to excrete body potassium even with little or no potassium intake.

Give at least two signs or symptoms of hypokalemia (Refer to Chapter 2, Table 6.)

a. ______________________________.

b. *______________________________.

— — — — — — — — — — — — — — — —

dizziness; muscular weakness; abdominal distention; decreased peristalsis; arrhythmia

---

50 Eighty to ninety percent of body potassium loss is excreted in the urine. If oliguria (decreased urine output) occurs, what type of potassium imbalance will occur? ______________________.

— — — — — — — — —

hyperkalemia or serum potassium excess

---

51 The infant's serum chloride level is 96–116 mEq/L. For the first few months of the infant's life the serum sodium level is ______________ and the serum potassium level is ______________. The child's serum chloride level is 98–105 mEq/l.

— — — — — — — — — — — — — — — —

134–150 mEq/L; 3.5–5.8 mEq/L

---

52 The calcium in the cord blood is higher than the maternal serum calcium, but after birth the infant's calcium level decreases to 3.8 mEq/L or 7.7 mg/dl. With low birth weight infants, the serum calcium tends to remain lower for a longer period of time. (Child's serum calcium level is 4.5–5.8 mEq/L or 9–11.5 mg/dl.)

Infants do not have calcium stored in the bones as do adults. If the infant is fed cow's milk, the body calcium level may remain low since cow's milk has higher phosphorus content, which lowers the calcium level.

Breast-fed infants receive more calcium and retain it since breast milk contains less phosphorus.

Which infant retains more body calcium—the infant receiving cow's milk or the breast-fed infant? *______________________.

— — — — — — — — — — — — — — — —

breast-fed infant

---

53 Calcium will be ionized in an acidotic state but not in an alkalotic state. Tetany symptoms occur when hypocalcemia (serum calcium deficit) is present in a normal acid-base balance or when an alkalotic state exists.

Symptoms of tetany are *______________ and *______________. (Refer to Chapter 2, Table 18.)

— — — — — — — — — — — — — — — —

tingling of fingers; twitching around mouth. Also, carpopedal spasm

---

54 Newborn infants tend to have a low pH, metabolic acidosis. This is the result of increased acid metabolites due to the infant's increased metabolic rate and physiologic changes from birth. The pH becomes closer to normal after the first few days or weeks of life. In low birth weight infants, the pH remains low for several weeks.

Cow's milk has a low pH; and with a low calcium level, tetany (could/could not) occur ______________________.

— — — — — — — — — — — — — — — — —

could not. Calcium is ionized in an acidotic state regardless of how low it is.

CLINICAL APPLICATIONS

Table 54 gives a simple method for calculating daily fluid and electrolyte requirements for infants and children, and this table is presented only as information for fluid and electrolyte maintenance in children. There are many tables and nomograms used for fluid calculations; however, Table 54 describes a method used by many pediatricians and hospital personnel. The calculations are according to milliliters times infant's weight for fluid, and milliequivalents times 100 ml of water for electrolytes.

TABLE 54. FLUID AND ELECTROLYTE DAILY REQUIREMENT FOR INFANTS AND CHILDREN

Body Weight (kg)	Fluid Requirement (ml/24 hr)	Electrolyte Requirement (mEq/L/24 hr)
1-10	100 ml/kg	
11-20	1000 ml +50 ml/kg for each kg above 10	3 mEq of sodium + 2 mEq of potassium for each 100 ml of water
21 and above	1500 ml +20 ml/kg for each kg above 20	

Adapted from Wilmington Medical Center, Pediatric Dept, and HI Hochman, et al.: Dehydration, diabetic ketoacidosis and shock in the pediatric patient. *Pediatr Clin of North Am* 26(4): 805, November 1979.

55 Daily fluid requirement for an infant weighing 6 kg is:
( ) a. 300 ml
( ) b. 600 ml
( ) c. 900 ml

— — — — — — — — — — — — — — — — —

a. –; b. X; c. –

---

56 A child weighs 15 kg. The daily fluid requirement is:

( ) a. 1150 ml
( ) b. 1250 ml
( ) c. 1500 ml

The daily sodium requirement according to the fluid requirement is:

( ) d. 15 mEq/L
( ) e. 25.5 mEq/L
( ) f. 37.5 mEq/L

The daily potassium requirement according to the fluid requirement is:

( ) g. 25 mEq/L
( ) h. 30 mEq/L
( ) i. 35 mEq/L

- - - - - - - - - - - - - - - - - -

a. –; b. X; c. –; d. –; e. –; f. X; f. X; g. X; h. –; i. –

---

57 Intravenous flow rate should be checked every 15 minutes on an infant and young child, and every 30 minutes on an older child (age 6 and above). A microdrip chamber set should be used, and as a safety precaution, approximately 2 hours of calculated solution should be in the solution container/set.

Intravenous fluid administered to a 2-year-old child should be checked every

__________________ minutes.

- - - - - - - - - - - - - - - - - -

15 minutes

NURSING ASSESSMENT

Table 55 lists the nursing assessment with rationale for assessing fluid and electrolyte imbalance in infants and children. This table can be used as an assessment tool in the hospital, clinic, or at home. In the nursing assessment column, you would either check or fill in the blanks. The rationale would be eliminated when used as a tool. Study the table and complete the frames related to the nursing assessment. Refer back to the table as needed.

TABLE 55. NURSING ASSESSMENT OF FLUID AND ELECTROLYTE IMBALANCE IN CHILDREN

Observation	Nursing Assessment		Rationale/Action
Changes in behavior and general appearance	Irritable	______	Early symptoms of fluid volume deficit are irritability, purposeless movement, and an unusual high-pitched or whining cry.
	Unusually quiet	______	As dehydration continues, lethargy and unconsciousness may occur.
	Lethargic	______	Gray or pallor color indicates a decrease in peripheral circulation from severe fluid loss—shock
	Purposeless movement	______	Flushed color could indicate sodium excess.
	Different cry	______	
	Won't eat	______	
	Color		
	Pale	______	
	Gray	______	
	Flushed	______	
	Unconsciousness	______	
Neurologic signs	Abdominal distention	______	Abdominal distention and weakness may indicate potassium deficits.
	Diminished reflexes	______	Tetany symptoms could indicate calcium and magnesium
	Weakness/paralysis	______	Confusion could be due to potassium deficit and/or fluid volume deficit.
	Tetany tremors	______	
	Twitching	______	
	Sensorium		
	Confusion	______	
	Comatose	______	
	Other	______	
Weight change	Preillness weight	______	Weight loss can indicate the degree of dehydration (fluid loss) Mild—2-5% loss of body weight Moderate—6-10% loss Severe—11% and above loss Routine weights should be taken on the same scale and at the same time each day.
	Present weight	______	
Skin (tissue) turgor	Elevated pinched skin (after 2 to 3 seconds)	______	Skin and subcutaneous tissue should be checked together to avoid misinterpretation of dehydration or normal skin turgor. To test skin turgor, pinch the skin over the abdominal and chest wall, and the medial aspect or the thigh. Skin turgor on obese infants or children can appear normal even with a fluid deficit. Undernourishment could cause poor tissue turgor with fluid balance. Abdominal distention may mask turgor loss. Poor skin turgor begins with 3% or more weight loss, considered as fluid loss.
	Lack of elasticity	______	
Dryness of mucous membrane	Dryness in oral cavity (cheeks and gums)	______	Mucous membrane and the tongue are dry with fluid deficit.
	Dry tongue	______	

TABLE 55. (Continued)

Observation	Nursing Assessment		Rationale/Action
			Dry tongue may indicate mouth breathing, so check to see if this is the cause of dryness. Some medications and vitamin deficiencies cause dryness. Dryness in the oral cavity (cheeks and gums) is a better indicator of fluid loss.
Sunken eyeballs and fontanels	Sunken eyeballs	______	Sunken eyes and dark skin around them can indicate severe fluid volume deficit.
	Sunken fontanels	______	Depression of the anterior fontanel is often an indicator of fluid deficit.
Absence of tearing and salivation	Absence of tearing Decrease in saliva Absence of saliva	______ ______ ______	Absence of tearing and salivation are indicators of fluid volume deficit. Tearing does not begin until approximately 4 months of age. This occurs with moderate dehydration (6–10% body weight loss).
Thirst	Thirst Mild Avid	 ______ ______	Thirst is an indicator of dehydration-fluid loss. Thirst may be difficult to determine when vomiting is present. If vomiting is present, offer carbonated drink (flattened Coca Cola or ginger ale) Avid thirst may indicate serum hyperosmolality and cellular dehydration.
Change in urine output	Number of voidings Amount ml/8 hr ml/25 hr Weight of saturated urine diaper (subtract dry diaper weight) Urine color Specific gravity	______ ______ ______ ______ ______ ______	Decrease in urine output and very concentrated urine (dark yellow color and ↑ specific gravity) indicate fluid volume deficit. With severe fluid deficit, the infant may not void for 16–24 hours and not have abdominal distention. Polyuria (increased urine output) with low specific gravity (dilute) could indicate kidney damage, excess fluid intake, extracellular shift from interstitial fluid to plasma, or ↓ ADH. Oliguria could be due to renal insufficiency or extracellular shift from the plasma to the interstitial space.

TABLE 55. (Continued)

Observation	Nursing Assessment		Rationale/Action
			Potassium excess or sodium excess could result. *Normal Range of Urine Output* 6 months: 12 ml/hour 1 year: 18–25 ml/hour 5 years: 20–30 ml/hour 12 years: 25–35 ml/hour Adult: 35–50 ml/hour *or* Newborn: 50 ml/kg/day All others: 25 ml/kg/day
Stools	Number Consistency Color Amount	______ ______ ______ ______	The consistency, color, and amount of the stool should be noted. If the stool is of liquid consistency, it should be measured. Frequent, liquid stools can lead to fluid volume deficit, potassium and sodium deficit, and bicarbonate deficit (acidosis).
Vomitus	Number Consistency Color Amount	______ ______ ______ ______	Vomitus needs to be described according to the amount, color, and consistency. Frequent vomiting with large quantity will lead to fluid loss, potassium and sodium loss, and hydrogen and chloride loss (alkalosis).
Vital Signs			Vital signs should be taken every 15 to 30 minutes while the infant or child is seriously ill.
Body temperature changes	Admission Temperature Time ______ Time ______	______ (1) ______ (2) ______	Fever will increase insensible water loss. The child's extremities may feel cold because of hypovolemia (fluid volume deficit), which decreases peripheral circulation. A subnormal temperature may be due to reduced energy output.
Changes in pulse	Admission pulse Pulse rate Time ______ Time ______ Pattern	______ (1) ______ (2) ______ ______	Increased pulse rate (↑ 160 infant and ↑ 120 child) may indicate a fluid volume deficit (hypovolemia) and the possibility of shock. Full, bounding pulse could mean fluid volume excess. Irregular pulse could be due to hypokalemia.
Changes in breathing	Admission rate Respiration Time ______	______ (1) ______	Note the rate, depth, and pattern of the infant's breathing.

TABLE 55. (Continued)

Observation	Nursing Assessment	Rationale/Action
	Time ______ (2) ______ Pattern ______	Rapid breathing increases insensible fluid loss. Rapid, deep, vigorous breathing (Kussmaul breathing) frequently indicates metabolic acidosis. Acidosis could be due to poor hydrogen excretion by the kidneys, diarrhea, salicylate poisoning, or diabetes mellitus. Shallow, irregular breathing could be due to metabolic alkalosis.
Blood pressure	Admission BP ______ Blood pressure Time ______ (1) ______ Time ______ (2) ______	Elasticity of young blood vessels may keep blood pressure stable even when a fluid volume deficit is present. Increased blood pressure may indicate fluid volume excess. Decreased blood pressure may indicate severe fluid volume deficit, extracellular shift from the plasma to the interstitial space, or sodium deficit.
Blood chemistry and hematology changes	Electrolytes Time / Time K ______ ______ Na ______ ______ Cl ______ ______ Ca ______ ______ Mg ______ ______ BUN ______ ______ Creatinine ______ ______ Hgb ______ Hct ______	One set of blood chemistry is not sufficient for assessment. Electrolytes should be frequently monitored when they are not in normal range. Norms are Na: Infant 134–150 mEq/L Child 134–146 mEq/L K: Infant 3.5–5.8 mEq/L Child 3.5–5.8 mEq/L Ca: Infant 3.7–6.0 mEq/L (7.3–12 mg/dl) Child 4.5–5.8 mEq/L (9–11.5 mg/dl) Cl: Infant 96–116 mEq/L Child 98–105 mEq/L BUN if elevated could indicate fluid volume deficit or kidney insufficiency. Creatinine frequently indicates kidney damage. Norms: Bun 10–25 mg/dl Creatinine 0.7–1.4 mg/dl Elevated hemoglobin and hematocrit could indicate hemoconcentration caused by fluid volume deficit. If

TABLE 55. (Continued)

Observation	Nursing Assessment	Rationale/Action
		anemia is present, the hemoglobin and hematocrit may appear falsely normal.
	Blood gases pH ____ PCO_2 ____ PO_2 ____ HCO_3 ____ BE ____	Normal range for pH is 7.35–7.45. pH 7.35 and less indicate acidosis. pH 7.45 and higher indicate alkalosis. Normal range for PCO_2 is 35–45 mm Hg. PCO_2 ↓ 35 means respiratory alkalosis or compensation (overbreathing-hyperventilating); PCO_2 ↑ 45 means respiratory acidosis (lung disorder). Normal range for HCO_3 is 24 to 28 mEq/L and BE -2 to +2. HCO_3 ↓ 24 and BE ↓ -2 metabolic acidosis (common in newborns). HCO_3 ↑ 28 and BE ↑ +2 metabolic alkalosis (occurs with vomiting—loss of HCl).
GI suction, drainage tubes, and fistula	GI suction ____ Amount ____ Drainage tube ____ Amount ____ Fistula ____ Color ____ Amount ____	Fluid loss from all sources should be measured. Fluid loss from GI suctioning, drainage tubes, and fistula can contribute to severe fluid and electrolyte imbalances.

58 The child in the hospital should be assessed for fluid balance every 8 hours; the child with an existing fluid imbalance should be assessed every hour.

The fluid status of a child in fluid imbalance should be closely monitored every ____________.

— — — — — — — — — — — — — — — —

59 Nursing assessment includes observing changes in behavior and general appearance. Irritability, lethargic, purposeless movement, high-pitched or whining cry, pallor or gray color could indicate:

( ) a. fluid volume deficit
( ) b. fluid volume excess
( ) c. cellular fluid excess

— — — — — — — — — — — — — — — —

a. X; b. –; c. –

60 Neurologic changes could indicate fluid and electrolyte imbalance. Match the neurologic assessment with the probable imbalance. (Refer to Chapter 2—potassium, calcium, and Chapter 5—dehydration and edema.)

______ 1. abdominal distention	a. potassium deficit
______ *2. confusion	b. calcium deficit
______ 3. muscle weakness	c. fluid volume excess
______ 4. tetany symptoms	d. fluid volume deficit

\- - - - - - - - - - - - - - - - -

1. a; 2. a, d; 3. a; 4. b. Good.

---

61 Twelve-percent weight loss would be comparable to what degree of dehydration (fluid loss)?
( ) a. mild dehydration
( ) b. moderate dehydration
( ) c. severe dehydration

\- - - - - - - - - - - - - - - - -

a. –; b. –; c. X

---

62 Tissue (skin) turgor can be misleading as an indicator of fluid volume loss. Explain why. *____

____.

\- - - - - - - - - - - - - - - - -

Subcutaneous fat can give the appearance of normal skin turgor, or abdominal distention may mask poor skin turgor

---

63 Indicate which areas of the body skin turgor should be checked.
( ) a. face
( ) b. chest wall
( ) c. abdomen
( ) d. top of thighs
( ) e. medial aspect of thighs

\- - - - - - - - - - - - - - - - -

a. –; b. X; c. X; d. –; e. X

---

64 To determine dryness of the mucous membrane, which part of the mouth should be assessed?
( ) a. cheeks and gums of the oral cavity
( ) b. teeth

— — — — — — — — — — — — — — — — —

a. X; b. –

---

65 Sunken eyeballs and fontanels frequently do not occur in infants until there is a 10% body weight loss (as fluid loss). Give the type of fluid imbalance that would be present with a 10% fluid loss. *______________________________

______________________________.

— — — — — — — — — — — — — — — — —

Moderate to severe dehydration, or the upper range of moderate dehydration

---

66 Thirst and absence of tearing and salivation are indicators of fluid volume

______________________________.

— — — — — — — — — — — — — — — — —

deficit

---

67 Normal specific gravity for a young infant is 1.002-1.010 and for a child is 1.005-1.030. If the child's urinary output is decreased and the specific gravity is > 1.030, the fluid imbalance would be *______________________________.
Usually neonates and young infants cannot concentrate urine even when hypo volemia is present.

— — — — — — — — — — — — — — — — —

fluid volume deficit or hypovolemia

---

68 Hyperkalemia (serum potassium excess) could result from (polyuria/oliguria)

______________________________.

— — — — — — — — — — — — — — — — —

oliguria

---

69 Frequent and increased quantities of vomitus and stools can lead to:
( ) a. hypokalemia
( ) b. hyperkalemia
( ) c. hyponatremia
( ) d. hypernatremia
( ) e. acidosis
( ) f. alkalosis
( ) g. dehydration

- - - - - - - - - - - - - - - - -

a. X; b. –; c. X; d. –; e. X(↑ stools due to loss of HCO_3); f. X (↑ vomitus due to loss of HCl); g. X

---

70 Vital signs should be monitored every *__________ minutes for the seriously ill infant or child. Check the vital signs that can indicate fluid volume loss.
( ) a. subnormal temperature
( ) b. rapid, weak pulse
( ) c. rapid respiration
( ) d. fever
( ) e. shallow breathing
( ) f. systolic pressure below 80

- - - - - - - - - - - - - - - - -

15–30; a. –(in some cases it could indicate loss); b. X; c. X; d. X; e. –; f. X

---

71 An irregular pulse (arrhythmia) could be caused by (potassium deficit/potassium excess). ____________________ .

A full bounding pulse could mean ____________________ .

Why are blood pressure readings in infants and young children a poor indicator of fluid imbalance? *______________________________

____ .

- - - - - - - - - - - - - - - - -

potassium deficit; fluid volume excess. Elasticity of young blood vessels keeps blood pressure stable

72 An elevated BUN (blood urea nitrogen) could indicate * ____________

or ____________; whereas an elevated serum creatinine indicates

* ____________________.

\- - - - - - - - - - - - - - - - - -

dehydration (fluid volume deficit) or kidney insufficiency (renal damage); kidney damage

---

73 Arterial blood gases (ABGs) should be part of the assessment tool for assessing acid-base balance.

a. pH of 7.27 would indicate ____________.

b. pH of 7.48 would indicate ____________.

c. Pco_2 of 28 would indicate * ____________.

d. Pco_2 of 60 would indicate * ____________.

e. HCO_3 of 34 and BE +8 would indicate * ____________.

f. HCO_3 of 18 and BE -6 would indicate * ____________.

\- - - - - - - - - - - - - - - - - -

a. acidosis; b. alkalosis; c. hyperventilation due to respiratory alkalosis or compensation for metabolic acidosis; d. respiratory acidosis; e. metabolic alkalosis; f. metabolic acidosis

---

74 Name the fluid imbalance that can result from gastrointestinal suction and from secretions from drainage tubes and fistula. ____________.

Name the two electrolytes that are lost from gastrointestinal suctioning.

____________ and ____________.

\- - - - - - - - - - - - - - - - - -

dehydration (fluid volume deficit); potassium and sodium (important electrolytes), also hydrogen, chloride, bicarbonate, and magnesium

75 From the following list of observations and nursing assessments, check the ones that indicate fluid volume deficit (hypovolemia or dehydration).

( ) a. BUN elevated
( ) b. Hemoglobin and hematocrit elevated
( ) c. Increased secretions from GI suction
( ) d. Increased blood pressure
( ) e. Irritability, high-pitched cry
( ) f. Confusion, disorientation
( ) g. Weight gain
( ) h. Decreased, concentrated urine
( ) i. Avid thirst
( ) j. Poor skin turgor
( ) k. Absence of tearing and salivation
( ) l. Sunken eyeballs and anterior fontanel
( ) m. Increased number and quantity of vomitus and stools
( ) n. Temperature 98.2°F or 36.8°C
( ) o. Rapid, weak pulse
( ) p. Rapid breathing

\- - - - - - - - - - - - - - - -

a. X; b. X; c. X; d.–; e. X; f. X; g. –; h. X; i. X; j. X; k. X; l. X; m. X; n. –; o. X; p. X

---

76 Diarrheal dehydration is the number-one villain causing fluid and electrolyte imbalance in children. When vomiting with diarrhea occurs, fluid and electrolyte loss is more severe.

Other clinical problems causing fluid and electrolyte imbalance are surgery, pyloric stenosis, renal disease, diabetic ketoacidosis, gastroenteritis, and syndrome of inappropriate ADH (SIADH).

The most common cause of fluid and electrolyte imbalance is *______________

______________________________.

\- - - - - - - - - - - - - - - -

diarrheal dehydration

77 The degree of body fluid loss can be determined by the weight loss. However, the weight before illness needs to be known. Parents should be encouraged to keep an accurate record of the child's weight. Body weight loss in excess of 1% per day represents loss of body water.

To determine the degree of dehydration:

1. Convert pounds to kilograms (2.2 lb = 1 kg)
2. Subtract the present weight from the preillness (wellness) weight
3. Divide the weight loss by the wellness weight.

Example: Mary weighed 30 pounds or 13.6 kg before she became ill. Now she weighs 26 pounds or 11.8 kg. She has lost 4 pounds or 1.8 kg.

$$1.8 \text{ kg} \div 13.6 \text{ kg} - 0.13 \text{ or } 13\%$$

The percentage of weight loss is ________________. Would the degree of dehydration be (mild/moderate/severe)? ________________________.
(Refer to Table 43.)

- - - - - - - - - - - - - - - -

13; severe dehydration

DEGREES OF DEHYDRATION

Table 56 gives the three degree categories of dehydration: mild, moderate, and severe, with clinical assessment. The table should help you identify degree of dehydration according to clinical symptoms. Double plus (++) means the clinical symptoms are more pronounced (greatly increased). Plus (+) means the symptoms are present; minus and plus (-+) means the symptoms may or may not be present; and minus (-) means they are not present. Study the table and refer to it as needed.

TABLE 56. ASSESSMENT OF DEGREES OF DEHYDRATION

Assessment (Clinical Symptoms)	Mild 1-5%	Moderate 6-10%	Severe 11-15%
Thirst	+	++	++
Irritable to lethargic	-+	+	+
Confusion	-	+	++
Convulsions	-	+	+
Comatose	-	-	+
Reduced skin turgor	-+	+	++
Dry mucous membranes	-+	+	++
Absence of tearing and salivation	-+	+	+
Sunken eyeballs and fontanels	-	-+	+
Skin color	Pale	Gray	Gray/Mottled
Vital Signs			
↑ Pulse	+	+	++
↑ Respiration	+	++	++
↓ BP	-	-+	+
↑ Temperature (except in young infant)	-+	+	+
Urine			
Volume	Decreased	Oliguria	Oliguria/Anuria
Osmolality (mOsm/L)	600	800	1000+
Specific gravity	1.025	1.030+	1.035+
Blood Chemistry			
↓ pH (acidosis)	-	-+	+
↑ BUN	-+	+	++
↑ Creatinine	-	-	-+
↑ K	-	- +	+

Adapted with permission, from HI Hochman et al.: Dehydration, diabetic ketoacidosis and shock in the pediatric patient. *Pediatr Clin North Am* 26(4): 810, November 1979.

---

78 Indicate which symptoms are present (+ or -+) during mild dehydration (1-5%). Refer to Table 56 as needed.

( ) a. thirst
( ) b. irritable
( ) c. confusion
( ) d. comatose
( ) e. reduced skin turgor
( ) f. dry mucous membranes
( ) g. sunken eyeballs
( ) h. increased pulse rate
( ) i. increased respiration
( ) j. decreased blood pressure
( ) k. oliguria
( ) l. increased BUN

- - - - - - - - - - - - - - - - - -

a. X; b. X; c. –; d. –; e. X; f. X; g. –; h. X; i. X; j. –; k. –; 1. X

79 Indicate which symptoms are present or markedly present (+ or ++) with moderate and/or severe dehydration:

( ) a. thirst
( ) b. confusion
( ) c. convulsions
( ) d. dry mucous membranes
( ) e. sunken fontanels
( ) f. reduced skin turgor
( ) g. increased pulse rate
( ) h. increased respiration
( ) i. decreased blood pressure
( ) j. oliguria

All of them (a–j). The thirst drive may no longer be present.

---

80 With severe dehydration, extracellular fluid and intracellular fluid are lost. With a slow, progressive fluid loss, the intracellular fluid (ICF) loss is equal to the extracellular fluid (ECF) loss. As ECF loss occurs, ICF shifts into the extracellular fluid compartment (vessels and tissue spaces).

What do you think could occur if dehydration developed rapidly? *________________

____.

Extracellular fluid loss would be greater than intracellular fluid loss. ICF could not quickly replace ECF loss

---

Table 57 estimates the percentage of extracellular and intracellular fluid loss caused from a slow onset to a rapid onset of dehydration. ECF is lost first. The nurse or doctor should determine the onset and severity of fluid loss. Then you can calculate the percentage of fluid loss from the ECF and ICF compartments with the use of this table. Refer back to it as needed.

TABLE 57. EXTRACELLULAR FLUID AND INTRACELLULAR FLUID LOSS

Onset of Dehydration	ECF % loss	ICF % loss
Rapid onset 2 or less days	44	25
Average onset 2 to 7 days	60	40
Slow onset 8 or more days	50	50

Adapted with permission, from HI Hochman, et al.: Dehydration, diabetic ketoacidosis and shock in the pediatric patient. *Pediatr Clin North Am* 26(4): 810, November 1979.

---

81 If a child developed dehydration over a period of 5 days, the percentage of ECF loss would be ______________ and of ICF loss would be ______________.
If dehydration occurs over a period of 2 weeks, the percentage of ECF loss would be ______________ and of ICF loss would be ______________.

\- - - - - - - - - - - - - - - - -

60%; 40%; 50%; 50%

---

82 Dehydration is classified according to the serum concentration of solutes (osmolality). Sodium is the primary contributor to the serum osmolality. Dehydration has three classifications in relation to osmolality and sodium concentration: (1) iso-osmolar dehydration (isonatremic dehydration); (2) hyperosmolar dehydration (hypernatremic dehydration); and (3) hypo-osmolar dehydration (hyponatremic dehydration).

Which dehydration has the highest osmolality (concentration)? *______________ ______________________________.

\- - - - - - - - - - - - - - - - -

hyperosmolar or hypernatremic dehydration

---

83 All degrees of dehydration are frequently associated with iso-osmolality or isonatremic dehydration. This is the most common type of dehydration, and fluid and sodium are proportionately lost.

With hypernatremic dehydration and hyponatremic dehydration, name the electrolyte involved. ______________________________.

\- - - - - - - - - - - - - - - - -

sodium

TYPES OF DEHYDRATION

Table 58 lists the three types of dehydration: isonatremic, hypernatremic, and hyponatremic. For each of the dehydrations, the water and sodium loss, serum sodium level, ECF and ICF loss, causes, symptoms and treatments are described. Study the table carefully, and refer to it as needed.

TABLE 58. TYPES OF DEHYDRATION: ISONATREMIC, HYPERNATREMIC, AND HYPONATREMIC

Types of Dehydration	Comments
Isonatremic dehydration (iso-osmolar or isotonic dehydration)	*Water and Sodium Loss:* Proportionately equal loss of water and sodium. *Serum Sodium Level:* 130 to 150 mEq/L *ECF and ICF Loss:* Extracellular fluid volume is markedly decreased (severe hypovolemia). Since sodium and water loss are approximately the same, there is no osmotic pull from ICF to ECF. The plasma volume is significantly reduced and shock occurs from decreased circulating blood volume. ICF volume remains virtually constant. *Causes:* Diarrhea, vomiting, and malnutrition (decrease in fluid and food intake) are the most common causes. *Symptoms:* With severe fluid loss, symptoms are characteristic of hypovolemic shock: rapid pulse rate; rapid respiration; and, later, a decreasing systolic blood pressure. Other symptoms are weight loss, lethargy, pale or gray skin color, dry mucous membranes, reduced skin turgor, sunken eyeballs, absence of tearing and salivation, and decreased urine output. *Treatment:* Fluid should be restored rapidly to correct hypovolemic shock. Iso-osmolar solutions, i.e., Ringer's lactate or 5% dextrose in 0.2% NaCl or 0.3% NaCl are some of the choices. Replacement should be calculated over 24 hours; and if dehydration is severe, half of the amount of solution should be given the first 8 hours and the remaining half over the next 16 hours.
Hypernatremic dehydration (hyperosmolar or hypertonic dehydration)	*Water and Sodium Loss:* Water loss is greater than sodium loss; sodium excess. *Serum Sodium Level:* ↑ 150 mEq/L *ECF and ICF Loss:* ECF and ICF volumes are both decreased. Increased ECF osmolality (solutes) results in a shift of fluid from the ICF to the ECF causing severe cellular dehydration. ECF depletion may not be as severe as ICF depletion. Loss of hypo-osmolar fluid raises the osmolality of ECF. *Causes:* This is the second leading type of dehydration in children. Severe diarrhea (water is lost in excess to solutes) and high solute intake with decreased water intake are the two most common causes. Others include fever, poor renal function, rapid breathing, or any combination of these conditions. *Symptoms:* Shock is less apparent since ECF loss is not as severe. Symptoms include weight loss, avid thirst, confusion, convulsions, tremors, thickened and firm skin turgor, sunken eyeballs and fontanels, absence of tearing, moderately rapid pulse, moderately rapid respirations, blood pressure frequently normal, urine output normal to decreased, and intracranial hemorrhage.

TABLE 58. (Continued)

Types of Dehydration	Comments
	Treatment: The goal is to increase the ICF and ICF volumes without causing water intoxication. Giving excessive hypo-osmolar solutions or only 5% dextrose in water would dilute ECF, causing water to shift to the ICF and water intoxication (intracellular fluid volume excess) to occur. A gradual reduction of sodium is safest. Five-percent dextrose with 0.2% NaCl may be ordered and, later, lactated Ringer's solution. Calcium gluconate is given to prevent tetany symptoms. With normal urinary flow, potassium can be added to the solutions (2–3 mEq/kg).
Hyponatremic dehydration (hypo-osmolar or hypotonic dehydration)	*Water and Sodium Loss:* Sodium loss is greater than water loss; excess water. *Serum Sodium Level:* ↓ 130 mEq/L *ECF and ICF Loss:* ECF is severely decreased, and ICF is increased. The osmolality of ECF is lower than the osmolality of ICF. Water shifts from the ECF to the ICF (lesser to the greater concentration). The cerebral cells are frequently affected first, and the excess water interferes with brain cell activity. *Causes:* Severe diarrhea (sodium is lost in excess to water), excessive water intake, electrolyte-free fluid infusions (5% dextrose in water), sodium-losing nephropathy, and diuretic therapy. *Symptoms:* Thirst, weight loss, lethargy, comatose, poor skin turgor, clammy skin, sunken and soft eyeballs, absence of tearing, shock symptoms (rapid pulse rate, rapid respirations, and low systolic blood pressure), and decreased urine output. *Treatment:* Ringer's lactate or 5% dextrose in 0.45% NaCl (½ NSS) would be helpful in correcting serum sodium level of 125–135 mEq/L. For serum sodium level of 120 mEq/L, normal saline could be used. For serum sodium 115 mEq/L or less, 3% saline may be indicated. Rapid fluid correction with electrolytes could cause an excessive shift of cellular fluid into the plasma. The result could be overhydration and congestive heart failure.

---

84 With isonatremic dehydration, there is a proportionate loss of the ions ________________ and ________________. The serum sodium level would be between ________________ and ________________. The ECF volume is (increased/decreased) ________________.

— — — — — — — — — — — — — — — —

sodium and water; 130 to 150 mEq/L; decreased

---

85 Two common causes of isonatremic dehydration are ________________ and ________________.

With severe dehydration, shock symptoms are common. Give three shock symptoms:

a. *______________________________.

b. *______________________________.

c. *______________________________.

- - - - - - - - - - - - - - - - - -

diarrhea and vomiting, also malnutrition;
a. rapid pulse rate; b. rapid respiration; c. decreasing systolic blood pressure. Others could be: gray skin color, lethargy

---

86 With hypernatremic dehydration, which is lost to greater degree: water or sodium? ____________. The serum sodium level would be ____________ mEq/L. ECF and ICF volumes are decreased. Which body fluid compartment has the largest fluid loss—ECF or ICF? ____________.

- - - - - - - - - - - - - - - - - -

water; 150 ↑; ICF

---

87 Give two causes of hypernatremic dehydration.

a. *______________________________.

b. *______________________________.

Shock symptoms (are/are not) ______________ common with this type of dehydration.

Check the symptoms found with hypernatremic dehydration:

( ) a. avid thirst
( ) b. convulsions
( ) c. tremors
( ) d. skin turgor firm and thickened
( ) e. absence of tearing
( ) f. excess urine output

- - - - - - - - - - - - - - - - - -

a. severe diarrhea; b. high solute intake; are not; a. X; b. X; c. X; d. X; e. X; f. –

---

88 With hyponatremic dehydration, the loss of ________________ is greater than the loss of ________________. The serum sodium level would be ______________ mEq/L. ECF is (increased/decreased) ________________ and ICF is (increased/decreased) ________________.

— — — — — — — — — — — — — — — — —

sodium; water; 130; decreased; increased

---

89 Give three causes of hyponatremic dehydration:

a. *____.

b. *____.

c. *____.

Check the symptoms found with hyponatremic dehydration:

( ) a. thirst
( ) b. weight gain
( ) c. poor skin turgor
( ) d. clammy skin
( ) e. Sunken, soft eyeballs
( ) f. rapid pulse rate
( ) g. rapid respiration
( ) h. low systolic blood pressure

— — — — — — — — — — — — — — — — —

a. severe diarrhea; b. excessive water intake; c. electrolyte-free fluid infusions. Others—sodium-losing nephropathy and diuretic therapy;
a. X; b. –; c. X; d. X; e. X; f. X; g. X; h. X

---

90 With hypovolemic shock, fluids should be restored (slowly/rapidly) __________. Replacement of fluids should be calculated over 24 hours and, if dehydration is severe, half is frequently administered in ______________ hours and the second half in ______________ hours.

What type of fluid imbalance could occur if intravenous fluids are given too rapidly? ______________________________.

— — — — — — — — — — — — — — — —

rapidly; 8, 16; overhydration or pulmonary edema (CHF)

91 The goal for correcting hypernatremic dehydration is to avoid causing what major type of fluid imbalance? * ________________.

What electrolytes should be replaced when correcting hypernatremic dehydration?

a. ________________.

b. ________________.

If the serum sodium level is below 115 mEq/L, what type of intravenous fluid may be indicated? * ________________.

- - - - - - - - - - - - - - - -

water intoxication; calcium and later potassium with normal kidney function; 3% saline

---

CASE REVIEW

Susan, age 4, has had diarrhea and anorexia for 3 days. She has only taken sips of fruit juices for the last 3 days. She weighed 38 lb or 17.3 kg preillness and now weighs 35 lb or 15.9 kg. Her cheeks and gums are dry and her skin turgor is reduced. She is irritable. Susan has voided once in the last 24 hours. Vital signs: T 99°F or 37.2°C, P 110, R 32, BP 90/60. Blood was drawn for serum electrolytes and BUN.

1. Because of diarrhea, anorexia, decreased fluid intake, and weight loss, the nurse would assume the fluid imbalance to be:
   ( ) a. intracellular fluid volume excess (water intoxication)
   ( ) b. extracellular fluid volume excess (edema or overhydration)
   ( ) c. extracellular fluid volume deficit (dehydration)
2. In kilograms, Susan has lost ________ kg. What degree of dehydration is present? ________. (Refer to Table 56 if needed.) The severity of her dehydration would be (mild/moderate/severe) ________.
3. The onset of fluid loss (dehydration) has been ________ days. Give the percentage of fluid loss from the ECF ________ and the ICF ________. (Refer to Table 57 if needed.)
4. List Susan's signs and symptoms of dehydration from your nursing assessment.

   a. * ________________.

   b. * ________________.

   c. ________________.

   d. * ________________.

   e. * ________________.

5. Is Susan's blood pressure indicative of shock? (Yes/No) ________.

   Explain. * ____.

6. The ranges of serum electrolytes for children are:

   a. K * ____________________

   b. Na * ____________________

   c. Cl * ____________________

   d. Ca * ____________________

   (Refer to Frames 42, 48, 51, 52, or to Table 55.)

7. Susan's serum potassium was (increased/decreased) ____________, serum sodium was (increased/decreased) ____________, serum chloride was (increased/decreased) ____________, and serum calcium was (increased/decreased) ____________.

8. Give some signs and symptoms of hypokalemia.

   a. ____.

   b. ____.

   c. ____.

9. With a serum calcium level of 8.4 mg/dl, the nurse should be observing for symptoms of ____________. Give two of the symptoms. (Refer to Frame 53 if needed.)

   a. * ____.

   b. * ____.

10. Susan's BUN was (increased/decreased) ____________. Most likely cause of her BUN is ____________. Explain. * ____________________

    ____.

11. Susan's preillness weight is 38 lb or 17.3 kg. What would her daily fluid requirements be? ____________ ml per 24 hr. (Refer to Table 54 if needed.)

    Her daily sodium requirement would be ____________ mEq/L and her daily potassium requirement would be ____________ mEq/L. (Refer to Table 54 as needed.)

12. Susan's type of dehydration according to her serum sodium level would be (isonatremic/hypernatremic/hyponatremic). ____________. Why? * ____________

    ____.

13. For the type of dehydration in question 12, fluids should be replaced (slowly/rapidly) ____________.

14. Name three shock symptoms that could occur.

    a. *________________________________.

    b. *________________________________.

    c. *________________________________.

15. Give Susan's vital signs that could be indicative of impending shock due to hypovolemia (fluid volume loss). *______________ and *______________.

16. With moderate dehydration, the changes in the urine output and concentration are: (Refer to Table 56 if needed.)

    a. Urine volume ______________

    b. Urine osmolality ______________ mOsm/L

    c. Urine specific gravity ______________

---

1. c
2. 1.4; 8%; moderate
3. 3; 60%; 40%
4. a. cheeks and gums dry; b. reduced skin turgor; c. irritable; d. voided once (decreased urine output); e. pulse rate and respiration increased
5. No. Blood pressure for that age group is normal; however, a baseline blood pressure from an office visit would be helpful
6. K 3.5-5.8 mEq/L; b. 134-146 mEq/L; c. 98-105 mEq/L; d. 4.5-5.8 mEq/l or 9-11.5 mg/dl
7. decreased; decreased; decreased; decreased
8. dizziness, soft muscles, abdominal distention, decreased peristalsis, arrhythmia, BP
9. tetany; a. tingling of fingers; b. twitching of mouth. Others—tremors and carpopedal spasms
10. increased; dehydration. With a decreased urine output, there will be an increase in serum solutes such as urea (due to hemoconcentration)
11. 1365 ml; Na 40.8 or 41 mEq/L; K 27.2 or 27 mEq/l
12. isonatremic. The serum sodium level for isonatremic is 130-150 mEq/L
13. rapidly
14. a. rapid pulse rate; b. rapid respiration; c. decreasing blood pressure. Others—clammy skin, pale or gray color
15. P 110 (rapid pulse rate) and R 32 (rapid respiration). Baseline vital signs would be helpful.
16. Oliguria or decreased; 800; 1.030+

NURSING DIAGNOSES

Potential for fluid volume deficit related to inadeuqate fluid intake, vomiting, diarrhea, GI suction as evidenced by poor skin turgor over the abdomen, chest wall, and medical aspect of the thigh, sunken and soft eyeballs, depressed anterior fontanels, weight loss, changes in vital signs, and/or behavioral changes.

Potential for fluid volume excess related to excessive fluid administration as evidenced by rapid weight gain, possible lung congestion, and hyponatremia.
Potential alteration in nutrition related to an insufficient intake of nutrients and GI suction.
Potential for electrolyte imbalance related to vomiting, diarrhea as evidenced by behavioral and neurological changes and abnormal serum electrolytes.
Potential alteration in bowel elimination (diarrhea) related to GI viral or bacterial infection as evidenced by an increased number of loose stools.
Potential for ineffective breathing patterns related to fluid and electrolyte imbalances as evidenced by rapid and/or irregular breathing rate.
Potential alteration in the family process related to separation of family members.
Potential alteration in parenting related to child separation due to hospitalization and care given by multiple caretakers.

NURSING ACTIONS

1. Check the serum electrolyte ranges in infants and children, i.e., Na—134-150 mEq/L (infant) and 134-146 mEq/L (children); K—3.5-5.8 mEq/L (infants and children); Cl—96-116 mEq/L (infant) and 98-105 mEq/L (children); and Ca—3.7-6 mEq/L or 7.3-12 mg/dl (infant) and 4.5-5.8 mEq/L or 9-11.5 mg/dl (children).
2. Recognize causes of hyponatremia in infants and children, i.e., overhydration, continuous administration of electrolyte-free solutions, and SIADH.
3. Determine the daily fluid and electrolyte requirement for infants and children and note imbalances resulting.
4. Observe for behavioral and neurologic changes resulting from fluid and electrolyte imbalances.
5. Obtain a data base on the infant or child from the parent(s), former physician, child, or record. Provide time for parent(s) to answer the questions about the child's health problems.
6. Weigh infant or child daily on the same scales and at the same time to determine the degree of dehydration according to weight loss.
7. Teach the parent to keep an accurate weight chart on the infant or child.
8. Assess for signs and symptoms of fluid loss, i.e., skin turgor, mucous membranes, eyeballs and fontanels, absences of tearing and salivation, and thirst of infants and children.
9. Check the cheeks and gums of the oral cavity for dryness.
10. Assess skin turgor by pinching the skin over the abdomen, chest wall, and medial aspect of the thigh.
11. Check for sunken and soft eyeballs and depressed anterior fontanels.
12. Observe for changes in urinary output, urine color, and specific gravity. Monitor the urinary output and withhold potassium with physician's permission if oliguria or anuria is present.
13. Assess the stools and vomitus for frequency, consistency, color, and amount.
14. Check for abnormal vital signs. Report temperature changes—slight elevation fluid loss); pulse changes—arrhythmia (↓ K), fast thready pulse (↓ ECF), full,

bounding pulse (↑ ECF); blood pressure change—decreased BP (fluid loss—↓ ECF); respiratory changes—rapid, deep, vigorous breathing (metabolic acidosis) and shallow, irregular breathing (metabolic alkalosis).

15. Observe for shock symptoms; i.e., rapid pulse rate, rapid respirations, confusion, lethargic, restlessness, gray or pale skin color, clammy skin, decreasing systolic blood pressure, and oliguria or anuria, especially when isonatremic dehydration and hyponatremic dehydration are present.
16. Monitor blood chemistry and hematology results and report changes.
17. Observe for signs and symptoms of hypo-hyperkalemia, hypo-hypernatremia, and hypo-hypercalcemia.
18. Check the serum BUN and serum creatinine. Know the significance of an elevated BUN and creatinine in relation to fluid volume loss and kidney function.
19. Identify abnormal arterial blood gases and be able to explain the acid-base imbalance occurring.
20. Measure secretions from gastrointestinal suction, drainage tubes, and fistulae.
21. Determine the degrees of dehydration and its severity, i.e., mild (1-5%), moderate (6-10%), and severe (11-15%).
22. Estimate the percentage of extracellular and intracellular fluid loss according to onset (time) and severity of fluid loss.
23. Identify the tvpe of dehydration present according to the serum sodium level, i.e., 130-150 mEq/L (isonatremic), ↑ 150 mEq/L (hypernatremic), and ↓ 130 mEq/L (hyponatremic).
24. Observe for symptoms of water intoxication (intracellular fluid volume excess) when using excessive hypo-osmolar solutions including 5% dextrose in water.
25. Assess for symptoms of hyperkalemia and hypocalcemia in infants and children with hypernatremic dehydration.
26. Observe for symptoms of overhydration when administering excessive hyperosmolar solutions (5% dextrose in normal saline or in 3% saline).

GASTROINTESTINAL SURGERY

92 People undergoing minor surgery freqently have little or no fluid and electrolyte changes.

In major surgery, there is a tendency for sodium and water to be retained and for potassium to be lost. Before potassium is administered, it is necessary to make certain that:

( ) a. the person can tolerate food
( ) b. renal function is adequate

— — — — — — — — — — — — — — — — —

a. —; b. X

---

93 After major surgery, there is a tendency for sodium ____________________,

water ____________________, and potassium ____________________.

— — — — — — — — — — — — — — — — — —

retention; retention; loss

---

94 Many people undergoing gastrointestinal surgery will have had a previous fluid and electrolyte imbalance along with concurrent losses. Replacement of losses is necessary before, during, and following surgery.

Frequently, these people are dehydrated and will need sufficient water to reestablish renal function. The following type of solution would be indicated in reestablishing renal function:

( ) a. Hydrating solutions
( ) b. Plasma expanders
( ) c. Replacement solutions with potassium replacement

— — — — — — — — — — — — — — — — — —

a. X; b. –; c. –

---

95 Many people undergoing gastrointestinal surgery will require fluid and electrolyte replacement therapy:

( ) a. before surgery
( ) b. during surgery
( ) c. after surgery

— — — — — — — — — — — — — — — — — —

a. X; b. X; c. X

96 Gastric or intestinal intubation (tube passed into the stomach or intestine) for suctioning purposes will frequently be inserted before surgery. This will be used to alleviate vomiting due to an obstruction in the gastrointestinal tract or to decompress the stomach or bowel, or both, before and after an operation.

For gastric intubation, a Levine tube is inserted via the nose into the stomach. For intestinal intubation, a Miller-Abbott tube or Cantor tube is inserted via the nose and stomach into the intestine. The intestinal tubes are longer than the gastric tube, and they contain, on the tip of their tubes, a small balloon filled with air or mercury which aids in stimulating peristalsis (bowel tone). A gastric or an intestinal tube is frequently used following abdominal surgery in order to remove secretions until peristalsis returns and to relieve abdominal distention.

Gastric or intestinal intubation before abdominal surgery is used to *________

____

____.

Gastric or intestinal intubation after abdominal surgery is employed to *______

____

and to *____.

— — — — — — — — — — — — — — — —

alleviate vomiting and decompress the bowel or stomach;
remove secretions until peristalsis returns; relieve abdominal distention

---

Table 59 lists the electrolytes that are in the stomach and intestine. Note which electrolytes are more concentrated in gastric and intestinal fluids. When the person is vomiting, having diarrhea, or is being intubated (gastric or intestinal), fluid and electrolytes are lost. Study the table and be able to state which electrolytes, with the highest concentration, are lost.

TABLE 59. CONCENTRATION OF ELECTROLYTES IN THE STOMACH AND INTESTINE (mEq/L)

Area	Body Fluid	Na^+	K^+	Cl^-	HCO_3^-	
Stomach	Gastric juice	60.4	9.2*	84*	0–14	H^+*
Small intestine	Intestinal juice	111.3*	4.6	104.2*	31*	

Note: *Electrolytes that are highly concentrated in these areas.

---

97 With vomiting and gastric intubation, which electrolytes are lost that are highly concentrated in the stomach? ______________, ______________ and ______________. Give the other electrolytes lost from the stomach. ______________ and ______________.

\- - - - - - - - - - - - - - - - - -

potassium; chloride; hydrogen;
sodium; bicarbonate (Chloride is in high concentration in the stomach and the intestine.)

---

98 If the person has diarrhea or intestinal intubation, what are the major electrolytes lost? ______________, ______________, and ______________.
Another electrolyte lost from the intestine is ______________.

\- - - - - - - - - - - - - - - - - -

sodium; chloride; bicarbonate;
potassium

---

99 Is the sodium ion more plentiful in the (stomach/intestine)? ______________.
Is the potassium ion more plentiful in the (stomach/intestine)? ______________.

\- - - - - - - - - - - - - - - - - -

intestine
stomach

---

100 Bicarbonate ion is more plentiful in the (stomach/intestine) ______________.
If a large amount of bicarbonate is lost, what type of acid-base imbalance could result? *______________.

\- - - - - - - - - - - - - - - - - -

intestine
metabolic acidosis

---

101 Hydrogen is more plentiful in the (stomach/intestine) ________________.

If hydrogen is lost, what type of acid-base imbalance can result? *________

________________________________.

- - - - - - - - - - - - - - - -

stomach
metabolic alkalosis

---

102 If potassium, chloride, and hydrogen are lost from the stomach, name the electrolyte–acid-base imbalance that can occur. *________________________.

- - - - - - - - - - - - - - - - -

hypokalemia alkalosis

CLINICAL APPLICATIONS

Mr. Drum, a 29-year-old man, was admitted to the hospital complaining of severe, persistent hiccups and abdominal pain. The patient notices a mass in the lower left quadrant of the abdomen for approximately 5–6 days before admission. According to Mr. Drum, he previously had several episodes of this left groin mass that he was able to reduce manually.

Mr. Drum stated he had not had a bowel movement for the past 3 days, nor was he able to "keep anything down" over the past 3 days. His skin was warm and dry and lacked elasticity. He was very weak.

---

103 The following signs and symptoms might indicate that Mr. Drum was dehydrated:

( ) a. Vomiting—unable to retain food for 3 days
( ) b. Skin warm, dry, and lacking elasticity
( ) c. Not having a bowel movement for 3 days
( ) d. Weakness
( ) e. Hernia could be manually reduced
( ) f. Severe abdominal pain in left lower quadrant

- - - - - - - - - - - - - - - -

a. X; b. X; c. –; d. X; e. –; f. –

TABLE 60. LABORATORY STUDIES OF MR. DRUM

Laboratory Tests	On Admission	First Day	Second Day	Third Day	Fourth Day
Hematology					
Hemoglobin (12.9–17.0 g)	21.2	18.4	13.1	13.2	
Hematocrit (40–46%)	58	54	38	39	
WBC (white blood count) (5000–10,000/cu mm)	10,700				
Biochemistry					
BUN (blood urea nitrogen) (10–25 mg/dl)**	85	68	68	19	19
Plasma/serum† CO_2 (50–70 vol %)	52	61	—	39	50
(22–32 mEq/L)	24	28		18	22
Plasma/serum chloride (98–108 mEq/L)	73	78	73	91	97
Plasma/serum sodium (135–146 mEq/L)	122	128	122	132	145
Plasma/serum potassium (3.5–5.3 mEq/L)	5.2	4.0	4.0	4.2	4.1

†*Plasma* and *serum* are used interchangeably.

**mg/100 ml = mg/dl

Table 60 gives the laboratory studies of Mr. Drum and shows how his results deviate from the norm at the time of his illness. It is important for you to memorize the "normal" laboratory ranges as given in Table 60 in the left-hand column. You will be using these ranges throughout the chapter. However, you may refer to Table 60 as needed.

---

104 Give "normal" ranges for hemoglobin ________________, for hematocrit ________________, and for white blood count ________________.

- - - - - - - - - - - - - - - - - -

12.9–17.0 g; 40–46%; 5000–10,000/cu mm

---

105 BUN is the abbreviation for blood urea nitrogen. Do you know how urea is formed? * ______________________.

And how it is excreted? * ______________________.

What is the "normal" BUN range? * ______________________.

— — — — — — — — — — — — — — — —

As a by-product of protein metabolism
Through the kidney
10–25 mg/dl

106 The "normal" serum CO_2 range is * ______________ vol % and * ______________ mEq/L.

Would a patient with a serum CO_2 of 18 mEq/L be in metabolic (acidosis/alkalosis)? ______________. Refer to Chapter 3 for further clarification.

— — — — — — — — — — — — — — — —

50–70
22–32 mEq/L
acidosis

107 The "normal" range for serum chloride is 98–108 mEq/L. Do you recall the "normal" range for serum potassium and sodium? Potassium * ______________ sodium * ______________. Refer to Chapter 2 for further clarification.

— — — — — — — — — — — — — — — —

K:3.5–5.3 mEq/L; Na: 135–146 mEq/L

---

108 Which of the following admission laboratory results of Mr. Drum would indicate fluid and electrolyte imbalance:

( ) a. Hemoglobin 21.2 g
( ) b. Hematocrit 58%
( ) c. Serum CO_2 52%
( ) d. Serum CO_2 24 mEq/L
( ) e. Serum chloride 73 mEq/L
( ) f. Serum sodium 122 mEq/L
( ) g. Serum potassium 5.2 mEq/L

\- - - - - - - - - - - - - - - - -

a. X; b. X; c. –; d. –; e. X; f. X; g. –

---

109 Mr. Drum's elevated hemoglobin and hematocrit on admission and the first day postoperatively would indicate ____________________.

\- - - - - - - - - - - - - - - - -

dehydration (Did you answer hemoconcentration? OK.)

---

110 A high BUN is indicative of renal impairment or dehydration, or both. Mr. Drum's elevated BUN would indicate:

( ) a. an increased urine output
( ) b. a retention of urea, the by-product of protein metabolism, in the circulating blood
( ) c. an abnormal excretion of urea, the by-product of protein metabolism

\- - - - - - - - - - - - - - - - -

a. –; b. X; c. –

---

111 The third day postoperatively, Mr. Drum's plasma CO_2 decreased. This would indicate:

( ) a. an increased bicarbonate ion in the plasma
( ) b. a decreased bicarbonate ion in the plasma
( ) c. metabolic acidosis
( ) d. metabolic alkalosis

\- - - - - - - - - - - - - - - - -

a. –; b. X; c. X; d. –

---

112 Below are some laboratory results for Mr. Drum. Label the imbalance that they might indicate, using:

D for dehydration
K for kidney dysfunction
E for electrolyte imbalance
M for metabolic acidosis
O for normal range or for those that do not pertain to the above four

Some results may be associated with more than one imbalance.

_______ a. Hematocrit 38%

_______ b. Hemoglobin 21.2 g

_______ c. BUN 68 mEq/L

_______ d. BUN 19 mEq/L

_______ e. Serum potassium 4.0 mEq/L

_______ f. Serum sodium 122 mEq/L

_______ g. Serum CO_2 18 mEq/L

_______ h. Serum CO_2 28 mEq/L

_______ i. Serum chloride 73 mEq/L

- - - - - - - - - - - - - - - - - -

a. O; b. D; c. K, D; d. O; e. O; f. E; g. M; h. O; i. E

CLINICAL MANAGEMENT–PREOPERATIVE

The preoperative management for Mr. Drum would include:

1. Hydrate rapidly utilizing 4 to 5 liters over the next 6–8 hours.
2. Insert a Levine tube and connect to low intermittent suction.
3. Prepare for OR for a left inguinal herniorrhaphy as soon as he is hydrated.
4. Check for renal function–urine output.

Solution for Hydration: 4500 cc 5% D /-1/2 NS (dextrose in 1/2 normal saline or 0.45%)

113 Due to vomiting, Mr. Drum had:

( ) a. severe dehydration
( ) b. water intoxication
( ) c. a loss of sodium and chloride
( ) d. a low serum bicarbonate level
( ) e. a low serum potassium level

He was hydrated (before/after) the herniorrhaphy.

A Levine tube was inserted to:

( ) a. relieve distention
( ) b. remove secretions from the stomach
( ) c. lessen vomiting
( ) d. provide nutrition

\- - - - - - - - - - - - - - - - -

a. X; b. –; c. X; d. X; e. –

before

On admission his potassium was high normal, probably because of the severe dehydration that caused hemoconcentration.

a. X; b. X; c. X; d. –

CLINICAL MANAGEMENT–POSTOPERATIVE

114 The postoperative management for Mr. Drum would include:

1. Connect Levine tube to low suction and check drainage hourly
2. Monitor parenteral therapy:
   1000 cc 5% D / 1/2 NS
   1000 cc 5% D/NS with one ampule of sodium bicarbonate
   1000 cc 5% lactated Ringer's
3. Check urine hourly and test for specific gravity and pH
4. Assess serum electrolyte findings
5. Administer penicillin for febrile condition
6. Encourage patient to breathe deeply and cough every 30 minutes for the first few hours
7. Other management procedures not related to fluid, electrolyte, and acid-base imbalances

Mr. Drum received gastric intubation following surgery to *________________ ____________________ and to *______________________________.

\- - - - - - - - - - - - - - - - -

relieve abdominal distention; remove gastric secretions

115 Gastrointestinal secretions contain solid particles that may accumulate and obstruct the tube. Irrigation of the tube will assure patency and proper drainage.

Frequent irrigations, using large amounts of water, should be avoided to prevent the loss of fluid and electrolytes.

Irrigation of Mr. Drum's tube will assure *____________________

______________________________________.

Name the "major" electrolytes lost through frequent gastric irrigation with large quantity of water. ________________, ________________

and ________________.

\- - - - - - - - - - - - - - - - - -

patency for proper drainage;
potassium; hydrogen; chloride

116 The tube should be irrigated at specific intervals with small amounts of saline to keep it patent.

Change of position helps in alleviating obstruction and aids in maintaining patency of the tube.

The use of small amounts of air to check the patency of the tube may be ordered instead of solution in order to prevent the loss of fluid and electrolytes. One would listen with the stethoscope for a "whoosh" sound.

The three methods that can be employed to check the patency of Mr. Drum's gastric tube are to:

1. *______________________________________
   ______________________________________.
2. *______________________________________.
3. *______________________________________
   ______________________________________.

\- - - - - - - - - - - - - - - - - -

irrigate at specific intervals with small amounts of saline;
change of position;
introduce small amount of air and listen with a stethoscope for a "whoosh" sound

---

117 Mr. Drum was allowed sips of water to alleviate the dryness in his mouth and lessen irritation in his throat.

Special attention should be taken to limit the amount of water by mouth, for water can dilute the electrolytes found in the stomach and the suction would remove them.

What might happen to the electrolytes if Mr. Drum drank a great deal of water during intubation? * ____

____.

- - - - - - - - - - - - - - - - -

The electrolytes in the stomach would be diluted, and suction would remove them

---

118 After Mr. Drum's Levine tube is removed, the nurse should observe him for:

1. feeling of fullness
2. vomiting
3. abdominal distention
4. diminished bowel sounds

This would indicate that Mr. Drum's gastrointestinal tract (is/is not) __________ functioning.

The signs and symptoms that would indicate that Mr. Drum's peristalsis had not returned would be:

1. * ______________________________.
2. ______________________________.
3. * ______________________________.
4. * ______________________________.

- - - - - - - - - - - - - - - - -

is not

1. feeling of fullness
2. vomiting
3. abdominal distention
4. diminished bowel sounds

---

119 Frequently, the tube is clamped for a period of time and then unclamped. The amount of fluid is measured. A large amount of fluid returned by unclamped tube would indicate that peristalsis has not returned.

Feeling of fullness, vomiting, and abdominal distention are signs and symptoms that ____________________ has not returned.

Using the clamping and unclamping method, how would you know if peristalsis was present? *____

____.

\- - - - - - - - - - - - - - - - - -

peristalsis

A small amount of fluid return or none

---

120 Because suction will remove fluids and electrolytes, oral fluid intake is restricted and parenteral therapy is used.

Mr. Drum received intravenous fluids containing dextrose, saline, and lactated Ringer's for:

( ) a. replacing sodium and chloride loss
( ) b. maintaining nutritional needs
( ) c. maintaining electrolyte balance
( ) d. replacing and maintaining fluids volume

\- - - - - - - - - - - - - - - - - -

a. X; b. X; c. X; d. X

---

121 If Mr. Drum received 3 to 4 (or more) liters of 5% dextrose in water with no other solutes, what type of fluid imbalance could occur?

( ) a. Dehydration
( ) b. Overhydration
( ) c. Water intoxication

Why? *____

____.

\- - - - - - - - - - - - - - - - - -

a. –; b. – (maybe); c. X. The diluted fluid in the vessels (vascular) will shift to the cells due to the process of osmosis, or a similar response. Osmosis causes fluid to diffuse from the lesser to the greater concentration.

---

122 Mr. Drum received an ampule of sodium bicarbonate in one of his intravenous fluids. This would:

( ) a. increase the plasma CO_2 or bicarbonate
( ) b. decrease the plasma CO_2 or bicarbonate
( ) c. reduce his metabolic acidotic state
( ) d. reduce his metabolic alkalotic state

— — — — — — — — — — — — — — — — —

a. X; b. –; c. X; d. –

---

123 Mr. Drum should be encouraged to breathe deeply and cough to inflate the lungs and promote effective gas exchange. Inadequate ventilation due to pain, narcotics, and anesthesia causes CO_2 retention and respiratory acidosis.

In respiratory acidosis, would Mr. Drum (hypoventilate/hyperventilate)?

________________________.

Indicate the type of breathing associated with CO_2 retention (respiratory acidosis). (Refer to Chapter 3 if needed.)

( ) a. Deep, rapid, vigorous breathing
( ) b. Dyspnea—difficult or labored breathing
( ) c. Overbreathing

— — — — — — — — — — — — — — — — —

hypoventilate; a. –; b. X; c. –

CASE REVIEW

Preoperatively the nurse would assess Mr. Drum's fluid and electrolyte imbalance and kidney function. To assess the fluid imbalance, the nurse would check all his laboratory findings.

1. Mr. Drum's elevated hemoglobin and hematocrit on admission and the first day postoperatively was indicative of ________________________.
2. His elevated BUN is also indicative of ________________________and possibly of *________________________.
3. Mr. Drum's serum sodium and chloride were low due to fluid loss. His serum potassium was 5.2 mEq/L on admission. Explain how Mr. Drum's serum potassium is at high average with vomiting and dehydration. *________________________

____

____.

4. What occurred to Mr. Drum's serum potassium level when he was hydrated?

*____.

5. Mr. Drum's decreased serum CO_2 would indicate *____________________
____________________.

6. When hydrating a debilitated individual with 4 to 5 liters at a rapid infusion rate, what type of fluid imbalance can occur? ____________________
____________________.

7. If Mr. Drum were hydrated with 4 to 5 liters of 5% dextrose in water, what type of fluid imbalance might occur? *____________________.
Explain why. *____________________
____________________.

8. Mr. Drum received intravenous fluids containing dextrose, saline, and lactated Ringer's for:
   a. replacing ____________ and ____________ loss
   b. maintaining ____________ needs
   c. maintaining ____________ balance
   d. replacing and maintaining *____________

9. Frequent irrigations of the gastric tube (Levine tube) with a large quantity of water might have what effect? *____________________
____________________.

10. The three methods that the nurse can use to check the patency of Mr. Drum's gastric tube are to:
    a. *____________________.
    b. *____________________.
    c. *____________________
    ____________________.

11. Why is it important that the nurse assess Mr. Drum's urine output pre- and post-operatively? *____________________
____
____________________.

12. Why is it important that Mr. Drum breathe deeply and cough after surgery?
*____________________.

– – – – – – – – – – – – – – – –

1. dehydration
2. dehydration; renal impairment
3. With dehydration, potassium is lost from cells and accumulates in ECF. If urine volume is low, the serum potassium level could be high

4. serum potassium level decreased, with hydration
5. metabolic acidosis
6. hypervolemia or overhydration
7. water intoxication. Dextrose is utilized by the body, leaving water (hypo-osmolar, which passes into cells)
8. a. sodium; chloride
   b. nutritional
   c. electrolyte
   d. fluid volume
9. depletion of electrolytes in the GI tract
10. a. irrigate with small amounts of saline
    b. change of position
    c. introduce small amount of air and listen with a stethoscope for a "whoosh" sound
11. to determine kidney function. Also, it can be an indication for overhydration. When fluids—orally and parenterally—are being pushed and urine output is low, overhydration can occur
12. to inflate the lungs and promote effective gas (O_2 and CO_2) exchange

NURSING DIAGNOSES

Fluid volume deficit related to vomiting, gastrointestinal loss, decreased fluid intake, and fluid volume shift.

Potential for electrolyte imbalances related to vomiting and gastrointestinal intubation.

Potential alteration in nutrition related to nothing by mouth and dextrose/saline solutions.

Potential ineffective airway clearance related to ineffective deep breathing and coughing, thick mucous secretions, and pain.

Potential for impaired gas exchange (CO_2 retention) related to retained mucous secretions and inadequate deep breathing and coughing.

Potential for impairment of urinary elimination related to fluid volume loss, SIADH.

Potential alteration in comfort related to pain, body position, and decreased ventilation.

NURSING ACTIONS

1. Recognize that patients undergoing gastrointestinal surgery may have a previous fluid and electrolyte imbalance along with concurrent losses.
2. Assess the patient for dehydration and decreased urine output at the time of admission and before surgery. Signs and symptoms of dehydration (hypovolemia) are thirst, poor skin turgor, dry mucous membranes and secretions, rapid pulse rate, slightly elevated temperature. Urine output could be decreased due to severe dehydration. Hydration should increase urine output.

3. Measure drainage from gastric or intestinal suction. The gastrointestinal secretions indicate fluid and electrolyte losses.
4. Assess for signs and symptoms of acid-base imbalance, such as, for metabolic acidosis, Kussmaul breathing (deep, rapid, vigorous breathing), flushed skin, restlessness, ↓ serum CO_2; for metabolic alkalosis, shallow breathing, ↑ serum CO_2; for respiratory acidosis, dyspnea (labored or difficult breathing, disorientation, and, if arterial blood gases are drawn, ↑ PCO_2.
5. Check laboratory results and report abnormal findings. Check serum electrolytes (K, Na, Cl), serum CO_2, BUN, creatinine (if ordered), hemoglobin, and hematocrit. Abnormal results could be due to hypovolemia (dehydration), renal insufficiency, or vomiting.
6. Monitor parenteral fluid and adjust flow rate as ordered. Assess for signs and symptoms of overhydration (hypervolemia) when administering IV fluids rapidly to hydrate. Symptoms include constant, irritated cough; dyspnea; neck and vein engorgement; and chest rales.
7. Encourage patient to breathe deeply and cough. This promotes good ventilation and prevents CO_2 retention or pneumonia.
8. Check urinary output every 8 hours and 24 hours. Urine output less than 600 ml per 24 hours should be reported. It could be due to a lack of fluid intake (IV or orally), renal insufficiency, or an overproduction of antidiuretic hormone (SIADH)—commonly seen after surgery.
9. Maintain patency of the gastric suction tube by changing patient's position, irrigating tube with small amounts of saline, or introducing small amounts of air and listening with a stethoscope for a "woosh" sound.
10. Check for bowel sounds (peristalsis) by listening at the four quadrants of the abdomen. There should be approximately six bowel sounds per minute. Peristalsis is diminished if bowel sounds are less than two per minute. If the gastric tube is removed, the nurse should observe patient for feeling of fullness, abdominal distention, vomiting, and diminished bowel sounds.
11. Observe for signs and symptoms of hypokalemia, i.e., dizziness, arrhythmia, muscular weakness, abdominal distention, diminished peristalsis. Following surgery there is cellular potassium loss; however, the serum potassium level may be normal due to hemoconcentration. Replacing fluid loss (IV fluids) dilutes the serum potassium, and hypokalemia results. Also, potassium will shift back into the cells, and a severe potassium deficit could occur.
12. Assess the types (tonicity) of intravenous solution to be administered. Constant use of 5% dextrose in water without any solutes (electrolytes) can cause water intoxication. Dextrose is metabolized rapidly, and water remains, which dilutes body fluids.

TRAUMA: THE ACUTELY INJURED

124 Fluid, electrolyte, and acid-base changes occur rapidly in the acutely traumatized patient. Quick medical and nursing assessments and actions are needed for survival.

In trauma (acute injury), the sodium shifts into cells, potassium shifts, and the fluid shifts from the vascular to the interstitial spaces and cells. These shifts can result in severe fluid and electrolyte imbalances.

The two electrolytes that change spaces during trauma are ________________

and ________________.

Explain the fluid shifts during trauma. *________________________________

____.

- - - - - - - - - - - - - - - - - -

potassium and sodium. Fluid shifts from the vascular to the interstitial spaces and cells

---

PHYSIOLOGIC FACTORS

Table 61 describes the physiologic factors occurring after trauma. Study the physiologic involvement, changes, and rationale. The information in this table should help you with assessment skills.

TABLE 61. PHYSIOLOGIC FACTORS ASSOCIATED WITH TRAUMA

Physiologic Factors	Rationale
Cellular electrolyte changes Potassium, sodium, chloride, bicarbonate	Potassium is lost from cells due to catabolism (cellular breakdown). As potassium leaves, sodium and chloride with water shift into the cells. The sodium pump does not function properly (see Chapter 2, Table 12).
Fluid changes	Fluids along with sodium shift into cells and to the third space (interstitial space—at the injured site). The increased cellular and third space fluids cause vascular fluid deficit (dehydration) and hyponatremia.
	Serum osmolality may be normal or increased due to fluid deficit and excess solutes other than sodium, such as potassium and urea. Remember, sodium influences the osmolality of plasma. (see Chapter 2, frame 121)
	The volume and composition of extracellular fluid (ECF) will fluctuate depending on the number of cells injured and the body's ability to restore balance. Two to three days following injury, fluid shifts from the third space at the injured site back to the vascular space.
Protein changes	With trauma, there is nitrogen loss due to increased protein catabolism, decreased protein anabolism, and/or protein shift with water to the interstitial space. The colloid osmotic pressure is decreased in the vascular fluid and increased in the interstitial

TABLE 61. (Continued)

Physiologic Factors	Rationale
	fluid (tissues), which cause fluid volume deficit (vascular) and edema.
Capillary permeability	Increased capillary permeability causes water to flow in and out of the cells and to tissue spaces. This contributes to hypovolemia (fluid volume deficit).
Hormonal influence	ADH and aldosterone help in restoring ECF. With a vascular fluid deficit and/or increased serum osmolality, more ADH is secreted, which causes water reabsorption from the distal tubules of the kidneys. In certain traumatic situations (surgery, trauma, pain), SIADH (syndrome of inappropriate ADH secretions) occurs and causes excess water reabsorption from the kidneys.
	Aldosterone is secreted from the adrenal cortex due to hyponatremia and stress.Aldosterone promotes sodium reabsorption from the renal tubules, and water accompanies the sodium. Potassium is excreted.
Kidney influence	Kidney activity is altered during and after a severely traumatic bodily insult. Sodium, chloride, and water shifts to the injured site, which causes hypovolemia. Decreased circulatory flow can decrease renal arterial flow, which could cause temporary or permanent kidney damage. With a decrease in kidney function, hyperkalemia results.
Acid-base changes	With cellular breakdown, acid metabolites, e.g., lactic acid, increase in the vascular fluid causing metabolic acidosis.
	Kidneys will conserve or excrete the hydrogen ion to maintain acid-base balance. Decreased kidney function can cause hydrogen retention and acidosis.
	The lungs will try to compensate for the acidotic state by blowing off excess CO_2—hyperventilation. Blowing off CO_2 will decrease the formation of carbonic acid.

125 Following a severe traumatic injury, there is cellular breakdown due to cell damage and hypoxia.

Explain what happens to the following electrolytes and water:

a. Potassium *____________________.

b. Sodium *____________________.

c. Chloride *____________________.

d. Water *____________________.

\- - - - - - - - - - - - - - - - -

a. Potassium is lost from the cells
b. Sodium shifts into the cells
c. Chloride shifts into the cells
d. Water shifts into the cells

---

126 The sodium pump is necessary for cellular activity (see Table 12).

Explain the sodium pump action. *___________________________

____.

- - - - - - - - - - - - - - - - -

To maintain cellular activity, sodium shifts into the cell and potassium shifts out repeatedly. Or, sodium shifts into the cell and depolarization occurs, and then potassium shifts back into the cell and repolarization occurs

---

127 During and after an acute injury, fluids shift to ______________ and *______________.

Do you know what is meant by fluids shifting to the third space? *__________

____.

- - - - - - - - - - - - - - - - -

cells and injured site(s) *or* cells and interstitial—third space.
Fluids shift to the interstitial space at the injured site. Fluid in the third space is considered physiologically useless or nonfunctional fluid

---

128 With vascular fluid deficit (loss), serum osmolality would be:

( ) a. decreased
( ) b. increased

What happens to the permeability of capillaries as a result of injury?

*____.

- - - - - - - - - - - - - - - - -

a. —; b. X; increase *or* increased capillary permeability

---

129 Explain how ADH and aldosterone restored water balance.

ADH *____.

Aldosterone *______________________________________.

- - - - - - - - - - - - - - - - -

ADH promoted water absorption from the distal tubules of the kidneys due to hypovolemia and/or increased serum osmolality, or a similar response

Aldosterone causes sodium to be reabsorbed from the distal tubules of the kidney when hyponatremia or stress is present, or a similar response

---

130 The syndrome of inappropriate antidiuretic hormone secretions (SIADH) frequently occurs following surgery, trauma, stress, pain, and CNS depressants (narcotics). The water reabsorption could be continuous for several days.

The nurse should assess for what type of fluid imbalance:

( ) a. Overhydration (hypervolemia)

( ) b. Dehydration (hypovolemia)

- - - - - - - - - - - - - - - - - -

a. X; b. –

---

131 Kidneys are the chief regulators of sodium and water balance. Kidneys conserve sodium when there is a sodium deficit. The hormone that is responsible for sodium reabsorption is ____________. This hormone also causes potassium (excretion/retention). ____________.

- - - - - - - - - - - - - - - - - -

aldosterone; excretion

---

132 A decrease in circulation from trauma, stress, or shock can cause a decrease in renal arterial blood flow. What effect does this have on the kidney?

* ____________________________.

For circulating blood to perfuse the kidneys, the systolic blood pressure should be ____________ or greater. (Refer to section on dehydration if needed.)

- - - - - - - - - - - - - - - - - -

temporary or permanent kidney damage; 60

---

133 Due to severe trauma, acid metabolites, such as lactic acid, are released from cells. What type of acid-base imbalance can occur? * ________________________. How do the lungs compensate for this imbalance? ____________________________.

- - - - - - - - - - - - - - - - - -

metabolic acidosis. By blowing off CO_2 (reducing carbonic acid)

---

134 From the following list of fluid, electrolyte, and acid-base changes, check those that apply to trauma, and correct the incorrect responses.

( ) a. Cellular loss of potassium
( ) b. Sodium shifts into cells
( ) c. Hypervolemia or overhydration
( ) d. Fluid shifts to the cells and injured site(s)
( ) e. Protein loss
( ) f. Decreased capillary permeability
( ) g. ADH secretion promoting reabsorption of water from the kidneys

— — — — — — — — — — — — — — — —

a. X; b. X; c. –; d. X; e. X; f. –; g. X
Corrections: c. Hypovolemia or dehydration; f. Increased capillary permeability

---

CLINICAL APPLICATIONS

---

135 The physician will assess the injured sites, order fluid replacements, and perform medical or surgical interventions as needed.

The nurse's responsibility is to assess and report fluid, electrolyte, and acid-base imbalances that may rapidly occur.

The imbalances that the nurse should assess are: ________________,

________________, and ________________.

— — — — — — — — — — — — — — — —

fluids, electrolytes, and acid-base

Table 62 lists the nursing assessments for ten observations: vital signs, behavioral changes, neurologic and neuromuscular signs, fluid loss, skin and mucous membranes, chest sounds and vein engorgements, ECG, fluid intake, previous drug regime, and laboratory results (chemistry, hematology, and ABG). This table could serve as a nursing assessment tool when providing nursing care to the acutely injured. Study the observations, nursing assessments, and rationale/comments. Refer to the table as needed.

TABLE 62. NURSING ASSESSMENT OF FLUIDS, ELECTROLYTE, AND ACID-BASE IMBALANCES IN THE ACUTELY INJURED PATIENT

Observation	Nursing Assessment		Rationale/Comments
1. Vital signs Pulse	Pulse rate Volume Pattern	______ ______ ______	Changes in vital signs (VS) are indicators of patient's physiologic status. Several VS should be taken and the first reading act as the baseline and for comparison. Pulse rate and pattern should be monitored frequently. Pulse rate ↑ 120 may indicate hypovolemia and possibility of shock. Full, bounding pulse could mean hypervolemia, and irregular pulse could mean hypokalemia.
Blood pressure	Admission BP Time Time	 ______ ______	Changes in BP (systolic and diastolic) may not occur until several hours after the injury. Several BP should be taken, and the first BP acts as the baseline and for comparison. Drop in systolic pressure could indicate hypovolemia. Pulse pressure (systolic minus diastolic) of < 20 could indicate shock.
Respiration	Respiration Pattern	______ ______	Note changes in rate, depth, and pattern. Rate > 32 could indicate hypovolemia. Deep, rapid, vigorous breathing could indicate acidosis as a result of cellular damage and shock. Hyperventilating (fast, shallow breathing) could be due to anxiety or hypoxia.
Temperature	Temperature on admission Date/Time	______ ______	Changes in temperature do not occur immediately after injury. If there is a slight elevation several hours or days later, it could indicate hypovolemia or impending shock.
2. Behavioral changes	Irritable Apprehensive Restless Confused Delirious Lethargic	______ ______ ______ ______ ______ ______	Irritability, apprehension, restlessness, and confusion, are indicators of hypoxia and later of fluid and electrolyte imbalances (hypovolemia, water intoxication, and potassium imbalance).
3. Neurologic and neuromuscular signs	Sensorium: Confused Semi-conscious Comatose Muscle weakness	 ______ ______ ______ ______	Changes in sensorium can be indicative of fluid imbalance. Muscle weakness and abdominal distention may be due to potassium deficit (may

TABLE 62. (Continued)

Observation	Nursing Assessment		Rationale/Comments
	Abdominal distention	________	not occur for hours or days later). Symptoms of tetany could indicate calcium and magnesium deficit.
	Pupil dilation	________	
	Tetany		
	Tremors	________	
	Twitching	________	
	Others	________	
4. Fluid loss	Wound(s)	________	Note the presence of an open draining wound.
	Urine		
	Number of voidings	________	Kidneys regulate fluids and electrolytes. Monitoring the urine output hourly is most important.
	Amount	________	
	ml/hr	________	
	ml/8 hr	________	
	ml/24 hr	________	
	Color	________	Oliguria could indicate lack of fluid intake or renal insufficiency due to decreased circulation/circulatory collapse or hypovolemia.
	Specific gravity	________	
	Vomitus		Frequent vomiting in large quantities leads to fluid, electrolyte (potassium, sodium, chloride), and hydrogen losses. Metabolic alkalosis can occur.
	Number	________	
	Consistency	________	
	Amount	________	
	Nasogastric tube		Gastrointestinal secretions should be measured. Large quantity loss of GI secretions could cause hypovolemia.
	Amount—ml/8 hr	________	
	Amount—ml/24 hr	________	
	Drain(s)		Excess drainage could contribute to fluid loss and should be measured if possible.
	Number	________	
	Amount	________	
5. Skin and mucous membrane	Skin color		Pale and/or gray-colored skin could indicate hypovolemia or shock. Flushed skin could be due to hypernatremia or metabolic acidosis.
	Pale	________	
	Gray	________	
	Flushed	________	
	Skin turgor		Poor skin turgor can result from hypovolemia/dehydration. This may not occur until 1–3 days after the injury.
	Normal	________	
	Poor	________	
	Edema—pitting peripheral		Edema indicates sodium and water retention. Sodium, chloride, and water could shift into the cells and to the injury site(s) (interstitial or third space).
	Feet	________	
	Legs	________	
	Dry mucous membranes	________	Dry, tenacious (sticky) secretions and dry membranes are indicative of dehydration or fluid loss. This may not occur until 1–3 days after the injury.
	Sticky secretions	________	
	Diaphoresis	________	Increased insensible fluid loss can result from diaphoresis

TABLE 62. (Continued)

Observation	Nursing Assessment		Rationale/Comments
			(excess perspiration). Amount of fluid loss from skin could double.
6. Chest sounds and vein engorgement	Chest rales	______	The chest should be checked for rales due to overhydration (pulmonary edema).
	Neck vein engorgement Hand vein engorgement	______ ______	Neck and hand vein engorgements are indicators of fluid excess. Rales and vein engorgements can occur from excess IV fluids or rapid IV administration.
7. ECG (EKG)	T wave Flat Inverted Peaked	 ______ ______ ______	Flat and inverted T wave indicates cardiac ischemia and/or potassium deficit. Peaked T wave indicates potassium excess.
8. Fluid intake	Oral fluid intake Amount ml/8 hr ml/24 hr	 ______ ______	Oral fluids should not be given until the injury(s) can be assessed. If surgery is indicated, the patient should be NPO.
	Types of IV fluids Crystalloids Colloids Blood Amount ml/8 hr ml/24 hr ml/hr	 ______ ______ ______ ______ ______ ______	Crystalloids, i.e., 5% dextrose in water, normal saline, lactated Ringer's, are normally ordered first to restore fluid loss, correct shocklike symptoms, restore or increase urine output, and serve as a life line to administer IV drugs. Five-percent dextrose in water given continuously can cause water intoxication (ICF volume excess).
9. Previous drug regime	Diuretics Digitalis Cortisone Antibiotics	______ ______ ______ ______	Drug history should be taken and reported to the physician. Potassium-wasting diuretics taken with digitalis preparation could cause digitalis toxicity. Cortisone causes sodium retention and potassium excretion. Antibiotics cause sodium retention.
10. Chemistry, hematology, and arterial blood gas changes	Electrolytes *Serum* K ______ Na ______ Cl ______ Ca ______ Mg ______	 *Urine/24 hr* K ______ Na ______ Cl ______	Electrolytes should be drawn immediately after severe injury and used as a baseline for future electrolyte results. Urine electrolytes are compared to serum electrolytes. Normal range for urine electrolytes are: K 25–120 mEq/24 hr Na 40–220 mEq/24 hr Cl 150–250 mEq/24 hr

TABLE 62. (Continued)

Observation	Nursing Assessment		Rationale/Comments
	Serum CO_2	______	Serum CO_2 > 32 mEq/L indicates metabolic alkalosis and < 22 mEq/L indicates metabolic acidosis.
	Osmolality Serum Urine	 ______ ______	Serum osmolality > 295 mOsm/L indicates hypovolemia/dehydration and < 280 mOsm/L indicates hypervolemia. Urine osmolality could be 100–1200 mOsm/L with a normal range 200–600 mOsm/L.
	BUN Creatinine	______ ______	Elevated BUN could indicate fluid volume deficit or kidney insufficiency. Elevated creatinine indicates kidney damage. Normal Range BUN 10–25 mg/dl Creatinine 0.7–1.4 mg/dl
	Blood glucose	______	Blood sugar increases during stress (up to 180 mg/dl).
	Hbg ______	Hct ____	Elevated hemoglobin and hematocrit could indicate hemoconcentration caused by fluid volume deficit (hypovolemia).
	Arterial blood gas (ABG) pH PCO_2 HCO_3 BE	 ______ ______ ______ ______	pH: < 7.35 indicates acidosis and > 7.45 indicates alkalosis. PCO_2 (respiratory component) Norms 35–45 mm Hg. Respiratory acidosis (pH ↓, PCO_2 ↑) may occur due to inadequate gas exchange. A ↓ PCO_2 and ↑ pH indicate respiratory alkalosis from hyperventilating due to anxiety and apprehension. HCO_3 (renal component) Norms 24–28 mEq/L. A ↓ HCO_3 and ↓ pH means metabolic acidosis, which is the most common acid-base imbalance following injury. Cellular catabolism occurs. (See Table 61.) A ↑ HCO_3 and ↑ pH means metabolic alkalosis. BE (base excess) Norms +2 to −2. Same as bicarbonate.

---

136 Vital signs should be constantly monitored during an acute injury. Tachycardia or pulse rate > 120 could indicate ______________ and should be reported.

- - - - - - - - - - - - - - - - -

hypovolemia or fluid volume deficit

---

137 Blood pressure does not immediately fall after an injury. With fluid volume deficit, the pulse rate increases first, and later the blood pressure will drop if fluid loss is not replaced.

A pulse pressure of less than 20 could indicate ______________.

- - - - - - - - - - - - - - - - -

shock

---

138 Respiration rate greater than 32 could indicate ______________.

Deep, rapid, vigorous breathing occurring after cellular damage or shock due to acute injury could indicate (metabolic acidosis/metabolic alkalosis).

* ______________ Why? * ______________

______________.

- - - - - - - - - - - - - - - - -

hypovolemia or fluid volume deficit; metabolic acidosis. Acid metabolites, such as lactic acid, are released from cells due to cellular breakdown. Good!

---

139 Temperature changes frequently do not occur immediately after injury. When there is a slight temperature elevation, this could indicate ______________.

If a high temperature elevation occurs 3 to 5 days after the injury, do you know what it might indicate? ______________.

- - - - - - - - - - - - - - - - -

hypovolemia *or* fluid volume deficit *or* dehydration; infection. Very good!

---

140 Irritability, apprehension, restlessness, and confusion are usually the result of hypoxia and of fluid and electrolyte imbalances. Name two fluid and electrolyte imbalances associated with the stated behavioral changes.

Fluid imbalances: 1. ______________________

2. *______________________

Electrolyte imbalance: 3. ______________________

- - - - - - - - - - - - - - - - -

1. hypovolemia *or* fluid volume deficit; 2. water intoxication; 3. hypokalemia

---

141 Match the following neurologic and neuromuscular signs with fluid and electrolyte imbalances:

____ 1. Decreased sensorium		
____ 2. Muscle weakness	a.	Hypokalemia
____ 3. Abdominal distention	b.	Hypovolemia
____ 4. Tetany—tremors and twitching	c.	Hypocalcemia

- - - - - - - - - - - - - - - - -

1. b; 2. a; 3. a; 4. c

---

142 Oliguria is not uncommon following an acute injury. Decreased urine output could be due to *______________________.
Do you recall what an elevated specific gravity (> 1.030) could indicate?

______________________.

- - - - - - - - - - - - - - - - -

hypovolemia, circulatory collapse, *or* a lack of fluid intake;
dehydration/hypovolemia *or* lack of fluid intake

143 Vomitus, nasogastric tubes, and diarrhea can cause what type of fluid imbalance?

______________________________.

Indicate which acid-base imbalance occurs with the following causes of fluid loss:

______ 1. Vomiting	a. Metabolic alkalosis
______ 2. Nasogastric tubes (loss of stomach secretions)	b. Metabolic acidosis
______ 3. Diarrhea	

\- - - - - - - - - - - - - - - - -

hypovolemia; 1. a; 2. a; 3. b

---

144 Pale or gray-colored skin could indicate ________________________.

Poor skin turgor, dry mucous membranes, and tenacious or sticky mucous secretions are indicative of ________________________.

\- - - - - - - - - - - - - - - - -

hypovolemia *or* shock; dehydration *or* fluid loss

---

145 When the patient is receiving intravenous therapy at an increased flow rate to correct fluid loss, the patient should be assessed for overhydration.

Excess and/or rapidly administered IV fluids can cause what type of fluid imbalance? ______________________________.

Name two symptoms associated with this imbalance. (Refer to Table 44 or to Chapter 5.)

1. * ______________________________.
2. * ______________________________.

\- - - - - - - - - - - - - - - - -

overhydration *or* ECFV excess (extracellular cellular fluid volume excess); 1. chest rales, 2. neck or hand vein engorgement. Also constant irritated cough or dyspnea

---

146 Flat or inverted T wave could indicate ______________________________,

and a peaked T wave could indicate ______________________________.

- - - - - - - - - - - - - - - - - -

hypokalemia *or* potassium deficit *or* cardiac ischemia; hyperkalemia *or* potassium excess

---

147 Immediately after an acute injury, oral fluids (should/should not) __________

be given. Why? *____

____.

- - - - - - - - - - - - - - - - - -

should not. The patient may need surgery and would be NPO (nothing by mouth)

---

148 Indicate which of the following are crystalloids used in intravenous therapy.

( ) a. 5% Dextrose in water
( ) b. Dextran 40–6%
( ) c. Plasmanate
( ) d. Normal saline (0.9% NaCl)
( ) e. Lactated Ringer's

Give two reasons for using crystalloids.

1. *____.
2. *____.

What type of fluid imbalance occurs when using 5% dextrose in water for

several days? *____.

- - - - - - - - - - - - - - - - - -

a. X; b. –; c. –; d. X; e. X; 1. restore fluid loss; 2. correct shocklike symptoms (correction may be temporary). Also, increase urine output; water intoxication (ICFV excess). In early shock/trauma massive infusions of 5% D/W could cause hyperglycemia and increased diuresis.

---

149 The drugs that cause sodium retention are:

( ) a. Diuretics
( ) b. Digitalis
( ) c. Cortisone
( ) d. Antibiotics

The drugs that cause potassium excretion are:

( ) e. Diuretics (potassium-wasting)
( ) f. Digitalis
( ) g. Cortisone
( ) h. Antibiotics

- - - - - - - - - - - - - - - - - -

a. –; b. –; c. X; d. X; e. X; f. –; g. X; h. –

---

150 After a severe injury, the serum potassium level is (increased/decreased) ________ with normal urine output.

Which type of sodium imbalance would be present? ________.

Lactic acid is released from the cells due to cellular breakdown. Would the serum CO_2 be (increased/decreased)? ________. Serum CO_2 is a bicarbonate determinant.

- - - - - - - - - - - - - - - - -

decreased (could be in normal range when there is severe cell damage—excess release of potassium. If urine output is poor, serum K would be increased.); hyponatremia; decreased

---

151 A serum osmolality of > 295 mOsm/L indicates (hypovolemia/hypervolemia). ________.

A serum osmolality of < 280 mOsm/L indicates (hypovolemia/hypervolemia). ________. Why? *________ ________.

- - - - - - - - - - - - - - - - -

hypovolemia; hypervolemia. It is caused from overhydration or hemodilution (excess water in proportion to solutes)

---

152 An elevated BUN could indicate *________ or *________.

After hydration, if the BUN does not return to normal, the elevated BUN would then indicate *________.

An elevated hemoglobin and hematocrit level could indicate hemoconcentration caused by ________.

- - - - - - - - - - - - - - - - -

fluid volume deficit or renal insufficiency; renal insufficiency; hypovolemia *or* fluid volume deficit

---

153 What type of acid-base imbalance is present if the patient's arterial blood gases are: pH 7.25; pCO_2 35 mm Hg; and HCO_3 18 mEq/L? *________.

- - - - - - - - - - - - - - - - -

metabolic acidosis

---

154 From the following list of observations and nursing assessments, check the ones that indicate fluid volume deficit (hypovolemia).

( ) a. Pulse 76
( ) b. Blood pressure 86/68
( ) c. Irritability, restlessness, confusion
( ) d. Specific gravity 1.034
( ) e. Excess GI drainage (> 2 liters)
( ) f. Dry mucous membrane and dry, tenacious mucous secretions
( ) g. Chest rales
( ) h. Peaked T waves
( ) i. Elevated BUN and Hgb

— — — — — — — — — — — — — — — — — —

a. –; b. X; c. X; d. X; e. X; f. X; g. –; h. –; i. X

CASE REVIEW

Marjorie Rockland, age 58, was in an automobile accident and was taken by ambulance to the emergency room of a large medical center. Her vital signs on admission were BP 134/88, P 106, R 30, T 98.8°F (37.1°C). She complained of pain in her abdomen and leg. A liter of 5% dextrose in water was started. Blood chemistry and x-rays (leg and abdomen) were ordered. Abdominal area appeared to be distended, and there were diminished bowel sounds. A nasogastric tube was inserted and attached to intermittent suction.

1. Ms. Rockland's pulse rate indicates tachycardia (mild to moderate) and could be indicative of ______________________________.
2. Her blood pressure is (normal/high/low) ______________ and could mean * ______________________________.
3. Name the solution category for 5% dextrose in water. ______________.
4. Distended abdomen and decreased peristalsis could indicate * ______________________________.
5. The purpose for the nasogastric tube connected to suction would be * ______________________________.

The x-rays showed a fractured right femur and possible abdominal fluid. Vital signs 2 hours later were BP 106/86, P 128, R 34. Ms. Rockland was apprehensive and restless, and had periods of confusion. Blood chemistry results were: K 3.7 mEq/L, Na 134 mEq/L, Cl 99 mEq/L, serum CO_2 24 mEq/L. A Foley catheter was inserted, and 350 ml of urine was obtained. The secretions from GI suction were "bloody."

6. Changes in the vital signs indicate:

   a. Pulse * ______________________________.

   b. Blood Pressure * ______________________________.

   c. Respiration * ______________________________.

7. Indicate whether the results from the blood chemistry were normal (N), low (L), or high (H).
   ( ) a. Potassium
   ( ) b. Sodium
   ( ) c. Chloride
   ( ) d. Serum CO_2

8. Why was a Foley catheter inserted? * ______________________________.

   Was the amount of urine obtained (adequate/inadequate)? ______________.

9. Bloody gastrointestinal secretions could indicate * ______________________.

10. Ms. Rockland's apprehension, restlessness, and bouts of confusion could indicate

    ______________________________.

Four hours after admission, Ms. Rockland's vital signs were BP 84/66, P 136, R 36. Her skin color was gray and she was diaphoretic. Urine output was averaging 15–20 ml/hr. Blood chemistry, type, and cross-match, and blood gases were ordered. A second liter of 5% dextrose in water to run for 4 hours was ordered.

11. Vital signs are indicative of * ______________________________.

12. Explain why there is a fluid volume deficit. * ______________________________

    ______________________________.

13. Is the hourly urine output adequate? ____________. Why? * ______________________

    ______________________________.

14. Is 5% dextrose in water an appropriate intravenous solution to be used continuously? ____________. Why? * ______________________________

    ______________________________.

An hour later, she was scheduled for OR. The second laboratory results were: K 5.0 mEq/L, Na 130 mEq/L, Cl 95 mEq/L, serum CO_2 18 mEq/L, blood sugar 166 mg/dl, BUN 32 mg/dl, ABG–pH 7.32, PCO_2 35 mm Hg, HCO_3 19 mEq/L.

15. Her lab results indicate:

    a. Potassium ______________________________.

    b. Sodium ______________________________.

    c. Chloride ______________________________.

    d. Serum CO_2 * ______________________________.

    e. Blood sugar ______________________________.

f. BUN ____.

g. pH ____.

h. PCO_2 *____.

i. HCO_3 *____.

- - - - - - - - - - - - - - - - - -

1. fluid loss or hypovolemia or impending shock
2. normal; heart rate (pulse) was compensating for fluid loss and in response to injury. Later, if heart rate does not compensate, blood pressure would fall.
3. crystalloids. This group of solutions increases fluid volume and acts as a lifeline for emergency IV drugs.
4. loss of bowel tone, fluid shift to the abdominal area (most likely a traumatized or injured area)
5. to remove accumulated stomach and intestinal fluid (secretions) that resulted from an abdominal injury
6. a. tachycardia from fluid volume deficit (hypovolemia)
   b. drop in BP and pulse pressure 20 indicate hypovolemia and shock (impending)
   c. hypovolemia and stress
7. a. N; b. L; c. N; d. N
8. to monitor urine output. (This is common practice following a traumatic injury.); adequate
9. GI injury *or* abdominal injury
10. hypovolemia and/or hypoxia
11. hypovolemia (severe) and shock
12. fluid shift from the vascular fluid to the cells and to the injured sites (third spaces—abdominal area and injured leg tissue area). Fluids are also lost from GI suction and from diaphoresis
13. No. It is less than 25 ml per hour
14. No. Five-percent D/W would not correct hyponatremia and could cause water intoxication if used continuously
15. a. normal
    b. hyponatremia
    c. hypochloremia
    d. metabolic acidosis
    e. stress
    f. hypovolemia/dehydration
    g. acidosis
    h. low normal
    i. metabolic acidosis

NURSING DIAGNOSES

Fluid volume deficit related to fluid loss at the injured site; fluid volume shift from vascular area to injured site.

Potential alteration in fluid volume excess related to excessive IV infusions, cardiac insufficiency, and renal dysfunction.
Impairment of urinary elimination related to fluid loss to the injured site, SIADH.
Alteration in acid-base balance related to acid metabolites and CO_2 retention.
Potential for electrolyte imbalance related to fluid and electrolyte shift to the injured site (third space), renal dysfunction, and cellular damage.
Anxiety related to the outcome of the trauma; fear of impending death.
Potential for infection related to traumatized area(s) and/or septic technique.
Potential for impaired gas exchange related to trauma; anxiety (tachypnea).
Potential for impaired verbal communication related to mechanical devices, e.g., respirators, and decreased level of consciousness.
Potential alteration in comfort related to the trauma; pain.
Potential for ineffective family action related to the crisis situation with the family member and the unknown outcome.

NURSING ACTIONS

1. Check the vital signs and report signs of tachycardia, arrhythmia, low blood pressure, low pulse pressure, tachypnea, dyspnea, and temperature elevation. Most of these signs and symptoms could indicate hypovolemia, hypoxia, and hypokalemia.
2. Assess the behavioral and neurologic status of the injured patient. Irritability, apprehension, restlessness, and confusion are symptoms of fluid volume deficit. Fluid and sodium frequently shift into cells, thus decreasing the vascular fluid.
3. Monitor urine output hourly and for 24 hours. Urine output should be at least 25 ml per hour and 600 ml per day.
4. Record amounts of vomitus, diarrhea, stools, secretions from nasogastric suction, and drains as fluid loss.
5. Assess the skin color and skin turgor. Pallor or gray color and poor skin turgor are symptoms of body fluid loss.
6. Check the mucous membranes for dryness and for dry, tenacious secretions.
7. Monitor intravenous fluid therapy. Regulate flow rate to prevent overhydration (hypervolemia). Report to physician if the patient is receiving 5% dextrose in water continuously without any other solutes such as saline.
8. Assess chest for rales. Overhydration from excess fluids and rapid administration of IV fluids can cause pulmonary edema. Heart failure could also be another result.
9. Teach the patient to cough and breathe deeply to expand the lungs and provide effective ventilation.
10. Check for neck vein engorgement and/or hand vein engorgement when overhydration is suspected. Check jugular vein at 45° angle, and hand veins by raising hand above heart level. Look for hand engorgement after 10 seconds.
11. Check for pitting edema in the feet and legs. Weigh patient to determine if there is fluid retention.
12. Monitor ECG readings and report flat, inverted, or peaked T waves, which are indicative of potassium imbalance.

13. Check for insensible fluid loss from diaphoresis and hyperventilation. Excessive sweating and overbreathing could double the normal fluid loss from the skin and lungs.
14. Take a drug history on the injured patient. Report if the patient is regularly taking potassium-wasting diuretics, digitalis, cortisone, or antibiotics.
15. Have blood chemistry, hematology, and blood gases drawn as ordered. Report abnormal findings to the physician (if he or she has not seen or been informed of lab results).
16. Assess for signs and symptoms of hypokalemia and hyperkalemia. Serum potassium levels can vary after trauma. The serum level could be low from excess urine output. The serum level could be in normal range even though there is a cellular potassium deficit. Also, hyperkalemia could be present due to excessive cellular breakdown and oliguria (decreased urine output). The serum potassium level should be known before administering potassium in IV fluids.
17. Check the BUN and creatinine. If both are highly elevated, it could be due to renal insufficiency. If they are slightly elevated and return to normal when the patient is hydrated, it could be due to fluid volume deficit.
18. Assess for signs and symptoms of acid-base imbalance. Check arterial blood gases. Deep, rapid, vigorous breathing, ↓ pH, and ↓ HCO_3 indicate metabolic acidosis. Dyspnea, ↓ pH, and ↑ P_{CO_2} indicate respiratory acidosis. Both types of acidosis are commonly seen in the acutely injured patient.
19. Explain to the patient the nursing care you plan to give. Answer patient's questions and, if answers are unknown, refer the question(s) to other health professionals.

RENAL FAILURE: HEMODIALYSIS AND PERITONEAL DIALYSIS (Sally Marshall)

155 Renal failure is the inability of the kidneys to excrete the by-products of cell metabolism and normal amounts of body water.

If the kidneys can no longer excrete waste products and normal amounts of body water *________________________ is indicated.

\- - - - - - - - - - - - - - - - - -

renal failure

ACUTE AND CHRONIC FAILURE

Renal failure is classified as acute or chronic. *Acute renal failure/ARF* results from an acute insult, primarily ischemia, toxicity, or obstruction. Table 63 lists examples of the causes, with contributing problems. Refer to the table as needed.

TABLE 63. CAUSES OF ACUTE RENAL FAILURE

Ischemia	Toxicity	Obstruction
Dehydration Shock: Distributive/sepsis Hypovolemic/hemorrhagic Cardiogenic	Antibiotics: Aminoglycosides, Penicillins, Cephalosporins. Nonsteroidal antiinflammatory agents Organic compounds: Carbon Tetrachloride Methyl alcohol Miscellaneous: Myoglobin, transfusion reactions	Prerenal: Arterial emboli Aneurysm Postrenal: Ureteral obstruction Bladder obstruction Catheter obstruction

---

156 Acute renal failure usually results from insult to the kidney. The three (3) major categoric causes of acute renal failure are ________________, ________________, and ________________.

- - - - - - - - - - - - - - - - - -

ischemia; toxicity; obstruction.

---

157 Examples of the major causes of renal failure are the following:

Ischemia: ________________.

Toxicity: ________________.

Obstruction: ________________.

- - - - - - - - - - - - - - - - - -

Ischemia: shock. Toxicity: antibiotics, e.g., aminoglycosides. Obstruction: arterial emboli (prerenal) or ureteral obstruction (postrenal).

Chronic renal failure/CRF often results from a disease process that affects the renal parenchyma and eventually causes cessation of function. Causative diseases are basically glomerular, renal vascular, and tubulointerstitial. Table 64 lists examples of causes and problems of chronic renal failure. Refer to the table as needed.

TABLE 64. CAUSES OF CHRONIC RENAL FAILURE

Glomerular	Renal Vascular	Tubulointerstitial
Glomerulonephritis	Malignant hypertension	Pyelonephritis
Acute poststreptococcal (glomerulonephritis)	Nephrosclerosis	Analgesic nephropathy
Systemic lupus erythematosus	Necrotizing vasculitis	Interstitial nephritis
Diabetes glomerulosclerosis		Oxlate nephropathy
		Papillary necrosis

---

158 The three (3) groups of causes associated with chronic renal failure are

________________, ________________, and ________________.

\- - - - - - - - - - - - - - - - - -

glomerular; renal vascular; tubulointerstitial.

---

159 Give an example of a renal problem that occurs in each of the three categories.

Glomerular: ________________________.

Renal vascular: ________________________.

Tubulointerstitial: ________________________.

\- - - - - - - - - - - - - - - - - -

Glomerular: glomerulonephritis. Renal vascular: malignant hypertension. Tubulointerstitial: pyelonephritis.

---

160 Two measurements of nitrogenous waste products, by-products of protein metabolism, are blood urea nitrogen/BUN and creatinine. A rise in the level of BUN and creatinine is known as azotemia.

The two by-products of protein metabolism or nitrogenous waste products are ________________ and ________________. Azotemia occurs when * ________________________.

\- - - - - - - - - - - - - - - - - -

BUN (blood urea nitrogen) and creatinine.
BUN and creatinine levels rise.

---

161 Urine output in renal failure varies. No-urine output is labeled *anuria.* In ARF the cause may be prerenal or postrenal obstruction. In CRF anuria indicates total loss of parenchymal function.

ARF means * ____________________.

CRF means * ____________________.

Anuria, which is * ________________ differs in ARF and CRF. In ARF anuria may be the result of * ________________ ________________ and in CRF anuria may cause * ________________ ________________.

\- - - - - - - - - - - - - - - - -

ARF: acute renal failure; CRF: chronic renal failure
no-urine output
prerenal or postrenal obstruction; total loss of parenchymal function.

---

162 Normal urine output is >30 ml/hr. Urine output of less than 400–500 ml/24 hr is *oliguria.* No-urine output is called ________________.

In ARF ischemia and nephrotoxicity are the causes of oliguria, whereas in CRF oliguria indicates a decline in parenchymal function.

\- - - - - - - - - - - - - - - - -

anuria

---

163 The earliest manifestations of a disease that may lead to CRF are hematuria and proteinuria. If undetected, renal disease progresses, renal function degenerates, and azotemia ensues.

In ARF oliguria is usually caused by ________________ and ________________. In CRF oliguria indicates * ________________ ________________.

In CRF early manifestations of renal disease are ________________, or ________________.

\- - - - - - - - - - - - - - - - -

ischemia and nephrotoxicity; a decline in parenchymal function; hematuria and proteinuria.

In ARF the earliest sign after the insult is a rise in BUN and serum creatinine accompanied by oliguria or nonoliguria. Acute renal failure has four characteristic phases. Refer to Table 65.

TABLE 65. CHARACTERISTIC PHASES OF ACUTE RENAL FAILURE

Phases	Rationale
Initial	This phase begins when insult occurs and continues until signs of azotemia and/or oliguria appear.
Oliguric	Urine output is less than 400ml a day. BUN and creatinine are markedly elevated. Dialysis is initiated in this phase.
Diuretic	It starts with a sudden increase in urine output, 1.5 liters (1500ml) or more per day. Nephrons do not concentrate urine sufficiently to conserve electrolytes and water. Risk of dehydration and death are high. This phase can continue for days to weeks.
Recovery	This phase is noted by decrease of azotemia with kidneys showing the ability to concentrate urine. It may last from weeks to months.

---

164 When the urinary volume is less than 400 ml per day and azotemia is present the patient is in the ____________________ phase.

If a sudden increase in urine output occurs (1.5 liters or more per day) the patient is in the ____________________ phase. In the diuretic phase the greatest risk is ____________________.

- - - - - - - - - - - - - - - - -

oliguric; diuretic; dehydration (if you answered electrolyte loss, true, but the greatest risk is ECFV deficit or dehydration).

165 Symptoms of ARF and CRF appear when renal function decreases to more than 75%. When renal function decreases to more than 90% death will result if dialysis treatment is not initiated.

Indicate which of the following problems contribute to acute and chronic renal failure by using ARF for acute renal failure and CRF for chronic renal failure.

_______ a. Burns

_______ b. Diabetes Mellitus

_______ c. Bladder obstruction

_______ d. Glomerulonephritis

_______ e. Malignant hypertension

_______ f. Severe fluid volume deficit

_______ g. Hemolytic blood transfusion

_______ h. Aminoglycoside therapy

_______ i. Pyelonephritis

_______ j. Nephrosclerosis

_______ k. Carbon tetrachloride

_______ l. sepsis

- - - - - - - - - - - - - - - - -

a. ARF; b. CRF; c. ARF; d. CRF; e. CRF; f. ARF; g. ARF; h. ARF; i. CRF; j. CRF; k. ARF; l. ARF

FLUID, ELECTROLYTE, AND ACID-BASE IMBALANCES IN RENAL FAILURE

Table 66 outlines the fluid, electrolyte, and acid-base changes that occur in renal failure and lists the rationale and signs and symptoms. Study the table carefully and refer to it as needed.

TABLE 66. FLUID, ELECTROLYTE, AND ACID-BASE IMBALANCES IN RENAL FAILURE

Imbalances	Rationale	Signs and Symptoms
Fluid Overload:		
Decreased urine output	Inability of the kidneys to concentrate, dilute, and excrete urine with normal or excessive intake of fluid.	Noninvasive: Elevated jugular venous pressure. Pitted edema: Preorbital, hands, feet, sacral, anasarca.
Sodium retention	Increased tubular reabsorption of sodium due to reduced renal perfusion and/or increased renin-angiotensin-aldosterone secretion. Occurs in ischemia and malignant hypertension.	Increased blood pressure. Weight gain. Moist rales. Dyspnea. Pulmonary edema.
Reduced oncotic pressure	Loss of intravascular protein through damaged glomeruli leads to decreased intravascular volume. ADH secretion increases water retention to maintain intravascular volume. Continued protein loss decreases oncotic pressure in capillaries and causes water to move into the interstitial space. Seen in nephrotic syndrome, glomerular diseases, and liver ascites.	Invasive: Increased central venous pressure. Elevated pulmonary artery and wedge pressure.
Potassium Excess:	**Hyperkalemia**	
Potassium retention	Inability of the kidneys to excrete potassium in severe oliguric and anuric states.	Noninvasive: Weakness, parathesia Nausea, vomiting ECG: elevated T wave Tachycardia Cardiac arrest.
Cellular injury	Massive tissue injury, acidosis, and protein catabolism cause potassium to leave cells.	Invasive: Serum K >5.3mEq/L
Potassium Deficit:	**Hypokalemia**	
Potassium loss	Excessive loss of gastrointestinal secretions or excessive loss in dialysis.	Non-invasive: Muscle weakness Abdominal distention Arrhythmia
Diuretic phase of ARF	Excessive loss of electrolytes and water due to the kidneys inability to concentrate urine.	Anorexia, N/V ECG: Flat or inverted T wave, prominent U wave, and A-V block.
Renal tubular acidosis	Nonoliguric azotemia causes excretion of K^+. Present in Fanconi syndrome, nephrotic syndrome, multiple myeloma, cirrhosis, and some drug toxicities.	Invasive: Serum K <3.5 mEq/L
Sodium Excess:	**Hypernatremia**	
Increased tubular sodium absorption	With decreased intravascular volume, aldosterone secretion increases	Noninvasive: Edema

TABLE 66. (Continued)

Imbalances	Rationale	Signs and Symptoms
	sodium retention to improve intravascular volume. May be seen in nephrotic syndrome and liver ascites.	Dry tongue Tachycardia Thirst Weight gain Increased BP
Dietary sodium ingestion	Increased dietary ingestion of sodium, especially in anuric states.	Invasive: Serum Na >145 mEq/L
Sodium Deficit:	**Hyponatremia**	
Sodium loss	Excessive loss of gastrointestinal secretion through suction, vomiting, and diarrhea.	Noninvasive: Decreased skin turgor Decreased BP Rapid pulse Dry mucuous membrane Muscle weakness
Diuretic phase of ARF	Excessive loss of electrolytes and water because of the kidneys inability to concentrate urine.	Invasive: Serum Na <130 mEq/L
Excessive fluid intake	Increased fluid intake with oliguria or anuria present will dilute serum sodium level.	
Metabolic acidosis	Sodium shifts into cells as potassium shifts to plasma during acidosis.	
Phosphorus Excess:	**Hyperphosphatemia**	
Phosphorus (phosphate) retention	Occurs because of decreased renal phosphate excretion which increases metabolic acidosis. Phosphorus affects serum calcium level by altering the balance of their reciprocal relationships. Parathyroid hormone/ PTH (parathormone) enhances phosphorus or phosphate excretion in the urine.	Noninvasive: Nausea and diarrhea Tachycardia Tetany with low Ca Hyperreflexia Muscle weakness Flaccid paralysis Invasive: Serum P >4.5 mg/dl >2.6 mEq/L
Calcium Deficit:	**Hypocalcemia**	
Increased phosphorus retention	Decreases the balance between calcium and phosphorus. PTH demineralizes the bone to increase serum calcium. Untreated hypo-calcemia and hyperphosphatemia lead to renal osteodystrophy and metastatic calcification (deposits of calcium phosphate crystals in soft tissues).	Noninvasive: Tetany: Muscle twitching Tingling Carpopedal spasm Laryngeal spasm Abdominal cramps Muscle cramps Decreased clotting Arrhythmias Positive Chvostek sign Positive Trousseau sign
Decreased absorption of calcium from intestines	Impaired vitamin D activity from renal impairment causes reduced calcium absorption.	Invasive Serum Ca <8.5 mg/dl or <4.5 mEq/L PTH >375 pg Eq/ml Ca and P product >70 mg/dl

TABLE 66. (Continued)

Imbalances	Rationale	Signs and Symptoms
Metabolic Acidosis:		
Hydrogen-ion retention	Inability of kidneys to excrete daily hydrogen ion load.	Noninvasive: Weakness Lassitude Increased respiration (rate and depth)
Reduced buffering mechanisms in tubules	Refer to Chapter 3, Table 32C, on renal regulatory mechanisms.	restlessness flushed skin
Ammonia	Reduced nephron function inhibits conversion of ammonia and HCl to NH_4Cl for excretion.	Invasive: pH <7.35 HCO_3 <24 mEq/L anion gap >16 mEq/L
Phosphate salts	Reduced nephron function inhibits the combination of hydrogen ion with $NaHPO_4$ to form NaH_2PO_4 for excretion.	
Bicarbonate/HCO_3	With damaged nephrons, less HCO_3 is regenerated and reabsorbed in the tubules.	
Retention of metabolic acids	Inability of the kidneys to excrete uric, sulfuric, phosphate, and other acids of metabolism.	
Lactic acid formation	Occurs from tissue hypoxemia from an ischemic insult.	
Increased fat breakdown	Malnutrition from decreased nutritional intake causes accumulation of ketone acids.	
Urine Sodium and Osmolality:		
Urine sodium	Ischemic state increases ADH and aldosterone secretions. Renal efforts to conserve sodium will show less Na ion in urine, <20 mEq. However, damaged nephrons cannot filter sufficiently; therefore the urine sodium level could be >20 mEq.	
Urine osmolality	Ischemic state increases renal filtration; thus urine osmolality is >500 mOsm/kg/H_2O. Injured nephrons with impaired filtration have urine osmolality similar to plasma, 290 mOsm/kg/H_2O.	

---

166 In renal failure fluid overload can occur from the following:

a. *________________________________.

b. *________________________________.

c. *________________________________.

\- - - - - - - - - - - - - - - - -

a. Decreased urine output. b. Sodium retention. c. Reduced oncotic pressure.

---

167 Name six (6) symptoms of fluid overload, both noninvasive and invasive:

a. __________________ d. __________________

b. __________________ e. __________________

c. __________________ f. __________________

\- - - - - - - - - - - - - - - - -

a. Elevated jugular venous pressure. b. Pitting edema. c. Weight gain. d. Increased blood pressure. e. Dyspnea. f. Increased central venous pressure. Others are: moist rales, pulmonary edema, elevated pulmonary artery, and wedge pressures.

---

168 Potassium excess occurs primarily in which urinary state?

________________________.

The effect of potassium excess on the cardiac muscle can cause

*________________________ and *________________________

____________________.

\- - - - - - - - - - - - - - - - -

Severe oligura and anuria
ventricular tachycardia; cardiac arrest

---

169 Potassium deficits may occur from the following:

a. *________________________________.

b. *________________________________.

c. *________________________________.

\- - - - - - - - - - - - - - - - -

a. Loss of GI secretions or from dialysis. b. Diuretic phase of ARF.
c. Renal tubular acidosis.

---

170 Sodium excess may occur from the following:

a. *______________________________.

b. *______________________________.

— — — — — — — — — — — — — — — —

a. Increased tubular absorption. b. Dietary ingestion

---

171 Sodium deficits may occur from the following:

a. *______________________________.

b. *______________________________.

c. *______________________________.

d. *______________________________.

— — — — — — — — — — — — — — — —

a. Loss of GI secretions. b. Diuretic phase of ARF. c. Excessive fluid intake. d. Metabolic acidosis

---

172 An excess serum phosphorus/phosphate in renal failure is caused by *______________________________. The retention of phosphorus/phosphate alters the balance between calcium and phosphorus. The result of the calcium imbalance is ______________________________.

— — — — — — — — — — — — — — — —

decreased excretion; hypocalcemia or calcium loss (deficit)

---

173 Calcium deficit results in symptoms of tetany which are ____________________, ____________________, and ____________________.

In which acid-base imbalance will the symptoms of tetany be decreased *______________________________.

— — — — — — — — — — — — — — — —

muscle twitching, tingling, laryngeal spasms. Also carpopedal spasm, positive chvostek, and Trousseau's signs. Metabolic acidosis

---

---

174 Calcium deficits will enhance the toxic effect of a high serum potassium by increasing cardiac muscle irritability.

When a low serum calcium exists serum potassium must be monitored to prevent * ______________________________.

– – – – – – – – – – – – – – – – –

cardiac muscle irritability

---

175 Metabolic acidosis occurs from the following renal failure:

a. * ______________________________.

b. * ______________________________.

c. * ______________________________.

d. * ______________________________.

e. * ______________________________.

– – – – – – – – – – – – – – – – –

a. Hydrogen ion retention. b. Reduction of buffering mechanisms. c. Retention of metabolic acids. d. Lactic acid formation. e. Increased breakdown of fats.

---

176 What respiratory symptom occurs in metabolic acidosis?

* ______________________________.

– – – – – – – – – – – – – – – – –

Increased respiratory rate and depth OR Kussmaul breathing

---

177 Two (2) urinary tests for assessing damaged nephrons are ______________________ and ______________________.

– – – – – – – – – – – – – – – – –

urine sodium; urine osmolality

---

178 A low urine sodium indicates increased ADH and aldosterone secretions. Nephrons damaged by ischemia or toxins cannot filter sufficiently and urine sodium could be ______________.

– – – – – – – – – – – – – – – – –

increased

---

179 A purpose for urine osmolality is to determine the kidneys' ability to dilute and concentrate urine. Damaged nephrons with impaired filtration are unable to concentrate urine. As a result of damaged nephrons, the urine would be

(diluted/concentrated) ______________________.

\- - - - - - - - - - - - - - - - -

diluted

Table 67 explains the effects of renal failure on body systems. Refer to the table as needed.

TABLE 67. SYSTEMIC EFFECTS OF RENAL FAILURE

Body Systems	Rationale
Neurological	Uremic waste products cause slow neural conduction. Changes in personality, thought processes, levels of consciousness, and seizures can occur.
Cardiovascular	Fluid retention causes fluid overload, hypertension, and cardiac hypertrophy. Electrolyte imbalance causes arrhythmias. Uremic waste products can irritate the pericardium and lead to pericarditis and cardiac tamponade.
Respiratory	Fluid retention causes pulmonary edema. Thick bronchial secretions and impaired immune response increase susceptibility to bacterial infections.
Gastrointestinal	Ulcerations can develop anywhere in the mucosa of GI tract from breakdown of urea to ammonia.
Hematologic	Failure of the kidneys to secrete erythropoietin results in decreased red cell production and anemia.
	Platelet survival is diminished and bleeding tendencies are increased. Immune deficiency develops from uremic waste products.
Musculoskeletal	Brittle bones and metastatic calcification result from bone demineralization in response to phosphorus and calcium imbalance.
Endocrine	Secondary hyperparathyroidism can develop from phosphorus and calcium imbalances. Sexual and menstrual dysfunctions are the result. Growth and mental retardation occur in children and carbohydrate and lipoprotein metabolism are altered.
Integumentary	Uremic waste products cause pruritis and dryness. Retained pigments and anemia give the skin a bronze cast.

180 Which of the following body systems are affected by renal failure?
( ) a. cardiovascular
( ) b. respiratory
( ) c. eye
( ) d. neurological
( ) e. gastrointestinal
( ) f. integumentary
( ) g. musculoskeletal
( ) h. endocrine
( ) i. hematological

– – – – – – – – – – – – – – – – –

a. X; b. X; c. –; d. X; e. X; f. X; g. X; h. X; i. X

DIALYSIS: HEMODIALYSIS AND PERITONEAL DIALYSIS

Dialysis is the process of filtrating uremic waste products and excess body fluid through a semipermeable membrane to restore body homeostasis. Refer to Chapter 1 as needed for the following definitions.

Diffusion is the movement of molecules/solutes in solution. In diffusion the rate of movement across the permeable membrane is greater from the higher to the lower concentration.

Osmosis is the movement of water molecules across a semipermeable membrane from an area of higher water concentration to an area of lower water concentration.

Ultrafiltration is the pressure gradient that enhances the movement of water molecules across the semipermeable membrane.

181 The movement of molecules across a semipermeable membrane from an area of higher concentration to an area of lower concentration is ______________.

The pressure gradient that enhances the movement of water molecules is called ______________.

The movement of water molecules across a semipermeable membrane from an area of higher water concentration to lower water concentration is called

______________.

– – – – – – – – – – – – – – – – –

diffusion; ultrafiltration; osmosis

---

182 The two types of dialysis therapy are hemodialysis and peritoneal dialysis.

The goals for both types are to restore electrolyte balance, to remove uremic waste products, and to restore the patient's dry weight. *Dry weight* is normal body weight without excess fluid.

Name two (2) goals of dialysis:

a. *_______________.

b. *_______________.

- - - - - - - - - - - - - - - - - -

a. To restore electrolyte balance. b. To remove uremic waste products. ALSO to restore patient's dry weight.

---

183 Hemodialysis and peritoneal dialysis are similar in the following ways: both use dialysate, which is a solution that contains electrolytes approximating normal plasma, and both use a semipermeable membrane.

List two ways in which hemodialysis and peritoneal dialysis are similar.

a. *_______________.

b. *_______________.

- - - - - - - - - - - - - - - - - -

a. They use dialysate similar in composition to normal plasma.
b. They use a semipermeable membrane.

---

184 In hemodialysis the artificial kidney/AK is the semipermeable membrane made of a cellophanelike material through which only molecules of a particular size can diffuse.

The semipermeable membrane of the artificial kidney through which only molecules of a *_______________.

- - - - - - - - - - - - - - - - - -

particular size can diffuse.

---

185 The dialysate solution is prepared in the delivery system, where it is mixed to the correct concentration, heated, and pumped to the artificial kidney.

List the functions of the delivery system:

a. * ________________________________.

b. * ________________________________.

c. * ________________________________.

— — — — — — — — — — — — — — — — —

a. Mixes dialysate in the correct concentration. b. Heats dialysate. c. Pumps dialysate

---

186 Dialysate for hemodialysis contains five (5) basic components: calcium chloride, magnesium chloride, potassium chloride, sodium chloride, and sodium acetate or bicarbonate. The concentration of these iso-osmolar/isotonic components resembles low plasma concentrations, but it can be prepared to correct an electrolyte or acid-base imbalance.

If a patient has a potassium excess dialysate can be prepared with low potassium. If a patient has metabolic acidosis acetate or bicarbonate in dialysate can be used.

Dialysate can be individualized for correcting imbalances of ________________

and ________________.

— — — — — — — — — — — — — — — — —

electrolytes or potassium; acid-base or acidosis

---

187 There are no uremic waste products in the dialysate. Therefore urea, creatinine, and other metabolic waste products diffuse rapidly from the blood across the membrane into the dialysate.

Uremic waste products rapidly diffuse from the ________________ into

the ________________.

— — — — — — — — — — — — — — — — —

blood; dialysate

---

188 Vascular access to the patient's bloodstream must be obtained to initiate hemodialysis. This access is surgically created or obtained by catheterizing a large vein.

To initiate hemodialysis vascular access to the ________________ must be obtained.

- - - - - - - - - - - - - - - - -

bloodstream

---

189 During hemodialysis blood is pumped through tubing to the membranes of the artificial kidney (AK). Two common types of AK design are the flatplate and hollowfiber. The flatplate is a stack of plastic plates with two membranes between each plate. Blood flows between these membranes. The hollowfiber has thousands of tiny hairlike fibers through which blood flows.

By using the flatplate AK blood flows *________________

________________.

With the hollowfiber AK blood flows *________________

________________.

- - - - - - - - - - - - - - - - -

between the membranes; through the tiny hairlike fibers

---

190 A complication of hemodialysis is blood clotting in the blood lines and AK. This can be prevented by administering heparin, an anticoagulant, during the procedure.

Heparin administered during hemodialysis prevents ________________.

- - - - - - - - - - - - - - - - -

clotting.

---

191 The delivery system pumps dialysate through the AK and around membranes. A negative pressure gradient, created by the delivery system, pulls excess water from the blood across the semipermeable membrane. A positive pressure gradient that pushes excess water across the membrane can occur. Ultrafiltration results and excess fluid is removed from the patient's bloodstream.

A negative pressure gradient results from *____

____________________________________. Excess fluid is removed from the patient

by ____.

\- - - - - - - - - - - - - - - - - -

the pull of water across semipermeable membrane; ultrafiltration

---

192 The goal of ultrafiltration is to obtain the patient's dry weight and prevent cardiovascular complications such as hypertension, pulmonary edema, and ventricular hypertrophy.

Dry weight is *____.

Maintaining the patient's dry weight prevents ____________________________ complication.

\- - - - - - - - - - - - - - - - - -

the normal body weight without excess fluid; cardiovascular

---

193 Hemodialysis treatment takes 3 to 5 hours to complete. The results should be the removal of *____, restoration of *____, and removal of *____.

Hemodialysis is done on a constant basis for patients with chronic renal failure and intermittently for those in acute renal failure until renal function improves.

\- - - - - - - - - - - - - - - - - -

uremic waste products; electrolyte balance; excess fluid

---

194 In peritoneal dialysis the peritoneum that surrounds the abdominal cavity is used as the semipermeable membrane.

The semipermeable membrane used in peritoneal dialysis is the ____________ ____________.

- - - - - - - - - - - - - - - - -

peritoneum

---

195 The dialysate for peritoneal dialysis is a sterile solution that contains similar levels of sodium, magnesium, calcium, and chloride as the plasma.

The dialysate electrolyte levels of sodium, magnesium, calcium, and chloride are similar to ____________.

The dialysate solution is ____________.

- - - - - - - - - - - - - - - - -

plasma; sterile

---

196 Potassium is not included in peritoneal dialysate solutions. The physician prescribes the amount of potassium that can be added to the dialysate according to the patient's serum potassium level.

Why do you think potassium is not added to the solution?

____

- - - - - - - - - - - - - - - - -

The patient's potassium level may be high. Adding too much potassium to the solution could increase the hyperkalemic state OR similar answer. VERY GOOD

---

197 Acid-base balance is corrected in peritoneal dialysis by adding acetate, a bicarbonate precursor, to the dialysate solution. This buffers metabolic acids.

Acetate can be added to peritoneal dialysate to correct what acid-base disorder? ____________.

- - - - - - - - - - - - - - - - -

metabolic acidosis

---

198 Ultrafiltration is accomplished in peritoneal dialysis by creating an osmotic pressure gradient with glucose. The glucose concentration in the dialysate creates a hyperosmolar solution that pulls water across the peritoneal membrane.

Glucose in peritoneal dialysate creates an * ______________________________ gradient. The dialysate solution in peritoneal dialysate is ______________________________.

- - - - - - - - - - - - - - - - - -

osmotic pressure; hyperosmolar

---

199 Peritoneal dialysate has three concentrations of glucose: 1.5, 2.5, and 4.5%. The physician prescribes the concentration needed on the basis of the patient's state of fluid overload. The higher the glucose concentration, the more hyperosmolar the solution. The result is more ultrafiltration.

More ultrafiltration will occur in peritoneal dialysis when the dialysate has a * ______________________________.

- - - - - - - - - - - - - - - - - -

high glucose concentration

---

200 Peritoneal dialysis begins by inserting a catheter into the peritoneal cavity. Capillary beds within the layers of peritoneum provide an indirect access to the bloodstream for the dialysate.

Access for peritoneal dialysis is obtained by a ____________________.

- - - - - - - - - - - - - - - - - -

catheter

---

201 A serious complication of peritoneal dialysis is peritonitis, the result of an infection caused by the peritoneal catheter, e.g., contamination.

Peritonitis is a serious complication of * ______________________________. It is caused by an * ______________________________.

- - - - - - - - - - - - - - - - - -

peritoneal dialysis; infection associated with the catheter

---

202 Two liters of dialysate is usually infused rapidly into the peritoneal cavity, where it remains for a short time. The excess fluid, electrolytes, and uremic waste products move through the peritoneal membrane into the dialysate. The solution is then drained from the abdomen by gravity.

---

203 Dialysate solution is infused rapidly into the *_______________

_______________. How is the dialysate removed from the abdomen? _______________.

- - - - - - - - - - - - - - - - -

peritoneal cavity. By gravity

---

204 Each infusion of fresh dialysate is referred to as an exchange. A complete peritoneal dialysis treatment involves about 40 exchanges and may last 48–72 hours.

One peritoneal dialysis treatment lasts _______________ hours and may take _______________ exchanges.

- - - - - - - - - - - - - - - - -

48–72; 40

---

205 Although the peritoneum is semipermeable, it does permit diffusion of proteins into the dialysate. Serum protein deficit may be compensated by increasing the dietary intake of protein.

Increasing dietary intake of protein could compensate for *_______________

_______________.

- - - - - - - - - - - - - - - - -

loss of proteins in dialysate

---

206 Heparin is sometimes added to dialysate to prevent the obstruction in the peritoneal catheter by fibrin or blood.

The purpose of heparin in the dialysate is to *_______________

_______________.

- - - - - - - - - - - - - - - - -

prevent fibrin and clot formation

---

207 The amount of dialysate return by gravity determines the amount of fluid loss from the body. The return must be accurately measured to maintain fluid balance. If the dialysate return is *more* than the infusion fluid is being removed by ultrafiltration, which is *______________________________________

_________________________. When the dialysate return is *less* than the infusion fluid is being (retained/excreted). ____________________.

\- - - - - - - - - - - - - - - - -

the movement of water molecules across semipermeable membrane. Retained

---

208 If 2000 ml of dialysate was infused and 2500 ml returned, 500 ml was

_________________________.

If 2000 ml of dialysate was infused and 1500 ml returned, 500 ml was

_________________________.

\- - - - - - - - - - - - - - - - -

ultrafiltrated or excreted; retained

---

209 Retained dialysate can be reabsorbed and will lead to fluid overload. Symptoms of fluid overload or overhydration are increased blood pressure, dyspnea, constant irritated cough, neck vein engorgement, chest rales, and edema.

A cause of fluid overload during peritoneal dialysis is the *__________________

____.

Name four (4) symptoms associated with fluid overload:

a. *___________________________________.

b. ____________________________.

c. *___________________________________.

d. *___________________________________.

\- - - - - - - - - - - - - - - - -

retention of dialysate fluid
a. increased blood pressure. b. dyspnea. c. neck vein engorgement. d. chest rales. Also edema and irritated cough.

---

210 Excessive ultrafiltration can result in dehydration from the use of 2.5% (398 mOsm/L) and 4.5% glucose dialysate (486 mOsm) or excessive exchanges. Symptoms of dehydration are decreased blood pressure, poor tissue turgor, tachycardia, dry mucous membranes, and hypernatremia.

Dehydration can result from ______________________ or

______________________.

Name three (3) symptoms of dehydration and fluid loss.

a. *______________________.

b. *______________________.

c. *______________________.

- - - - - - - - - - - - - - - - -

4.5% glucose dialysate; excessive exchanges.
a. decreased blood pressure. b. poor tissue turgor. c. tachycardia. Also dry mucous membranes.

Table 68 is a comparison of hemodialysis and peritoneal dialysis and the advantages and disadvantages of their use. This table can be used to summarize the effects of the two types of dialysis. Refer to the table as needed.

TABLE 68. COMPARISON OF HEMODIALYSIS VS PERITONEAL DIALYSIS

Hemodialysis	Peritoneal Dialysis
Rapid removal of fluid.	Fluid is removed slowly.
Potassium is lowered at a faster rate.	Potassium is removed slowly.
Waste products are removed at a faster rate.	Waste products are removed at a slower rate.
Rapid removal of poisonous drugs.	Inefficient for the removal of poisonous drugs.
Treatment time is 3 to 5 hours.	Treatment time is about 48–72 hours.
Requires complex equipment and specialized training.	Uses less complex equipment and less specialized personnel.
Needs vascular access.	Needs no direct access to the bloodstream; causes no blood loss; use for patients with poor vascular access, i.e., children and the elderly.
Requires large doses of heparin.	Requires very small amounts of heparin.
Poorly tolerated by patients with cardiovascular disease.	Minimal stress to patients with cardiovascular disease.
Contraindicated for patients in shock or hypotension.	Can be used for patients with unstable cardiovascular status.
Can be used for patients with abdominal trauma.	Should be avoided for patients who have a colostomy, abdominal adhesions, ruptured diaphragm, or recent surgery.
Cost is high.	Cost effective.
Risk of clotting vascular access.	Risk of peritonitis.

211 Enter HD for hemodialysis and PD for peritoneal dialysis opposite the effect of the appropriate dialysis procedure:

_______ a. Removes fluid rapidly.

_______ b. Removes excess potassium slowly.

_______ c. Removes waste products quickly.

_______ d. Effective for the removal of poisonous drugs.

_______ e. Treatment takes about 48–72 hours.

_______ f. Needs vascular access.

_______ g. Requires large doses of heparin.

_______ h. Poorly tolerated by patients with cardiovascular disease.

_______ i. Cost effective.

_______ k. Risk of peritonitis.

_______ l. Risk of clotting vascular access.

a. HD; b. PD; c. HD; d. HD; e. PD; f. HD; g. HD; h. HD; i. PD; j. HD; k. PD; l. HD

CLINICAL APPLICATIONS

Mr. Tom Smith, age 36, sustained critical injuries in a motor vehicle accident. Assessment in the emergency room indicated that he had suffered blunt trauma to the chest and abdomen and two fractured femurs. He was also in shock. He went to surgery immediately for an exploratory laporatomy in which a spleenectomy, aspiration of a large retroperitoneal hematoma, and repair of a ruptured diaphragm were performed. It was noted that he had bilateral contusion of both kidneys. An open reduction of his fractures was done and he was placed in balanced traction.

His urine after surgery was grossly bloody and the output ranged from 10–20 ml/hr. Mr. Smith experienced hypotension after his injury and surgery. He was transfused with eight (8) units of whole blood. Postoperatively, Mr. Smith was sent to the SICU on ventilatory support with a subclavian line for fluids, nutritional support and hemodynamic monitoring. Laboratory studies were done daily.

212 The intravascular fluid volume lost from Mr. Smith's injuries would cause increased ADH and aldosterone secretion.

Two days after his accident Mr. Smith's serum sodium was increased. Which hormone prevents excretion of sodium? ________________.

\- - - - - - - - - - - - - - - - - -

Aldosterone

Table 69 lists the lab results, urine output, weight, and urinary sodium for Mr. Smith during admission, surgery, and the first four days of hospitalization. Refer to the table as you respond to the frames.

TABLE 69. LABORATORY STUDIES: MR. SMITH

Tests	Admission	Surgery	Day 1	Day 2	Day 3	Day 4
Potassium (serum) (3.5–5.3 mEq/L)	3.6	4.2	6.5	5.1	5.9	6.8
Sodium (serum) (135–145 mEq/L)	138	142	144	143	144	143
Chloride (serum) (98–108 mEq/L)	110	110	110	108	103	104
CO_2 (serum) (22–32 mEq/L)	17	24	27	29	25	21
BUN (10–25 mg/dl)			30	34	57	84
Creatinine (serum) (0.6–1.2 mg/dl)			1.6	4.7	7.2	10.2
Urine output (ml/24 hours)			580	440	320	290
Weight (lb)		185	189	191	196	198
Urine sodium (mEq/L)						93

---

213 Which factors in Mr. Smith's case contributed to the development of the initial phase of ARF? ________________ and ________________.

\- - - - - - - - - - - - - - - - - -

hypotension and hemorrhagic shock

214 A rise in serum potassium was noted the day after surgery. An increase in potassium in this situation was caused by * ______________________.

\- - - - - - - - - - - - - - - - -

massive tissue damage

215 Mr. Smith showed signs of oliguric azotemia on the second and third days. What are the clinical indicators of decreased renal function? * ______________________ ______________________ and * ______________________ ______________________.

\- - - - - - - - - - - - - - - - -

urine output less than 400 ml in 24 hours; elevated BUN and creatinine

216 The nephrologist who cared for Mr. Smith diagnosed his renal problem as acute renal failure secondary to shock and myoglobinuria. Shock and myoglobinuria are listed under which causes of renal failure? ______________________ and ______________________.

\- - - - - - - - - - - - - - - - -

Ischemia; toxicity

217 On the fourth day Mr. Smith's serum potassium measured 6.8 mEq/L. His ECG showed peaked T waves and signs of cardiac irritability. These are symptoms of (hypokalemia/hyperkalemia) ______________________.

\- - - - - - - - - - - - - - - - -

hyperkalemia

---

218 Hyperkalemia can be treated temporarily by methods that will decrease serum potassium. (Refer to Chapter 2 on potassium if needed.)

List four (4) methods used on the treatment of hyperkalemia:

a. *______________________________.

b. *______________________________.

c. *______________________________.

d. *______________________________.

\- - - - - - - - - - - - - - - - -

a. Kayesalate and sorbitol. b. IV sodium bicarbonate. c. 10% calcium gluconate. d. Insulin and glucose

---

219 Because of ECG changes, rapid lowering of Mr. Smith's serum potassium level was necessary. Name two (2) methods that can be used to shift potassium back into the cells.

a. *______________________________.

b. *______________________________.

---

220 Another treatment prescribed for Mr. Smith was a kayexalate-retention enema. Kayexalate, a cation exchange resin, is mixed with sorbitol and given orally or rectally to induce an "osmotic diarrhes." The sodium in kayexalate is exchanged with potassium in the intestines to lower the serum potassium level.

The resin used in excreting potassium from the intestine is

____________________.

\- - - - - - - - - - - - - - - - -

kayexalate

---

221 It was noted that Mr. Smith had had a 13-pound weight increase since admission. Edema was evident in hands, feet, and face. His blood pressure was 160/80 and his central venous pressure/CVP and pulmonary artery wedge pressure/PAWP were elevated. Auscultation of lung fields revealed coarse rales bilaterally. These symptoms indicated what fluid imbalance? ____________________.

\- - - - - - - - - - - - - - - - -

fluid overload OR overhydration

---

222 Mr. Smith's electrolytes for day 4 showed a serum Na 143, Cl 104, and CO_2 21. He had an anion gap of 18. An anion gap greater than 16 mEq/L is indicative of what condition? __________________ (refer to Chapter 3 on anion gap if necessary).

\- - - - - - - - - - - - - - - - - -

acidosis. If you answered metabolic acidosis—OK.

---

223 Once the urine was cleared of blood a random urine sodium test was done. The result of 93 mEq/L means that the kidneys are unable to concentrate urine.

What does this test indicate about the nephrons? * __________________

__________________.

\- - - - - - - - - - - - - - - - - -

they are damaged

---

224 The decision was made that Mr. Smith would have hemodialysis for 3 hours every day. Explain why hemodialysis was chosen instead of peritoneal dialysis.

* ____________________________________

What type of access is need for hemodialysis? __________________.

____________________________________.

\- - - - - - - - - - - - - - - - - -

Hemodialysis removes fluid, corrects acidosis, and lowers potassium rapidly. Peritoneal dialysis is contraindicated after abdominal surgery. Hemodialysis also takes less time.
Vascular access.

---

225 During hemodialysis what method is used to remove excess fluid?

__________________.

What can be done in hemodialysis to lower serum potassium? * __________________

__________________.

\- - - - - - - - - - - - - - - - - -

ultrafiltration; use a low potassium dialysate.

---

226 In hemodialysis the rapid shift of fluids and electrolytes can cause central nervous system disturbances such as agitation, twitching, and seizures. This is known as *disequilibrium syndrome*.

This condition is caused by a *________________.

— — — — — — — — — — — — — — — —

rapid shift of fluids and electrolytes during hemodialysis

---

227 The nurse must observe for signs of disequilibrium syndrome such as

________________, ________________, and ________________.

— — — — — — — — — — — — — — — —

agitation; twitching; seizures

Table 70 lists Mr. Smith's test results on days 6, 14, 35, 37, 40, and 47. He was given hemodialysis until the 49th day. Refer to the table as needed.

TABLE 70. LABORATORY STUDIES II: MR. SMITH

Tests	Day 6	Day 14	Day 35	Day 37	Day 40	Day 47
WBCC (5000–10,000 mm^3)		38,000				
Potassium (serum) (3.5–5.3 mEq/L)	5.3	5.2	3.8	3.9	3.8	4.8
Sodium (serum) (135–145 mEq/L)	142	135	134	128	133	147
Calcium (serum) (8.5–10.5 mEq/L)		8.5		8.6		
Chloride (serum) (98–108 mEq/L)	102	93	99	90	97	112
Phosphorus (serum) (2.5–4.5 mg/dl)		9.7		4.2		
CO_2 (serum) (22–32 mEq/L)	25	25	21	21	22	17
BUN (10–25 mg/dl)	71	110	86	89	86	140
Creatinine (serum) (0.6–1.2 mg/dl)	9.2	11.1	7.1	6.5	5.8	4.3
Weight (lb)	193	180	171	169	165	164
Intake (ml/24 hr)	600	600	600	600	3000	4955
Output (ml/24 hr)	170	0	250	550	3000	4650

---

228 On day 14 Mr. Smith became anuric. He also developed a severe infection that caused catabolism and increased BUN, creatinine, and WBC. Hemodialysis time was increased to 5 hours per day. From Table 70 what problems were controlled with dialysis on day 14?

a. ______________________.

b. ______________________.

c. ______________________.

- - - - - - - - - - - - - - - - - -

a. acidosis. b. hyperkalemia. c. fluid overload

---

229 Mr. Smith's serum phosphorus was very high (9.7 mg/dl). When the serum calcium and phosphate are multiplied the Ca X P product is 83.1. This is an indicator of * ______________________.

- - - - - - - - - - - - - - - - - -

metastatic calcification

---

230 The nephrologist ordered a phosphate binding agent to lower serum phosphorus. This agent contains aluminum which attracts and binds phosphorus compounds in the intestines for excretion in the stool.

The drug/agent that lowers serum phosphate is ______________________ ______________________.

- - - - - - - - - - - - - - - - - -

a phosphate binding agent OR an agent that contains aluminum; e.g. Amphojel or Basaljel

---

231 On day 35 Mr. Smith began to have urine output and hemodialysis was decreased to 4 hours per day. By day 37 his urine output had increased further and dialysis times were reduced again.

Note: Electrolytes stabilize to low normal levels in daily dialysis therapy.

His intake and output were about the ______________. His serum phosphorus was within high normal limits because of the ______________ ______________________.

- - - - - - - - - - - - - - - - - -

same; phosphate binding agent (phosphate binders)

---

232 On day 40 Mr. Smith's urinary output was 3000 ml/24 hr. This indicated that he was in which phase of ARF? ______________. What is the most serious complication of this phase? ______________.

— — — — — — — — — — — — — — — — —

diuretic; dehydration

---

233 By day 47 Mr. Smith's urine output continued to be high, but his serum sodium and BUN were also high. Skin turgor was poor, mucous membranes dry, and he complained of thirst. His blood pressure was 120/60 and his pulse, 92. These symptoms are indicators of what type of fluid imbalance? ______________.

— — — — — — — — — — — — — — — — —

dehydration

---

234 By day 55 Mr. Smith's was no longer azotemic and his urine output and electrolytes were in normal range. This phase of ARF is ______________.

— — — — — — — — — — — — — — — — —

recovery

CASE REVIEW

Mrs. Alice Grady, age 68, had a 3-year history of renal insufficiency from glomerulonephritis and a history of several myocardial infarctions. Mrs. Grady was admitted to the hospital for shortness of breath but no dyspnea. Her laboratory results on admission were Hgb 6.1 g/dl, Hct 19%, BUN 78 mg/dl, creatinine, 5.2 mg/dl, serum CO_2, 13 mEq/L, serum potassium, 5.6 mEq/L, serum sodium, 124 mEq/L, serum calcium, 8.4 mg/dl, and serum phosphorus, 5.2 mg/dl. Physical assessment revealed edema of face, hands, and lower legs. Lung sounds indicated bilateral course rales but no frothy sputum. Blood pressure was 170/98 and her jugular venous pressure was elevated. She had noted a weight gain of 5 pounds in the preceding 3 days and her urine output was approximately 500 ml per day. Mrs. Grady complained of thirst.

1. Chronic renal failure usually develops over ______________. A progressive loss of ______________ is the result.

2. What is the cause of Mrs. Grady's chronic renal failure? ____________________.
3. Mrs. Grady's low hemoglobin and hematocrit indicate what clinical condition?

   ____________________. Explain why? * ____________________

   ____________________.

4. Her BUN and serum creatinine were (elevated/decreased) ____________________,

   which are indicative of ____________________.

5. Mrs. Grady's serum CO_2 was (elevated/decreased) ____________________,

   which is indicative of ____________________.

6. Her serum potassium was 5.6 mEq/L. Normal range is ____________________. Give two (2) reasons why hyperkalemia occurs in renal failure.

   a. ____________________.

   b. ____________________.

7. Her serum sodium was (elevated/decreased) ____________________. Give two (2) reasons for her hyponatremia.

   a. ____________________.

   b. ____________________.

8. Peritoneal dialysis was chosen to treat Mrs. Grady's uremia. Why would peritoneal

   dialysis be best for her? Explain. * ____________________

   ____________________.

9. What symptoms indicate signs of fluid overload?
   a.
   b.
   c.
   d.
10. After trocar insertion, the first exchange of peritoneal dialysate was a 4.25% dextrose dialysate with no potassium. The 4.25% dextrose dialysate is

    ____________________. It is used for ____________________.

    Dialysate without potassium increases * ____________________

    ____________________.

11. When dialysate infuses into the peritoneum, it can push the diaphragm upward.

    The nurse must assess the patient for signs of * ____________________

    ____________________.

12. Excessive use of 4.25% dextrose dialysate can lead to ____________________.

13. An essential intervention during peritoneal dialysis to maintain fluid balance is * ______________________________.

14. Retention of dialysate during exchanges can lead to * ______________________________.

15. On the second day of peritoneal dialysis Mrs. Grady's potassium was 2.8 mEq/L, which indicates ______________. Potassium must be added to the ______________.

16. Mrs. Grady's phosphorus was slightly elevated. The drug given to decrease serum phosphorus was ______________.

17. To prevent fluid overload in the future it was important to determine Mrs. Grady's * ______________________________.

- - - - - - - - - - - - - - - - -

1. periods of time; nephrons
2. Glomerulonephritis
3. Anemia. Kidneys are not producing erythropoietin to stimulate the bone marrow to build red blood cells
4. elevated. Azotemia
5. decreased. Acidosis
6. 3.5–5.3 mEq/L a. Potassium retention b. Massive tissue destruction
7. decreased. a. excessive fluid intake; b. metabolic acidosis
8. Minimal stress to patients with cardiovascular disease; frequently used for patients with unstable cardiovascular disease.
9. a. Weight gain. b. Edema. c. Increased blood pressure. d. Bilateral rales. Also elevated jugular venous pressure.
10. hyperosmolar; ultrafiltration; diffusion of potassium into the dialysate.
11. respiratory distress
12. dehydration
13. strict measurement of dialysate from each exchange
14. fluid overload
15. hypokalemia; dialysate
16. phosphate binding agents; e.g., Amphojel
17. dry weight

NURSING DIAGNOSES AND NURSING ACTIONS

1. *Alteration in renal tissue perfusion related to ischemia, or obstruction.*

   Monitor patient for conditions that predispose to alterations in renal perfusion. Monitor vital signs, especially B/P, for hypo-hypertension.

Monitor urine output; if it is less than 30 ml/hr call the physician.
Check specific gravity/SG; SG detects changes in the concentration that result from decreased intravascular volume and may lead to decreased renal perfusion.
Monitor CVP and PAWP for early signs of decreased cardiac output and decreased perfusion.
Check abdominal girth. Monitor for fluid shifts by noting rigidity or girth changes of the abdomen.
Check for signs of edema.

2. *Potential for fluid volume excess related to decrease urine output.*

Obtain daily weights and pre- and postdialysis for determining fluid volume changes. One liter of fluid weighs approximately 2.2 pounds.
Monitor accurate measurements of peritoneal dialysate return to determine gain in body fluid.
Check blood pressure and jugular venous pressure, which are effective indicators of fluid volume excess.
Assess for edema by checking the preorbital, hands, feet, and sacral for swelling.
Ausculate the lungs for chest rales (signs of pulmonary edema).
Monitor strict intake and output.
Give medications with a minimum of allowable fluid to prevent fluid excess.
Encourage patient to follow a restricted sodium diet and fluid limitation.

3. *Potential for fluid volume deficit related to excess ultrafiltration or diuresis.*

Monitor vital signs, especially pulse and blood pressure.
Excessive ultrafiltration during hemodialysis causes hypotension. If blood pressure postdialysis is low it may cause decreased blood flow through the vascular access and result in clotting. Excessive ultrafiltration from peritoneal dialysis can produce hypovolemia and hypernatremia because of the rapid movement of water across the peritoneum.
Monitor urinary output in the diuretic phase of ARF. Excessive urine output can deplete intravascular volume.
Intake must be increased to match output.

4. *Potential for electrolyte imbalance: a. Hyperkalemia related to potassium retention or massive tissue damage. b. Hypocalcemia and hyperphosphatemia related to retention of phosphate.*

4a. Monitor conditions tnat may lead to hyperkalemia. Any condition that causes tissue destruction, burns, crushing injuries, or infections can liberate potassium from the cells into the extracellular fluid. Blood transfusions that contain some lysed cells (red blood cell breakdown) can increase the potassium level in the blood.
Check ECG for peaked T waves, an early sign of hyperkemia.
Note patient's complaint of tingling and numbness in extremities.

4b. Check serum calcium and phosphorus levels.

Administer phosphate binding agent as ordered to control decrease in serum phosphorus level. A low serum calcium level could be caused by hyperphosphatemia, restricted calcium intake in the diet, and/or alterations in Vitamin D metabolism needed for intestinal absorption of calcium. Too much aluminum-phosphate binding drug could cause hypophosphatemia.

Assess the symptoms of tetany when the serum calcium level is low and acidosis is corrected. Checking for serum calcium deficits will indicate when Vitamin D supplements should be initiated.

5. *Potential for clotting of access site related to inflammation, mechanical factors, decreased perfusion, and temperature.*

Check for the patency of the vascular access site used for hemodialysis tid. A surgically recreated vascular access should have bruits and pulsation on auscultation. Blood can be aspirated and infused through femoral and subclavian catheters.

Use an aseptic technique in caring for the access site.

Check for swelling, redness, and drainage at the access site.

Check for mechanical factors, such as kinking of the tubing, displacement of the catheter, and lying on the access site which could cause poor blood flow.

Decreased perfusion can cause stasis in the vascular access which leads to clotting. Extreme cold may produce enough vasoconstriction to decrease blood flow and cause clotting of the vascular access.

Observe the peritoneal catheter site for crusting or redness.

Check the catheter for smooth flow.

Teach the patient how to care for the access site.

6. *Potential alteration in patterns of urinary elimination related to reduced glomerular filtration rate and decreased functioning nephrons.*

Teach patients with renal insufficiency to monitor their fluid intake and urine output. A decreased output may signify decreased function. Urinary output should be at a minimum of 30 ml/hr. The physician must be notified if the urine output drops.

7. *Potential for injury related to disequilibrium syndrome.*

Assess new dialysis patients and catabolic pateints in ARF for signs of central nervous system dysfunction that may lead to disequilibrium syndrome. Observe for tremors, irritability, confusion, and seizures.

Protect the patient by restraint when appropriate.

Keep an oral airway at bedside in anticipation of seizures.

Provide a quiet environment and limit noxious stimuli.

Explain all procedures clearly and concisely.

8. *Potential for alteration in bowel elimination and constipation related to intake of phosphate binding agents (Amphojel) and fluid limitation.*

Monitor for bowel frequency and record output.
Assess for bowel sounds and signs of nausea, vomiting, or distention. Notify the physician and suggest a stool softener if patient continues to have "hard stools."
Encourage activities that promote peristalsis.

9. *Potential for infection related to a depressed response of the immune system.*

Monitor temperature and WBCs for elevation.
Note any discharge from access sites or wounds. Culture potentially infected sites.
Encourage patient to breathe deeply and cough, especially if on bedrest.
Teach patient good hygiene, to avoid crowds, and to obtain yearly innoculation for flu viruses.

10. *Potential for bleeding tendency, related to a decreased platelet count, heparin administration, and ulcer formation.*

Monitor all stools or vomitus for signs of occult blood.
Teach patient to avoid straining if constipated. Encourage use of lubrications for bowel movements.
Instruct patient to use a soft toothbrush.
Encourage patient to avoid activities that may cause bruising. Capillary fragility in renal disease causes increased bleeding in body tissues.
Monitor signs of bleeding postdialysis.

11. *Potential alteration in respiratory pattern related to dialysate infusion.*

Assess for dyspnea when administering solutions for peritoneal dialysis. Rapid infusion of dialysate can push the diaphragm upward, thus decreasing the area for lung expansion.

BURNS AND BURN SHOCK

235 Following a burn, there is an extracellular fluid volume shift in which fluid and electrolytes shift from the intravascular space (plasma) to the interstitial spaces of the burned areas. (See Chapter 5 on ECF shift, if necessary.) This would result in a(n) (increase/decrease) ______________ of circulating plasma.

- - - - - - - - - - - - - - - - - -

decrease

---

236 With a decrease in circulating plasma volume, burn shock occurs. It is characterized by restlessness, confusion, tachycardia, decreased blood pressure, decreased urine output, metabolic acidosis, and paralytic ileus.

Burn shock results from *______________________________.

\- - - - - - - - - - - - - - - -

decrease in circulating plasma volume

PHYSIOLOGIC FACTORS

Table 71 describes the physiologic factors associated with burns. There is an increased capillary permeability, increased serum osmolality, increased circulatory resistance, decreased cardiac output, decreased renal function, increased hemolysis, electrolyte imbalances, acidosis, increased hematocrit, and decreased protein level. Study the table carefully, and refer to it as needed.

TABLE 71. PHYSIOLOGIC FACTORS ASSOCIATED WITH BURNS AND BURN SHOCK

Physiologic Factors		Rationale
Capillary permeability	Increased ↑	There is a rapid shift of fluid and protein from the intravascular space (vessels) to the burned site. If more than 25% of the total body surface is burned, fluid (edema) will accumulate in burned and unburned tissue spaces. Fluid shift to the burned site and tissue spaces is referred to as *fluid shift to the third space.* The fluid is nonfunctional, which causes vascular fluid deficit (hypovolemia). This is referred to as *burn shock.* Most of the fluid shift occurs during the first 18 hours, but could persist for 48 hours postburn. Approximately40–50% of vascular fluid can be lost to burned site and tissue spaces within the first 18 hours.
Serum osmolality	Increased ↑	Hemoconcentration results from loss of vascular fluid. The serum osmolality is > 295 mOsm/L since the proportion of solutes is greater than water.
Circulatory resistance	Increased ↑	Hypovolemia and decreased blood pressure are sensed by pressoreceptors in the aorta and carotid bodies and in the sympathetic nervous system to cause vasoconstriction in order to increase blood flow to the vital organs, i.e., heart, brain, and lungs.

TABLE 71. (Continued)

Physiologic Factors		Rationale
Cardiac output	Decreased ↓	With more than 40% of the total body surface area burned, cardiac output could be decreased 50% or more due to hypovolemia. Cardiac output = stroke volume × heart rate. Tachycardia is a compensatory response. Beta receptors in the myocardium increase heart rate.
Hematocrit	Increased ↑	Hematocrit level is elevated due to hemoconcentration from hypovolemia. Anemia is present postburn due to blood loss at burned site and hemolysis but it is not assessed until the patient is adequately hydrated.
Hemolysis (destruction of cells)	Increased ↑	Hemolysis causes a liberation of hemoglobin (free hemoglobin).
Renal function	Decreased ↓	Severe decreased blood volume (hypovolemia) causes a fall in blood pressure and oliguria or anuria. Systolic blood pressure < 60 mm Hg can cause renal insufficiency. Excess ADH (SIADH) is secreted during the first 48 hours, which causes water to be reabsorbed from the rneal tubules, and urine output to be decreased. Free hemoglobin from hemolysis is excreted by the kidneys as red-color urine and can cause renal damage.
Electrolyte imbalances Serum Sodium (hyponatremia)	Decreased ↓	Sodium enters the edema fluid in the burned area, lowering the sodium content of the vascular fluid. Hyponatremia may continue for days to several weeks because of sodium loss to edema fluid, sodium shifting into cells, and later, due to diuresis. After 48 hours, fluid shifts from the burned and interstitial spaces to the vascular space. Sodium and excess fluid are excreted by the kidneys.
Cellular potassium	Decreased ↓	Potassium is lost from the cells.

TABLE 71. (Continued)

Physiologic Factors		Rationale
Serum potassium (hyperkalemia)	Increased ↑	Potassium leaves the cells as sodium shifts in, and hyperkalemia can occur if urine output is decreased. Serum potassium value may vary from normal to a deficit or an excess depending on the urine output.
Serum calcium (hypocalcemia)	Decreased ↓	Hypocalcemia occurs because of calcium loss to edema fluid at the burned site. Multiple infusions of citrated blood decreases the serum calcium level.
Metabolic acidosis	Increased ↑	Burns cause cellular breakdown, and the cells release acid metabolites (lactic acid). Bicarbonate loss accompanies loss of sodium.
Serum protein	Decreased ↓	Protein is lost to burned site due to increased capillary permeability. Serum protein level will remain low until healing occurs.

---

237 Capillary permeability is (increased/decreased) __________ during the first 48 hours postburn. If *less* than 25% of the total body surface is burned, fluid accumulates in the:

( ) a. burned site
( ) b. unburned tissue spaces

- - - - - - - - - - - - - - - -

increased; a. X; b. –
If more than 25% of body surface area is burned, fluid shifts to the burned site and to unburned tissue spaces.

---

238 What is meant by fluid shift to the third space? * ________________________________

____.

Most of the fluid shift can occur during the first ______________ hours.
If 50% of the vascular fluid is lost to burned and unburned tissue spaces, severe

(hypervolemia/hypovolemia) ____________________ occurs.

— — — — — — — — — — — — — — — — —

Fluid shifts from the vascular to the tissue (burned or unburned) spaces; 18; hypovolemia

---

239 Erythrocytes or red blood cells are destroyed by hemolysis (destruction of red blood cells). The free hemoglobin released from the red blood cells may

cause * ____.

Erythrocytes are destroyed as a result of:
( ) a. increased plasma protein
( ) b. hemolysis

— — — — — — — — — — — — — — — — —

renal damage; a. –; b. X

---

240 Hemoconcentration is present in early burns. Explain why.

* ____.

The serum osmolality would be (increased/decreased) ____________________

and the hematocrit would be (increased/decreased) ____________________.

— — — — — — — — — — — — — — — — —

Fluid leaves the vascular (intravascular) space causing hypovolemia or dehydration; increased; increased

---

241 As a result of burned shock, there is (increased/decreased) ________________

circulatory resistance and (increased/decreased) ________________ cardiac output.

— — — — — — — — — — — — — — — — —

increased; decreased

---

242 After rehydration is established, the diminished number of red blood cells (erythrocytes) becomes apparent; thus anemia (does/does not) ______ occur. Hemoconcentration is present (after/before) ______ rehydration.

- - - - - - - - - - - - - - - - - -

does; before

---

243 Renal function is decreased because of:
( ) a. Hypervolemia
( ) b. Hypovolemia
( ) c. Systolic BP $<$ 60 mm Hg
( ) d. SIADH (syndrome of inappropriate antidiuretic hormone secretion)

- - - - - - - - - - - - - - - - - -

a. –; b. X; c. X; d. X

---

244 Electrolyte imbalances occur postburn. Hyponatremia is common because:
( ) a. Sodium is lost to edema fluid at the burned site
( ) b. Sodium shifts into cells
( ) c. Diuresis 48 hours postburn

- - - - - - - - - - - - - - - - - -

a. X; b. X; c. X

---

245 Intracellular potassium is lost from the cells and is replaced by sodium. Decreased urine output (oliguria) usually occurs in early postburn. The serum potassium level would be ______. Why? * ______ ______.

- - - - - - - - - - - - - - - - - -

increased; Kidneys excrete 80–90% of potassium loss

---

246 Serum calcium deficit may result from *______________________________

____.

Multiple infusions of citrated blood could cause (hypocalcemia/hypercalcemia)

______________________.

- - - - - - - - - - - - - - - - - -

calcium loss to edema fluid at the burned site; hypocalcemia

---

247 Serum protein is (increased/decreased). ______________________ Why?

*____

____.

- - - - - - - - - - - - - - - - - -

decreased. Protein leak from vascular to the burned site because of increased capillary permeability

---

248 Metabolic acidosis can result from:

( ) a. loss of serum bicarbonate
( ) b. increase in acid metabolites
( ) c. excess vascular fluid

- - - - - - - - - - - - - - - - - -

a. X; b. X; c. –

---

249 Indicate which of the following may occur as the result of burns. Correct the wrong statements.

( ) a. Increased capillary permeability
( ) b. Elevated serum osmolality in early postburn
( ) c. Hypovolemia
( ) d. Hemolysis
( ) e. Hemoconcentration before hydration
( ) f. Metabolic alkalosis
( ) g. Increased serum protein
( ) h. Hyperkalemia with oliguria
( ) i. Hyponatremia
( ) j. Hypercalcemia

- - - - - - - - - - - - - - - - - -

a. X; b. X; c. X; d. X; e. X; f. – (Metabolic acidosis); g. – (Decreased serum protein); h. X; i. X; j. – (Hypocalcemia)

DEGREES OF BURNS

Table 72 names the three degrees of burns, the affected tissues, and the characteristics of the burns. With first-degree burns, the epidermis—or the outer layer of skin—is involved. With second-degree burns, the dermis—or the true skin—is involved; and with third-degree burns, the subcutaneous tissues—or fatty tissues—are involved.

After studying this table carefully, proceed to the frames that follow. Refer to this table as needed.

TABLE 72. DEGREE OF BURNS AND THEIR CHARACTERISTICS

	Characteristics		
Degree	**Surface**	**Color**	**Pain**
First degree (epidermis)	Dry, no blisters	Erythema (redness of skin)	Painful, hyperesthetic (very sensitive)
Second degree (dermis)	Moist, blisters	Mottled red	Painful, hypesthetic (less sensitive)
Third degree (subcutaneous tissues)	Dry	Pearly white or charred	Little pain; anesthetic (not sensitive)

Reference: *Therapeutic Notes*, October 1963, p 236.

250 The three degrees of burns are

1. * ______________________.
2. * ______________________.
3. * ______________________.

— — — — — — — — — — — — — — —

1. first degree
2. second degree
3. third degree

---

251 Define the following terms:

Epidermis. *____________________________.

Dermis. *______________________________.

Subcutaneous tissue. *______________________.

Give the names of the tissue involved according to the degree of the burn.

First degree. _________________________.

Third degree. *________________________________.

Second degree. ___________________________.

\- - - - - - - - - - - - - - - - -

Epidermis. The outer layer of skin
Dermis. True skin
Subcutaneous tissue. Fatty tissue
epidermis; subcutaneous tissue; dermis

---

252 The characteristics commonly seen with first-degree burns are: the skin surface is __________, the color is ______________________, and the pain is ____.

The characteristics commonly seen with second-degree burns are: the skin surface is *________________________, the color is *_______________, and the pain is ____.

The characteristics commonly seen with third-degree burns are: the skin surface is __________, the color is *____________________, and the pain is ________________________________.

\- - - - - - - - - - - - - - - - -

dry
red (erythema)
hyperesthetic (very sensitive)
moist with blisters
mottled red
hypesthetic (less sensitive)
dry
pearly white or charred
anesthetic (not sensitive)

---

253 Place FD for first-degree, SD for second-degree, and TD for third degree burns beside their characteristics:

_______ Skin surface is dry and pearly white or charred

_______ Skin surface is red and dry

_______ Skin surface is moist with blisters

_______ Painful and hypesthetic

_______ Little pain and anesthetic

_______ Painful and hyperesthetic

_______ Skin surface mottled red

- - - - - - - - - - - - - - - -

TD; FD; SD; SD; TD; FD; SD

METHODS USED TO DETERMINE PERCENTAGE OF TOTAL BODY SURFACE BURNS

254 There are several methods used for determining the percentage of total body surface burns. The Berkow formula determines the percentage according to ages and 19 surface body areas. Another is the Lund and Browder chart, which estimates the body surface areas in smaller proportion, i.e., upper arm 2%, forearm 1½%, and hand 1½%. The Rule of Nines is frequently used as a quick method of estimation since it can be easily recalled.

The three methods used to determine the percentage of total body surface area that has been burned are:

1. *____.

2. *____.

3. *____.

- - - - - - - - - - - - - - - -

1. Berkow formula; 2. Lund and Browder chart; 3. Rule of Nines

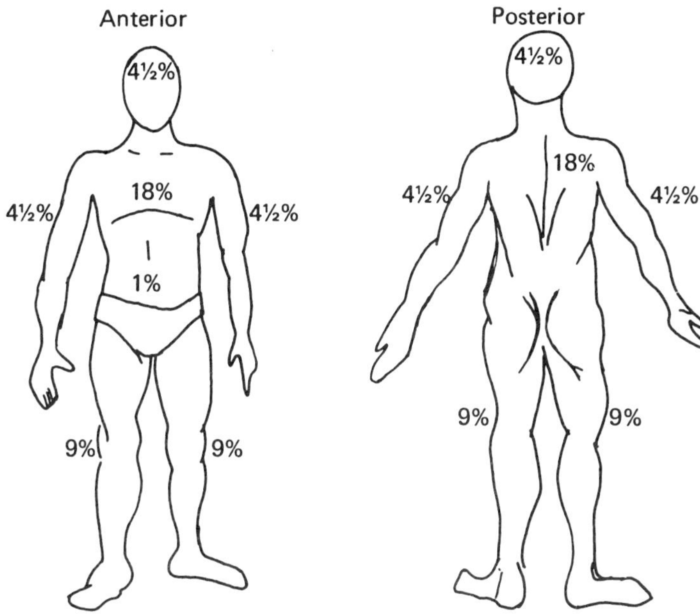

Diagram 11. Rule of Nines for estimation of body surface.

Diagram 11 explains the Rule of Nines in the estimation of amount and areas of body burns. The Rule of Nines uses 9% or multiples thereof in calculating the burned body surface. The five main regions in the estimation of burned surface are underlined. Be sure to know the five regions and the percentages of each. Refer to this diagram as needed.

Region	Percentage of Body Surface (%)
Head and neck	9
1. Anterior head and neck	
2. Posterior head and neck	
Upper extremities	18
3. Right arm—anterior and posterior	
4. Left arm—anterior and posterior	
Trunk and buttocks	36
5. Anterior surface	
6. Posterior surface	
Lower extremities	36
7. Right leg and thigh—anterior and posterior	
8. Left leg and thigh—anterior and posterior	
9. *Perineum and genitalia*	1

---

255 The Rule of Nines and the degree of burns are used in planning parenteral therapy for burned persons.

What are the five main regions of the body used in the Rule of Nines?

1. *____.
2. *____.
3. *____.
4. *____.
5. *____.

\- - - - - - - - - - - - - - - - -

head and neck
upper extremities
trunk and buttocks
lower extremities
perineum and genitalia

---

256 In planning parenteral therapy for burns, the *________________________ and *________________________ are used.

The estimated percentage used for the anterior and posterior surfaces of *each* arm is ____________ %.

The estimated percentage used for the anterior and posterior surfaces of the head and neck is ____________ %.

\- - - - - - - - - - - - - - - - -

Rule of Nines
percentage of burned surface
9
9

---

257 The estimated percentage used for the anterior surface of the trunk is ________ %.

The estimated percentage used for the anterior and posterior surfaces of *each* thigh and leg is ________ %.

\- - - - - - - - - - - - - - - - -

18; 18

Complete the percentages on the following chart:

Region	Percentage of Body Surface (%)
Anterior head and neck	________
Left anterior and posterior arm	________
Anterior and posterior surfaces of trunk and buttocks	________
Anterior surface of right thigh and leg	________
Perineum and genitalia	________
Anterior right arm	________
Posterior trunk and buttocks	________
Anterior and posterior surfaces of lower extremities	________

— — — — — — — — — — — — — — — —

4 1/2; 9; 36; 9; 1; 4 1/2; 18; 36

CLASSIFICATION OF BURNS

Table 73 differentiates between minor, moderate, and critical burns according to the degrees and percentage of burns. Know the three classifications oт burns, the degrees, and the percentages. Refer to the table as needed.

TABLE 73. CLASSIFICATION OF BURNS

Minor Burns	Moderate Burns	Critical Burns
First degree	Second degree of 15–30% of body surface	Second degree of over 30% of body
Second degree of less than 15% of body surface	Third degree of less than 10% of body surface except hands, face, feet, or genitalia	Third degree of more than 10% of body surface or of hands, face, feet, or genitalia
Third degree of less than 2% of body surface		

Reference: *Therapeutic Notes,* October 1963, p 236.

---

258 The three classification of burns are:

1. *____________________.
2. *____________________.
3. *____________________.

\- - - - - - - - - - - - - - - - - -

1. minor burns
2. moderate burns
3. critical burns

---

259 To be classified as *minor burns,* the skin surface involved must be first degree, or second degree of less than __________ % of body surface, or third degree of less than __________ % of body surface.

\- - - - - - - - - - - - - - - - - -

15; 2

---

260 To be classified as *moderate burns,* the skin surface involved must be second degree of __________ to __________ % of body surface, or third degree of less than __________ % of body surface, provided that the hands, face, feet, or genitalia are not burned.

\- - - - - - - - - - - - - - - - - -

15; 30;
10

---

261 To be classified as *critical burns,* the skin surface involved must be second degree of over __________ % of body surface, or third degree of more than __________ % of body surface, or of the hands, face, __________, or __________.

\- - - - - - - - - - - - - - - - - -

30; 10
feet; genitalia

---

262 Place the word *minor* for minor burns, *moderate* for moderate burns, and *critical* for critical burns beside the following statements:

__________ Burns of the hands, face, feet, or genitalia

__________ Third-degree burns of less than 10% of body surface

__________ Third-degree burns of more than 10% of body surface

__________ Second-degree burns of 15-30% of body surface

__________ Second-degree burns of over 30% of body surface

__________ Second-degree burns of less than 15% of body surface

__________ Third-degree burns of less than 2% of body surface

- - - - - - - - - - - - - - - - - -

critical; moderate; critical; moderate; critical; minor; minor

---

CLINICAL APPLICATIONS

---

263 Minor burns are generally treated in the physician's office or emergency room, and the person seldom needs hospitalization.

Moderate burns and burns (first and second degree) greater than 20% of total body area should receive intravenous fluids.

Minor burns and burns less than 20% of total body surface area frequently (do/do not) __________ need intravenous fluids unless the person is a child. Children with burns greater than 10% of total body surface area do need intravenous fluids.

- - - - - - - - - - - - - - - - - -

do not

---

264 Persons with burns involving 20-40% of body surface require careful and proper intravenous replacement therapy for survival.

In those with over 50% of the body surface involved, the mortality rate is high regardless of careful and proper intravenous therapy.

Individuals with 20-40% of body surface burns, having received careful and proper fluid replacement, have a (good/poor) __________ prognosis, whereas those with 50% or more have a (good/poor) __________ prognosis.

- - - - - - - - - - - - - - - - - -

good; poor

---

265 Various formulas have been devised and used as a basis for initiating therapy in the treatment of burns.

Cope and Moore state that accurately measured hourly urine flow is an important index for determination of adequate intravenous therapy. For all severely burned persons, an indwelling catheter would be advisable to obtain * ______________________________.

- - - - - - - - - - - - - - - - -

hourly urine output

---

266 The desired rate of urine flow is 30–50 ml (cc) per hour. Less than 25 ml of urine output per hour for an adult would indicate insufficient fluid intake or kidney dysfunction.

After 48 to 72 hours, the urine output could be 100 ml or more per hour from diuresis due to fluid shift from the burned and unburned tissue areas to the intravascular space (vessels). For a burned person, the desired urine output per hour is * ____________. Less than 25 ml per hour would indicate:

( ) a. too much fluids
( ) b. not enough fluids
( ) c. kidney dysfunction

- - - - - - - - - - - - - - - - -

30–50 ml; a. –; b. X; c. X

---

267 During the first 48 hours, fluid replacement should be at least three times the urine output because of fluid shift to the burned and unburned tissue areas.

When intravenous flow rate is increased to correct hypovolemia and to increase urine output, the nurse should observe for what type of fluid imbalance?

______________________________.

Give three symptoms of overhydration.

1. ______________________________.
2. ______________________________.
3. ______________________________.

- - - - - - - - - - - - - - - - -

overhydration; 1. constant, irritated cough; 2. dyspnea; 3. neck vein engorgement, and also chest rales

---

268 To determine whether poor urine output is due to renal damage or inadequate fluid intake, a Water Tolerance Test can be used. This test consists of giving 1000 ml of fluid in ½ hour.

A failure to note an increase in urine output would indicate *____.

- - - - - - - - - - - - - - - - -

renal damage

---

269 Persons who are burned complain frequently of thirst. In order not to hydrate orally with plain water, the following oral liquid can be prepared in quenching thirst.

1 teaspoon of $NaHCO_3$, commonly known as soda
1 teaspoon of NaCl, commonly known as salt
} in 1 quart of cold water

One quart of this oral solution may be given per day on physician's orders. For mild burns, Gatorade may be used for thirst.

Massive amounts of plain water should not be given to prevent:

( ) a. edema
( ) b. water intoxication
( ) c. dehydration

The desired oral solution to be given for thirst consists of *____.

- - - - - - - - - - - - - - - - -

a. –; b. X; c. –
1 teaspoon of soda and 1 teaspoon of salt in 1 quart of water

---

270 Later, high protein liquids, between meals or at mealtime, are helpful for cell reconstruction. The following protein liquids can be used:

( ) a. eggnog
( ) b. ginger ale
( ) c. milkshake
( ) d. Coca Cola

- - - - - - - - - - - - - - - - -

a. X; b. –; c. X; d. –

---

271 A very high hematocrit reading indicates *________________.

Many physicians prefer to maintain the hematocrit (Hct) at 45 or above for the first 48 hours after the burn so that in rehydration, the hematocrit will:

( ) a. drop very low
( ) b. return to a normal range
( ) c. show a marked increase

- - - - - - - - - - - - - - - - -

dehydration or hemoconcentration
a. –; b. X; c. –

---

272 The greatest fluid shift occurs during the first 18 hours after a burn and reaches its peak in 48 hours. Therefore, the critical period for fluid and electrolyte replacement is:

( ) a. the first 36 hours
( ) b. the first 48 hours
( ) c. the first 72 hours

- - - - - - - - - - - - - - - - -

a. –; b. X; c. –

---

273 A hazard in burn cases is infection. This will frequently delay the reabsorption of edema fluid from the site of the burn.

Surgical aseptic (sterile) techniques should be employed to reduce the possibility of ________________.

Infection will cause the edematous fluid to be reabsorbed (more slowly/more quickly) ________________.

- - - - - - - - - - - - - - - - -

infection
more slowly

274 After 48 hours, capillary permeability lessens, fluid reabsorption begins, and edema starts to subside. This is considered to be the *stage* of *diuresis.* Stage of diuresis generally begins after two days; however, it may take as long as two weeks before the stage develops because of the severity of the burns. Explain what happens when the stage of diuresis begins.

a. Capillary permeability ______________________.

b. Fluid ______________________ begins.

c. Edema starts to ______________________.

- - - - - - - - - - - - - - - -

a. lessens or decreases
b. reabsorption
c. subside

---

275 Extracellular fluid volume shift from the interstitial space to the intravascular space occurs during the stage of diuresis. This most likely occurs ____________ hours after being burned.

- - - - - - - - - - - - - - - -

48. This may not occur for several days or may take up to two weeks, according to the severity of the burn.

---

276 After 48 hours, IV therapy is frequently restricted or decreased, providing the serum sodium and potassium levels are near normal.

Continuous IV therapy may result in overhydration. This can be hazardous since it can overload the circulation, causing pulmonary edema and cardiac failure.

After 48 hours, intravenous administration is ______________________.

Overloading the circulation can result in:

( ) a. pulmonary edema
( ) b. gastritis
( ) c. cardiac failure
( ) d. pancreatitis

- - - - - - - - - - - - - - - -

decreased
a. X; b. –; c. X; d. –

FLUID CORRECTION FOR BURN SHOCK

There are many formulas for calculating fluid replacement for the first 48 hours after burns. Table 74 gives three types of formulas. Brooke and Evans are similar except for the amount of colloid and electrolyte replacement. Parkland's formula does not call for colloid and free water the first 24 hours. Parkland supporters believe that early colloid replacement shifts to the burned site and that by giving lactated Ringer's infusions cardiac output will increase and cell function will be restored. Refer to Table 74 as needed.

TABLE 74. FORMULAS FOR FLUID REPLACEMENT—48 HOURS FOR BURN SHOCK

Name	First 24 hours	Second 24 hours
Brooke Army Hospital	Colloid[a]: 0.5 ml per kg X % of burned area Electrolyte [b]: 1.5 ml per kg X % of burned area Water: 2000 ml 5% D/W	One-half (½) amount of colloid and electrolyte of the first 24 hours Water: 2000 ml 5% D/W
Evans	Colloid: 1 ml per kg X % of burned area Electrolyte: 1 ml per kg X % of burned area Water: 2000 ml 5% D/W	One-half (½) amount of colloid and electrolyte of the first 24 hours Water: 2000 ml 5% D/W
Parkland	Colloid: None Electrolyte: 4 ml per kg X % of burned area Water: None	Colloid: 20-60% of calculated plasma volume Water: 2000 ml 5% D/W

Note: Over 50% of body surface burns are calculated at 50% burns for fluid replacement purposes.

[a]Colloid used: blood, dextran, plasma, albumin.

[b]Electrolyte used: lactated Ringer's or normal saline.

1 kg = 2.2 pounds.

277 The three formulas used for fluid replacement 48 hours postburn are:

1. * ______________________________.

2. ______________________________.

3. ______________________________.

— — — — — — — — — — — — — — —

1. Brooke Army Hospital; 2. Evans; 3. Parkland

---

278 Solutions used for colloid replacement are ____________, ____________, ____________, or ____________.

Electrolyte solutions used are ________________ and ________________.

- - - - - - - - - - - - - - - -

blood; dextran; plasma, albumin;
lactated Ringer's; normal saline

CLINICAL EXAMPLE I

---

279 Mr. Greene, who weighed 154 pounds, or 70 kg, had 30% of his body surface burned.

To estimate his fluid needs for the first 24 hours, one would calculate his fluid needs according to the Brooke's formula as:

a. 0.5 ml colloid X 70 (kg of body weight) X 30 (% of burned body surface) = ____________ ml of colloid to be given. [*Note:* When multiplying by 30, do not use the decimal point.]

b. 1.5 ml electrolyte X 70 (kg of body weight) X 30 (% of burned body surface) = ____________ of lactated Ringer's to be given.

c. Plus 2000 ml of dextrose in water.

The total amount of parenteral fluids that Mr. Greene should receive in the first 24 hours following his burns would be ________________.

- - - - - - - - - - - - - - - - - -

1050 ml; 3150 ml; 6200 ml

---

280 For the second 24 hours, according to Brooke's formula, Mr. Greene should receive:

____________ ml of colloid

____________ ml of lactated Ringer's

____________ ml of dextrose in water

The total amount of parenteral fluid for the second 24 hours would be ____________ ml.

- - - - - - - - - - - - - - - -

525; 1575; 2000; 4100

---

281 After 48 hours of parenteral therapy, Mr. Greene should receive (more/less) ____________ intravenous fluid.

After 48 hours, fluid shifts from the burned and unburned tissue spaces to the vascular area, and if the same quantity of IV fluids is given, what type of fluid imbalance could occur? ______________________.

– – – – – – – – – – – – – – – – – –

less; overhydration or hypervolemia

---

282 Over 50% of body surface burns are calculated as ____________ % burns for fluid replacement purposes.

– – – – – – – – – – – – – – – – – –

50

---

283 During the first 24 hours, one-half of the fluids is given in the first 8 hours and the other half is given in the remaining 16 hours.

Mr. Greene should receive ____________ ml of intravenous fluids in the first 8 hours and ____________ ml in the remaining 16 hours of the first day.

– – – – – – – – – – – – – – – – –

3100; 3100

CLINICAL EXAMPLE II

---

284 Mrs. Silver, age 35, received 25% of second- and third-degree body surface burns when her farmhouse caught fire.

The following areas of her body were burned:

Face, 5%
Right arm and hand, 9%
Left arm, 5%
Back and upper chest, 5%

Since the face, hand, and upper chest were burned, Mrs. Silver would be considered to have:

( ) a. Minor burns
( ) b. Moderate burns
( ) c. Critical burns

– – – – – – – – – – – – – – – – –

a. –; b. –; c. X

---

285 Mrs. Silver's laboratory studies were:

Hemoglobin	13.5 g
Hematocrit	44%
White blood count	20,300 cells per cu mm
Polymorphonuclear cells (polys)	65%

Venous section (cutdown) was performed. In the emergency room she received:

Two injections of morphine sulfate: 1. grain 1/6 (IM)
2. later grain 1/6 (IV)
1000 ml normal saline with 2 million units of aq. penicillin IV
1000 ml normal saline with 5 million units of aq. penicillin IV
tetanus toxoid: 0.5 ml

Mrs. Silver received tetanus toxoid since she was subject to infection by anaerobic microbes, such as *Clostridium tetani.*

She received morphine sulfate for *______________________________.

She received aq. penicillin since her

( ) a. hemoglobin was elevated
( ) b. hematocrit was elevated
( ) c. WBC was elevated

– – – – – – – – – – – – – – – – –

relief of pain
a. –; b. –; c. X

TABLE 75. LABORATORY STUDIES FOR MRS. SILVER

Laboratory Tests	On Admission	1st Day	2nd Day	3rd Day	4th Day	5th Day	6th Day	7th Day
Hematology								
Hemoglobin (12.9–17.0 g)	13.5	17.6						
Hematocrit (40–46%)	44	49 54	56 52 61 64	55	51	46	36	35
WBC (white blood count) (5000–10,000/cu mm)	23,000		11,685					
Biochemistry								
BUN (blood urea nitrogen) (10–25 mg/dl)*	11	14						
Plasma/serum[†] CO_2 (50–70 vol %) / (22–32 mEq/L)	30/14	44/20	44/20		57/26		57/26	
Plasma/serum chloride (98–108 mEq/L)	105	105	105		97		103	
Plasma/serum sodium (135–146 mEq/L)	141	137	137		134		141	
Plasma/serum potassium (3.5–5.3 mEq/L)	4.3	4.8	4.8		4.2		4.4	

[†] *Plasma* and *serum* are used interchangeably.

*mg/100 ml = mg/dl.

The laboratory studies of Mrs. Silver, given in Table 75, show how her results deviated from the normal values at the time of her illness.

---

286 Mrs. Silver's serum chloride, sodium, and potassium (were/were not) ____________ in normal range.

Due to her low serum CO_2 on admissions, she would be in a state of metabolic ____________.

- - - - - - - - - - - - - - - - -

were; acidosis

---

287 In the preceding frame you noted that Mrs. Silver's serum electrolytes were in normal range. Can you recall the normal range for the following electrolytes without referring to Table 75?

Plasma chloride. *____________.

Plasma sodium. *____________.

Plasma potassium. *____________.

- - - - - - - - - - - - - - - - -

Cl, 98-108 mEq/L
Na, 135-146 mEq/L
K, 3.5-5.3 mEq/L

---

288 Her elevated hematocrit during the first 48 hours following admission would be an Indication of ____________.

- - - - - - - - - - - - - - - - -

dehydration (You may have answered hemoconcentration, O.K.)

---

289 Her elevated WBC indicated:

( ) a. an increased number of white blood cells
( ) b. infection
( ) c. head cold

- - - - - - - - - - - - - - - - -

a. X; b. X; c. –

CLINICAL MANAGEMENT

---

290 Mrs. Silver weighed 65 kg and had a total of 25% body burns. Calculate the following for fluid replacement, using the Brooke Army Hospital formula.

0.5 X 65 X 25 = ________________ ml colloid

1.5 X 65 X 25 = ________________ ml electrolyte

______2000______ ml dextrose/water

Total ________________ ml for first 24 hours

During the first 8 hours Mrs. Silver received one-half of the intravenous fluid.

This amount would be ________________ ml.

She received:

1000 ml normal saline
350 ml plasma
1000 ml normal saline
250 ml blood
2600 ml for the first 8 hours

Note: Mrs. Silver did not receive lactated Ringer's because her electrolytes were not low. You will notice in this program and in the clinical area that the figures are rounded off.

Then she received:

500 ml saline
250 ml plasma
400 ml 5% dextrose in water
1150 ml for the second 8 hours

Later she received 1500 ml 5% dextrose in water for the third 8 hours.

The total amount of colloid she received was ________________.

The total amount of electrolyte she received was ________________.

The total amount of dextrose in water she received was ________________.

— — — — — — — — — — — — — — — —

812.5 ml; 2437.5 ml; 5250 ml; 2600 ml; 850 ml; 2500 ml; 1900 ml

291 Check the following amounts of fluids that she should receive during the *second* 24 hours according to the Brooke's formula.

( ) a. 800 ml colloid
( ) b. 1250 ml electrolyte
( ) c. 1000 ml 5% dextrose in water
( ) d. 425 ml colloid
( ) e. 1500 ml electrolyte
( ) f. 2000 ml 5% dextrose in water
( ) g. 2600 ml total for second 24 hours
( ) h. 3675 ml total for second 24 hours
( ) i. 4225 ml total for second 24 hours

– – – – – – – – – – – – – – – – –

a. –; b. X; c. –; d. X; e. –; f. X; g. –; h. X; i. –

---

292 In Mrs. Silver's case, urine output was measured hourly and tested for specific gravity.

After the first 8 hours, Mrs. Silver's urine output was 250 ml per hour, so the IV fluid flow rate was cut back.

During the third 8 hours, her urine output had fallen 5–10–15 ml per hour.

The fluid flow rate should be (increased/decreased) ________________.

In a case like Mrs. Silver's, which of these intravenous fluids would be the preferred ones to give when there is a decrease in urine output?

( ) a. blood
( ) b. normal saline
( ) c. 5% dextrose in water
( ) d. 10% dextrose in saline

– – – – – – – – – – – – – – – – –

increased
a. –; b. X; c. X; d. –

---

293 The specific gravity of Mrs. Silver's urine ranged from 1.005–1.017. Specific gravity of urine is the weight (waste products) in relationship to water, 1.000. Specific gravity norm for urine is (1.010–1.030).

When Mrs. Silver's specific gravity was 1.005 and 1.008, there would be (more/less) ________ waste products in her urine than the norm.

The waste products would be more concentrated in her ________________.

– – – – – – – – – – – – – – – – –

less; plasma

CASE REVIEW

Mrs. Silver, age 35, received 25% second- and third-degree body surface burns. Using the Rule of Nines, her face received 5%, right arm and hand 9%, left arm 5%, and back and upper chest 5%. On admission, her hematocrit was 44%; on the first day it rose to 49% and 54%. By the second day, it was 56% to 64%. Her serum CO_2 on admission was 14 mEq/L, and on the second and third days it was 20 mEq/L. Her serum electrolytes were in normal range.

1. After a burn, there is extracellular fluid volume shift from *________________ to *________________.
2. Mrs. Silver's hematocrit was ________________ the first and second days following the burn. Explain why. *________________________________.
3. What type of acid-base imbalance did Mrs. Silver have the first three days? *________________________________. Explain the reason for this imbalance. *________________________________.
4. Besides fluid and electrolyte imbalance, another hazard in burn cases is ________________.
5. Name two methods used to estimate body burn surface. *________________ and *________________________________.
6. Mrs. Silver's burns were classified as (minor/moderate/critical). ________________. Explain why. *________________________________.

Mrs. Silver's fluid replacement was based on the Brooke Army Hospital formula. Review the Brooke's formula and Mrs. Silver's fluid replacement the first 24 hours.

7. In planning Mrs. Silver's parenteral therapy, the ________________ formula was used based on the Rule of Nines. Name two other formulas for determining fluid needs. ________________ and ________________.
8. Cope and Moore state that an important index for the determination of adequate parenteral therapy is an accurate record of *________________.
9. Urine output should be between ________ and ________ ml per hour.

Less than 25 ml of urine output can indicate *____________________

or *__________________________________.

10. Mrs. silver's colloid replacements were ________________________

and ____________________. What other colloids can be used?

____________________ and ________________________.

11. Name two types of electrolyte solutions used in parenteral therapy due to burns.

*______________________________________ and

*______________________________.

Give the type of electrolyte solution that Mrs. Silver received and state why.

*____

____.

12. After 2 days or, at the longest, 2 weeks, the stage of diuresis occurs. This is when the extracellular fluid volume shifts from *____________________

to *______________________________.

13. Overhydration of hypervolemia is a hazard when large quantities of intravenous fluids are administered for fluid replacement and for increasing urinary output. Give four symptoms of overhydration.

*____________________, ____________________,

*____________________, and *____________________.

1. the intravascular space; the interstitial space
2. elevated. Hemoconcentration was present due to ECF volume shift to the interstitial space (burned area), or you could answer dehydration due to ECF volume shift
3. metabolic acidosis. Due to a loss of serum bicarbonate and an increase in body acid metabolites (cellular breakdown)
4. infection
5. Rule of Nines; Lund and Browder's; also Berkow
6. critical. Mrs. Silver had burns to the hand and face plus a total of 25% second- and third-degree burns
7. Brooke's; Evans and Parkland
8. measured hourly urine output
9. 30; 50; insufficient fluid intake; kidney dysfunction
10. plasma; blood; albumin; dextran
11. normal saline; lactated Ringer's. Mrs. Silver's serum electrolytes were in normal range, so saline was given. Also, saline (NaCl) will replace the sodium loss to the burn area and to diuresis
12. interstitial spaces; the intravascular space
13. irritating cough; dyspnea; engorged neck and hand veins; moist-rales.

NURSING DIAGNOSES

Fluid volume deficit related to the fluid volume shift to the burned site.

Potential for alteration in fluid volume excess related to the fluid shift from the burned site and tissues to the vascular system; large quantities of administered IV fluids.

Potential for impairment of urinary elimination related to the fluid volume deficit, SIADH.

Potential for electrolyte imbalances related to the cellular damage (potassium loss) and the fluid shift to the burned site (sodium loss).

Potential for infection related to contamination at the burned site.

Anxiety and fear related to the outcome from burns, impending death, and dependence.

Potential for injury related to tissue damage and infection.

Potential alteration in nutrition related to inadequate nutrient intake.

NURSING ACTIONS: BURNS

1. Recognize the physiologic factors associated with burns and burn shock. The factors include increased capillary permeability, increased circulatory resistance, decreased cardiac output, hemoconcentration (increased serum osmolality and increased hematocrit), hemolysis, decreased renal function, electrolyte imbalances (hyponatremia, cellular potassium loss, hyperkalemia with oliguria, and hypocalcemia), decreased serum protein, and metabolic acidosis.
2. Assess the degrees of burns (first, second, or third) and recognize characteristics of skin surface, color, and pain related to the degrees.
3. Calculate the percentages of total burned body surface areas using the Rule of Nines. Know there are other methods used for calculating percentage, such as Berkow and Lund and Browder. Rule of Nines is easy to recall and can be used for quick estimation.
4. Know the classification of burns (mild, moderate, or critical) that is calculated according to the degree and percentage of burn.
5. Know that parenteral therapy is usually not indicated if the percentage of burn is less than 20% for an adult and less than 10% for a child.
6. Monitor hourly urine output, especially during the first 48 hours, and if it is less than 25 ml, increase the IV flow rate with the physician's permission.
7. Check the laboratory tests and report abnormal results. If the patient is receiving intravenous fluids with potassium and his serum potassium level is elevated and urine output is decreased, the physician most likely would change the IV order. IV fluids *without* potassium should be run rapidly to improve renal function and to decrease potassium level.
8. Calculate the flow rate for the first and second 24 hours. One-half of the daily IV fluid order should be administered in 8 hours, and the second one-half in 16 hours.
9. Assess the patient for overhydration (hypervolemia) when administering large quantities of IV fluids, and 48 hours postburn (fluid shifts from the burned

and unburned sites to vascular area). Symptoms of overhydration are constant, irritated cough; dyspnea; neck and hand vein engorgement; and chest rales.

10. Monitor vital signs and report changes associated with hypovolemia and burned shock, i.e., tachycardia (pulse > 100), increased respirations, and decreased blood pressure.
11. Observe for signs and symptoms of metabolic acidosis, i.e., deep, rapid, vigorous breathing, restlessness, low serum CO_2 (< 22 mEq/L).
12. Encourage the burned patient to drink and eat foods rich in protein. Protein replacement is needed for protein loss especially during the healing stage.
13. Use aseptic technique when changing dressings at the burned site. Infections will delay reabsorption of edema fluid.
14. Explain to patient the care given and why. Explanation could help to alleviate most of patient's anxiety.

CANCER (Julie Waterhouse)

---

294 Cancer is a group of diseases characterized by abnormal and uncontrolled cell grwoth. Cancer cells are malignant (capable of invading normal tissues and spreading to distant sites).

Cancer is characterized by *____. Cancer cells are considered to be (malignant/benign).________________________.

- - - - - - - - - - - - - - - - - -

abnormal and uncontrolled cell growth. Malignant

---

295 Fluid and electrolyte disturbances occur frequently in individuals with cancer because of the nature of the malignant cell growth and the effects of therapies used to control it.

Give two reasons why fluid and electrolyte disturbances occur frequently in individuals with cancer.

1. *____.
2. *____.

- - - - - - - - - - - - - - - - - -

1. nature of the malignant cell growth
2. effects of therapies used to control this cell growth

---

296 Cancer may begin as an individual solid tumor (carcinoma or sarcoma) or may arise throughout the body in the blood forming cells of bone marrow or lymph nodes (leukemia and lymphoma).

Match the types of cancer with their tissues of origin:

a. lymphoma
b. leukemia
c. carcinoma
d. sarcoma

____ ____ 1. epithelium or supporting tissue

____ 2. bone marrow

____ 3. lymphoid tissue

- - - - - - - - - - - - - - - - -

1. c, d; 2. b; 3. a

---

297 Malignant cells are more primitive (anaplastic) than normal cells. They may be undifferentiated or may differentiate in abnormal and bizarre ways.

This lack of normal differentiation causes malignant cells to produce unusual proteins, hormones, enzymes, and other chemicals. Because the proteins produced by malignant cells are abnormal, they do not respond to normal regulatory mechanisms such as diet and hormonal and metabolic controls.

Cancer of the colon may produce carcinoembryonic antigen/CEA, a fetal antigen.

Why? * ____

____.

- - - - - - - - - - - - - - - - -

malignant cells lack normal differentiation; thus a more primitive embryonic protein may be produced

298 One major consequence of these biochemical abnormalities is cachexia, a complex process manifested by anorexia, weight loss, wasting, weakness, anemia, fluid and electrolyte disturbances, and increased basal metabolic rate.

A major problem associated with cancer is ____________________. This serious problem is manifested by ________________, ________________, * __________________________________, and * ____________________ _______________________________.

\- - - - - - - - - - - - - - - - -

cachexia; anorexia, weakness, weight loss, fluid and electrolyte disturbances. Others are wasting, anemia, and increased basal metabolic rate.

---

299 Cachexia involves changes in the metabolism of all the major nutrients. Glucose utilization is higher than in normal cells and anaerobic metabolic pathways are used more often than the aerobic. This produces higher than usual concentrations of lactic acid and may result in lactic acidosis.

Glucose utilization is higher in malignant cells. What specific acid-base imbalance is caused by anaerobic metabolism? ___________________________.

\- - - - - - - - - - - - - - - - -

Lactic acidosis

---

300 Nitrogen transferred from body tissues to the tumor often leaves the patient in negative nitrogen balance. Similarly, cancer patients retain sodium, with a total of 120% that of healthy individuals. Sodium, however, is concentrated in the tumor and the serum and may be low (hyponatremia).

Often in patients with cancer the nitrogen balance is (positive/negative) ___________________________. Why? * ____.

\- - - - - - - - - - - - - - - - -

negative; nitrogen is transferred from the body tissues to the tumor, thus causing a negative nitrogen balance.

---

301 Would the cancer patient be (retaining/excreting) sodium? ________________.

Concentration of sodium is in the ______________. Because the sodium is not in the vascular fluid, would (hyponatremia/hypernatremia) result?

________________.

— — — — — — — — — — — — — — — —

retaining; tumor; hyponatremia

---

302 The patient with cancer may also experience malabsorption syndrome, which involves inflammation, ulceration, decreased patency, and decreased secretions of the GI tract. The problems that result in protein and fat absorption may compound fluid and electrolyte disturbances of cachexia.

Malabsorption syndrome may occur in cancer patients. It can involve

________________, ________________, and

*________________________________.

— — — — — — — — — — — — — — — —

inflammation; ulceration; decreased patency OR decreased secretions of the GI tract

---

303 A second major consequence/problem of the primitive biochemical function of malignant cells is the secretion of abnormal hormones (ectopic hormone secretion).

An example is bronchiogenic cancer which may secrete antidiuretic hormone (ADH), parathyroid hormone (PTH), or adrenocorticotropic hormone (ACTH). The resulting hormonal abnormalities may lead to fluid and electrolyte problems such as water intoxication (ICFV excess), hyponatremia, hypokalemia, and hypophosphatemia.

The first major consequence of biochemical abnormalities in progressive

cancer is ________________. The second major consequence is

*________________________________.

— — — — — — — — — — — — — — — —

cachexia; secretion of abnormal hormones OR ectopic hormone secretion.

304 Abnormal hormonal secretions may occur in bronchiogenic cancer. Examples are (use abbreviations) ________________, ________________, and ________________.

What fluid and electrolyte problems can result from these abnormal hormonal secretions? * ________________, ________________, ________________ and ________________.

— — — — — — — — — — — — — — — — —

ADH; PTH; ACTH
water intoxication; hyponatremia; hypokalemia; hypophosphatemia.

---

305 A third problem which commonly occurs in cancer patients in whom large numbers of malignant and normal cells are destroyed by radiation or chemotherapy is catabolism (breakdown) of purine nucleic acids in cells. The result is an increase in serum uric acid.

Three (3) problems that may result in fluid and electrolyte disturbances in cancer patients are the following:

1. ________________.
2. * ________________.
3. * ________________.

— — — — — — — — — — — — — — — — —

1. Cachexia
2. Secretion of abnormal hormones, i.e., ADH, PTH, ACTH
3. Catabolism of purine nucleic acids

---

306 Uric acid is poorly soluble in body fluids and is excreted primarily through the kidneys. Small increases above normal serum concentrations can cause uric acid precipitation in the renal tubules and collecting ducts.

Do you know what could happen if uric acid precipitated in the renal tubules?

* ________________.

— — — — — — — — — — — — — — — — —

renal disorders and possible renal failure. VERY GOOD.

FLUID AND ELECTROLYTE DISTURBANCES

Table 76 lists the fluid and electrolyte disturbances commonly associated with cancer and cancer therapy. Also given are abnormal serum levels and the rationale for their occurrence. Study the table carefully and refer to it as needed.

TABLE 76. FLUID/ELECTROLYTE DISTURBANCE IN CANCER/CANCER THERAPY

Fluid/Electrolyte Disturbance	Commonly Associated Cancer and Cancer Therapy	Defining Characteristics	Rationale/Comments
Hypercalcemia	Breast cancer, multiple myeloma, ovarian cancer, pancreatic cancer, leukemia, lymphoma, lung cancer, bladder cancer, kidney cancer, head and neck cancer, and prostate cancer. Mithramycin chemotherapy	Serum calcium >10.5 mg/dl or >5.8 mEq/L	Hypercalcemia occurs in 10–20% of all cancer patients and in 40–50% of those with metastatic breast cancer or multiple myeloma. It is caused by bone destruction by metastatic tumors, elevated parathyroid hormone/PTH levels related to some tumors, and elevated prostaglandin and osteoclast activating factor (OAF). Prolonged immobility is also a causative factor.
Hyponatremia (usually associated with dehydration)	Lung cancer, pancreatic cancer, multiple myeloma, head and neck cancer, stomach cancer, brain cancer, colon cancer, ovarian cancer, and prostate cancer. Aggressive diuretic therapy. High-dose cyclophosphamide/Cytoxan therapy. Daumorubicin or cytosine chemotherapy (decreases blast cell count).	Serum sodium <135 mEq/L	Hyponatremia is caused by liver, thyroid and adrenal insufficiencies, renal failure, and congestive heart failure. A condition known as cerebral salt wasting is caused by some intracranial neoplasms. In this condition the brain releases a postulated natriuretic factor or the neural innervation to the brain is altered. The result is the kidney's inability to conserve sodium.
Secretion of inappropriate anti-diuretic hormone (SIADH)	Lung cancer, pancreatic cancer, brain cancer, ovarian cancer, colon cancer, sarcoma, leukemia, prostate cancer, Hodgkin's disease, and other lymphomas. Vincristine, cyclophosphamide chemotherapy.	Serum sodium <130 mEq/L Serum osmolality <280 mOsm/Kg	SIADH occurs because of increased release of ADH from posterior pituitary or ectopically from neoplastic tumors. The posterior pituitary then becomes impervious to the usual feedback control mechanism.

Hyperuricemia	Leukemias, lymphomas, multiple myeloma. Any cancer treated aggressively with chemotherapy or radiation.	Serum uric acid >7.0 mg/dl Uric acid crystals in urine	Breakdown of large numbers of cells causes release of uric acid into the bloodstream (tumor lysis syndrome). Precipitation of uric acid in the kidneys results in gouty nephropathy, acute hyperuricemic nephropathy, and eventual renal failure. First signs of hyperuricemic renal failure may be nausea, vomiting, and lethargy. This type of renal failure may or may not be reversible. Symptoms of hyperuricemia include hematuria, flank pain, nausea, vomiting, and symptoms of renal failure.
Hypokalemia	Colon cancer, multiple myeloma, Hodgkins disease, pancreatic cancer, stomach cancer, thyroid cancer, adrenal adenoma, adrenal hyperplasia tumors, and cancers that secrete ACTH ectopically.	Serum potassium <3.5 mEq/L	Dietary intake of potassium is deficient when the patient is anorexic, vomiting, or NPO. Excessive diarrhea that leads to rapid potassium depletion occurs with many GI tumors, chemotherapy, radiation therapy to the lower abdomen, and antibiotic therapy. Excessive urinary excretion may be caused by diuretics, hypercalcemia, hypomagnesemia, antibiotic therapy, ectopic ACTH secretion, nephrotoxicity due to chemotherapy or radiation, and renal tubular necrosis due to Hodgkins disease, multiple myeloma, and acute blast crisis. Ileostomy, colostomy, fistulas, and the diuretic phase of renal failure also contribute to hypokalemia.
Hypomagnesemia	Lung cancer, espeically oat cell, ovarian cancer, and testicular cancer. Total parenteral nutrition/TPN. *Cis*-platinum chemotherapy.	Serum magnesium <1.5 mEq/L	Low magnesium level occurs most often in patients with severe diarrhea, vomting, malabsorption syndrome, cachexia, ADH secretion, or renal disease. *Cis*-platinum and nephrotoxic antibiotics also contribute to hypomagnesemia.

TABLE 76. (Continued)

Fluid/Electrolyte Disturbance	Commonly Associated Cancer and Cancer Therapy	Defining Characteristics	Rationale/Comments
Lactic acidosis	Hodgkins disease, lymphoma, leukemia, lymphosarcoma, and lung cancer (especially oat cell with liver metastasis).	Arterial blood pH <7.35 HCO_3 <24 mEq/L Serum CO_2 <22 mEq/L	Lactic acidosis (metabolic acidosis) occurs because rapidly growing malignant cells utilize large amounts of glucose. When the glucose is metabolized by the anaerobic pathway (glycolysis) pyruvic acid is the end product. When hypoxia exists pyruvate is converted to lactic acid. Elevated serum lactic acid concentrations may exceed the liver's ability to metabolize and the kidney's to excrete.
Hyperkalemia	Hodgkins disease, lymphoma, leukemia, lung cancer (especially oat cell), and liver metastasis. Aggressive chemotherapy.	Serum potassium >5.3 mEq/L	Intracellular-extracellular redistribution occurs during respiratory and metabolic acidosis (including lactic acidosis). Extracellular hydrogen ions shift into the cell in an attempt to raise serum pH. Intracellular potassium ions then shift out of the cell to compensate. Lysis of large numbers of malignant and normal cells during radiation and chemotherapy causes the release of massive amounts of potassium from destroyed cells (tumor lysis syndrome). Renal failure and hypoaldosteronism can cause renal retention of potassium.
Hypophosphatemia	Leukemia, multiple myeloma, PTH-secreting tumors. Total parenteral nutrition/TPN (hyperalimentation).	Serum phosphorus <1.7 mEq/L or <2.5 mg/dl	Hypophosphatemia occurs with cancers that contain and secrete PTH (PTH normally regulates the rate of phosphorus reabsorption by the kidneys). Aggressive hyperalimentation/parenteral nutrition often induces hypophosphatemia because the phosphorus influx into cells is accelerated during carbohydrate metabolism. Malabsorption, sepsis, diuretics, corticosteroids, and thrombocytopenia are other contributing factors. Symptoms of hypophosphatemia are fatigue, weakness, anorexia, irritability, paresthesia, seizures, and coma.

Decreased vascular volume (shift to the third space)	Liver cancer, including liver metastasis, stomach cancer, pancreatic cancer, colon cancer, and head and neck cancer.	Serum albumin <3/2 g/dl Decreased BP Increased H & H Increased BUN	Decreased vascular volume occurs when serum protein is decreased, when tumor cells exude fluids, or when vascular permeability is increased by infection. Protein depletion occurs with anorexia/cachexia, nausea, and vomiting due to disease or therapy or to decreased protein synthesis in cancer patients. Some individuals with cancer have increased loss of protein via the GI tract; elevated basal metabolic rate due to disease or infection results in accelerated protein loss. Decreased serum protein leads to decreased blood volume and a drop in blood pressure. The patient may have ample or excess extracellular fluid but is unable to retain it within the vascular space. Without treatment, cardiovascular failure and death result.
Hypocalcemia/ hyperphosphatemia	Leukemia, lymphoma, and multiple myeloma. Aggressive chemotherapy and radiation.	Serum calcium <8.5 mg/dl or <4.5 mEq/L Serum phosphorus >4.5 mg/dl >2.6 mEq/L	Rapid cell lysis causes the release of large amounts of phosphate. Immature blast cells contain up to four times more phosphate than mature lymphocytes. The rise in serum phosphorus then causes a drop in serum calcium. Renal failure may result from precipitation of calcium phosphate in the kidneys. Symptoms include oliguria, anuria, azotemia, and tetany.

---

307 Bone destruction which results from metastatic tumors causes what type of calcium imbalance? ______________. The serum level would be ______________.

— — — — — — — — — — — — — — — —

hypercalcemia or calcium excess; >10.5 mg/dl or >5.8 mEq/L

---

308 Could hyponatremia be associated with cancer? ______________.

Explain how? *______________________________.

Could hypernatremia be associated with cancer? ______________.

— — — — — — — — — — — — — — — —

Yes. Liver, thyroid, and adrenal insufficiencies are followed by a loss of sodium. Also intracranial neoplasms could result in cerebral salt wasting and the inability of the kidneys to conserve sodium.
Not usually.

---

309 Secretion of inappropriate antidiuretic hormone/SIADH can be associated with cancer. As a result more water would be (reabsorbed/excreted) by the kidneys.

______________.

The drug therapies that contribute to SIADH are ______________ and ______________.

— — — — — — — — — — — — — — — —

reabsorbed; vincristine; cyclophosphamide

---

310 Hyperuricemia can occur in any cancer treated aggressively with ______________ and ______________.

Explain the effect of a high-serum uric acid on kidney function. *______________________________.

— — — — — — — — — — — — — — — —

chemotherapy; radiation.
It could cause renal failure by precipitation of uric acid crystals in the kidneys.

---

311 Name three (3) symptoms of hyperuricemia:

1. ________________.
2. ________________.
3. ________________.

- - - - - - - - - - - - - - - - -

1. Hematuria. 2. Flank pain. 3. Nausea/vomiting. Also symptoms of renal failure.

---

312 Could hypokalemia occur as the result of cancer? ________________

Give three (3) reasons why there could be a low potassium level. ____________, ________________, and ________________.

The serum potassium level would be ________________.

- - - - - - - - - - - - - - - - -

Yes. Poor dietary intake of potassium; excessive diarrhea; excessive urinary secretion. Also vomiting, colostomy, and excess adrenal gland secretion from tumor.
<3.5 mEq/L

---

313 Could hyperkalemia be induced by cancer? ________________.

Explain why? * ________________________________

________________________________.

The serum potassium level would be ________________.

- - - - - - - - - - - - - - - - -

Yes. The breakdown of malignant and normal cells causes potassium to shift from cells to vascular fluid (or a similar answer).
>5.3 mEq/L

---

314 Why does hypomagnesemia develop? * ________________________________

________________________________.

- - - - - - - - - - - - - - - - -

Because of vomiting, malabsorption syndrome, severe diarrhea, cachexia, and total parenteral nutrition/TPN

---

315 Hypophosphatemia is usually associated with cancer in * ______________________

______________________ and * ______________________

______________________.

- - - - - - - - - - - - - - - - - -

PTH secreting tumors; total parenteral nutrition/TPN

---

316 A large amount of glucose is utilized by rapidly growing malignant cells. Pyruvic acid is the end product of anaerobic metabolism of glucose. When hypoxia exists pyruvate is converted to lactic acid. The specific acid-base imbalance

would be * ______________________.

The arterial blood pH would be ______________________.

The arterial bicarbonate would be ______________________.

- - - - - - - - - - - - - - - - - -

lactic acidosis
<7.35
<24 mEq/L

---

317 Calcium and phosphorus imbalance may be present in leukemia, lymphoma, multiple myleoma, and aggressive chemotherapy and radiation.

Identify the imbalances that occur together:

( ) a. Hypocalcemia
( ) b. Hypercalcemia
( ) c. Hypophosphatemia
( ) d. Hyperphosphatemia

- - - - - - - - - - - - - - - - - -

a. X; b. –; c. –; d. X

---

318 When protein depletion occurs because of anorexia/cachexia and nausea/ vomiting what type of fluid imbalance may result? * ______________________ ______________________.

Explain why * ______________________ ______________________.

— — — — — — — — — — — — — — — — —

Decreased vascular volume or ECFV deficit (vascular).
Protein loss decreases osmotic pressure and *less* fluid is held in the vascular space. When protein shifts to an injured or damaged site and permeability is increased, fluid shifts to the third space (to the injured or damaged site).

---

319 Indicate which of the electrolyte imbalances are frequently associated with cancer and cancer therapy:

( ) a. hypercalcemia
( ) b. hypernatremia
( ) c. hypokalemia
( ) d. hyperkalemia
( ) e. hypomagnesemia
( ) f. hypermagnesemia
( ) g. hypophosphatemia
( ) h. hypocalcemia/hyperphosphatemia

— — — — — — — — — — — — — — — — —

a. X; b. –; c. X; d. X; e. X; f. –; g. X; h. X

CLINICAL APPLICATIONS

The fluid and electrolyte disturbances listed in Table 76 can develop in almost any individual with cancer at any time during diagnosis, treatment, recovery, or terminal stages of the disease.

Fluid and electrolyte problems are *most common*, however, with the following clinical conditions:

1. *Cachexia.* Severe anorexia, nausea, vomiting, and/or diarrhea are present.
2. *Tumor lysis syndrome.* Large numbers of cells are destroyed by chemotherapy or radiation.

3. *Uncontrolled cell growth.* Rapid, widespread cell growth with multiple metastasis or multiple organ infiltration.
4. *Ectopic hormone production.* Ectopic hormones are secreted by the tumor(s).

Cachexia

Cachexia may occur because of the effects of the malignancy itself and/or be caused by radiation and chemotherapy. Contributing problems include anorexia, nausea, vomiting, diarrhea, draining wounds, and fistulas. The most frequently encountered fluid and electrolyte disturbances are hypomagnesia, hypokalemia, and decreased vascular volume.

---

320 Name the two (2) electrolytes that are most commonly lost due to cachexia.

______________ and ______________.

Do you recall the methods/routes for replacing potassium and magnesium? See Chapter 2 on potassium and magnesium replacement.

Potassium ______________ and ______________.

Magnesium ______________, ______________, and ______________.

- - - - - - - - - - - - - - - - - -

potassium; magnesium
intravenously; orally; intravenously; intramuscularly; orally

---

321 Anorexia, nausea, and vomiting decrease protein intake; diarrhea, malabsorption syndrome, and wound drainage increase protein loss in the cancer patient.

Decreased vascular volume occurs in cachexic patients because of *______________ ______________ (see Table 76).

Protein synthesis is (increased/decreased) ______________ in many cancer patients. Metabolic changes and infections accelerate *______________ ______________.

- - - - - - - - - - - - - - - - - -

protein depletion or loss; decreased; protein loss

---

322 The basic goal of therapy in cancer patients with decreased vascular volume is to maintain blood pressure. Whole blood, packed red blood cells (RBCs), or albumin may be given to increase plasma oncotic pressure (colloid osmotic pressure) to restore fluid balance in the vascular space. Whole blood, packed RBCs, and albumin restore the fluid balance in the vessels by * ________________ ____.

\- - - - - - - - - - - - - - - -

increasing plasma oncotic pressure OR increasing colloid osmotic pressure in vascular space

---

323 Carefully prescribed and monitored hyperalimentation/TPN can correct hypokalemia and hypomagnesemia and improve vascular volume.

What happens to the serum phosphorus level when TPN is aggressively administered? * ____ ____. (Refer to Table 76)

Prolonged parenteral nutrition without magnesium supplement can cause ________________________________.

\- - - - - - - - - - - - - - - -

Hypophosphatemia occurs because of phosphorus influx into cells during carbohydrate metabolism

hypomagnesemia

Tumor Lysis Syndrome

Following the destruction of large numbers of cells by chemotherapy, usually in leukemia or lymphoma, vast numbers of intracellular electrolytes enter the bloodstream. The cancer patient may then develop hyperuricemia, hyperkalemia, hyperphosphatemia, and/or hypocalcemia. Renal failure or cardiac arrest may result.

---

324 When a large number of cells in the body is destroyed by chemotherapy what two (2) life-threatening situations can result? *____________________ and *____________________.

Indentify the imbalances that occur during massive cell destruction.

( ) a. hypokalemia
( ) b. hyperkalemia
( ) c. hypocalcemia
( ) d. hypercalcemia
( ) e. hyperphosphatemia
( ) f. hyperuricemia

– – – – – – – – – – – – – – – – – –

renal failure; cardiac arrest
a. –; b. X; c. X; d. –; e. X; f. X

---

325 The purpose of therapy for cancer patients who are undergoing the destruction of large numbers of cells is the prevention of renal failure and

*______________________________.

Management includes aggressive hydration (3000 ml/day) to increase urinary volume and excretion of ________________, ________________, and ________________.

– – – – – – – – – – – – – – – – – –

electrolyte imbalance. If your answer was cardiac arrest, true, but that usually results from severe electrolyte imbalance. uric acid; potassium; phosphorus (phosphate).

---

326 Drugs used for the management of fluid and electrolytes include

potent diuretics such as furosemide/Lasix when there is fluid retention;
allopurinol to decrease uric acid;
calcium gluconate IV infusion if hypocalcemia develops;
Sodium bicarbonate to alkalinize the urine if hyperuricemia occurs (uric acid is less soluble in acid urine).

The imbalances found with massive cell destruction are ________________, ________________, ________________, and ________________.

– – – – – – – – – – – – – – – – – –

hyperkalemia; hyperphosphatemia; hyperuricemia; hypocalcemia

Uncontrolled Cell Growth

Individuals with cancer may experience severe electrolyte disturbances whenever rapid and widespread malignant cell growth occurs. This uncontrolled cell growth is marked by multiple metastatic lesions (metastatic carcinoma or sarcoma) or by multiple organ infiltration (leukemias and lymphomas).

---

327 Hypercalcemia is usually present when *____ (refer to Table 76).

Hypercalcemia in individuals with cancer develops more rapidly and becomes more severe than hypercalcemia from other causes. When acute hypercalcemic crisis occurs the mortality rate is extremely high (up to 50%).

\- - - - - - - - - - - - - - - - - -

metastatic tumors cause bone destruction OR metastatic cell growth or infiltration destroys bone.

---

328 Management of mild and moderate hypercalcemia involves IV normal saline (NaCl 0.9%) to achieve adequate hydration and promote calcium excretion.

For severe hypercalcemia the following drugs are indicated:

Furosemide/Lasix to decrease tubular reabsorption of calcium.
Steroids to increase calcium excretion.
Calcitonin to inhibit bone resorption.
Mithramycin to inhibit bone resorption.
IV inorganic phosphates (severe side effects could be calcium precipitation in lung, kidney, or heart tissues).

With mild and moderate hypercalcemia effective management includes

*____.

The drugs frequently prescribed for severe hypercalcemia are

____________________________ and ____________________________. A diuretic that promotes kidney excretion of calcium is ____________________________.

\- - - - - - - - - - - - - - - - - -

IV normal saline; Mithramycin and Calcitonin; furosemide

329 Lactic acidosis and hyperkalemia may occur during periods of uncontrolled malignant cell growth.

Lactic acidosis is the result of * ____

____. (Refer to Table 76).

Hyperkalemia is due to * ____

____. (Refer to Table 76).

— — — — — — — — — — — — — — — — —

rapidly growing malignant cells that utilize excess glucose with anaerobic metabolism and cause lactic acid as a by-product.
Acidosis causes a compensatory shift of extracellular hydrogen ions and intracellular potassium.

Ectopic Hormone Production

Abnormal hormones may be secreted by any malignant cells. The most commonly involved cancers and hormones are those listed in Table 77. Study the contents in the table and refer back to it as needed.

TABLE 77. HORMONES THAT ARE COMMONLY SECRETED ECTOPICALLY BY MALIGNANT CELLS

Hormones	Type of Cancer	Common Associated Problems
Antidiuretic hormone (ADH)	Lung (oat cell) Pancreas Hodgkins disease Prostate gland Sarcoma	SIADH Hyponatremia Hypomagnesemia
Parathyroid hormone (PTH)	Lung Leukemia Multiple myeloma Breast	Hypercalcemia Hypophosphatemia Hypomagnesemia
Adrenocorticotropic hormone (ACTH)	Lung (oat cell and non-oat cell)	Hypokalemia
Osteoclast activating factor (OAF)	Multiple myeloma Lymphoma	Hypercalcemia
Prostaglandins (E series)	Breast Kidney Pancreas	Hypercalcemia

---

330 Name three (3) ectopic hormones that can be secreted by malignant cells:

1. ______________________.
2. ______________________.
3. ______________________.

Ectopic hormone secretions occur most commonly with what type of cancer?

______________________.

- - - - - - - - - - - - - - - - -

1. ADH or antidiuretic hormone. 2. PTH or parathyroid hormone. 3. ACTH or adrenocorticotropic hormone.
Lung

---

331 Acute complications caused by ectopic hormone secretions in cancer patients include fluid and electrolyte imbalances. If the imbalances are severe cardiac arrest may occur.

According to Table 77, common electrolyte imbalances that result from ectopic hormone secretions are ______________, ______________, ______________, and ______________.

- - - - - - - - - - - - - - - - -

hypercalcemia; hypomagnesemia; hyponatremia; hyperphosphatemia. Also hypokalemia.

---

332 The primary goal of therapy in cancer patients with ectopic hormone secretion is the eradication or reduction of the hormone-secreting tumor. If surgery, radiation, and/or chemotherapy do not eliminate the tumor and control the symptoms long-term pharmacologic therapy may be ordered.

What are the three (3) methods that can be used to reduce or eradicate the hormone-secreting tumor?

a. ______________________, b. ______________________,

c. ______________________.

- - - - - - - - - - - - - - - - -

a. surgery; b. radiation; c. chemotherapy

---

333 Treatment of SIADH varies with the severity of the symptoms.

Mild SIADH: Restriction of water and fluid intake to 500–1000 ml per day.
Severe SIADH: 3–5% saline infusion to restore serum sodium; Furosemide/Lasix to increase water excretion; Demeclocycline or lithium carbonate to interfere with the action of ADH on renal tubules.

Extreme care should be taken when hyperosmolar saline is administered (3–5% saline solution) because it could raise the serum sodium level too rapidly and cause shrinkage of CNS neurons and neurologic dysfunction.

SIADH is the abbreviation for * ____

______________________________.

Treatment for mild SIADH is * ____

______________________________.

---

secretion (syndrome) of inappropriate antidiuretic hormone; restriction of fluid to 500–1000 ml daily.

---

334 Indicate which of the following treatments may be used for managing severe SIADH:

( ) a. 0.9% saline (normal saline solution)
( ) b. 3–5% saline infusion
( ) c. Radiation to tumor
( ) d. Chemotherapy to tumor
( ) e. Surgical removal of ADH–secreting tumor
( ) f. Lithium carbonate
( ) g. Ampicillin
( ) h. Demeclocycline

---

a. –; b. X; c. X; d. X; e. X; f. X; g. –; h. X

335 What could happen if excessive amounts of 3–5% saline are administered to correct SIADH? * ________________________

____________________.

\- - - - - - - - - - - - - - - - -

An elevated serum sodium level (hypernatremia) causes shrinkage of CNS neurons and neurologic dysfunction.

---

336 Name the four (4) major problems found in cancer patients that are commonly associated with fluid and electrolyte disorders:

1. ____________________.
2. * ____________________.
3. * ____________________.
4. * ____________________.

\- - - - - - - - - - - - - - - - -

1. cachexia 2. tumor lysis syndrome 3. uncontrolled cell growth 4. ectopic hormone production

---

CLINICAL EXAMPLE I

Ralph Peterson, a 53-year-old auto mechanic, complained to his doctor of progressive dyspnea, a persistent, productive cough, fatigue, anorexia, and weight loss.

A chest x-ray, sputum cytology, and bronchoscopy were performed and Mr. Peterson was diagnosed as having STAGE II squamous-cell lung cancer. The primary tumor was removed by lobectomy and radiation therapy was given a month later to reduce the risk of metastasis.

Fourteen months after his surgery Mr. Peterson was readmitted because of severe weight loss (30 pounds in 3 months), fatigue, and dyspnea. CT scans revealed that Mr. Peterson had a large metastatic lesion in the liver and two smaller tumors in the (R) lung.

Laboratory studies were ordered for Mr. Peterson on admission, on day 3, on day 10, on day 11, and a month later. The results of his laboratory tests are given in Table 78. Complete the frames related to his laboratory studies by following the table.

TABLE 78. LABORATORY STUDIES: MR. PETERSON

Laboratory Tests	On Admission	Day 3	Day 10	Day 11	One Month
Hematology:					
Hemoglobin/Hgb (M: 13.5–18 g)	18.5	14			
Hematocrit/Hct (M: 40–54%)	54	46			
Biochemistry:					
BUN (10–25 mg/dl)	34	21			
Creatinine/Cr (0.6–1.2 mg/dl)	2.1	1.5			
Uric Acid (M: 3.5–7 mg/dl)	7.8				
Lactate (serum) (6–16 mg/dl)	17				
Albumin (serum) (3.5–5.0 g/dl)	2.8	4.3			
Potassium (K) (3.5–5.3 mEq/L)	3.1	4.2	3.0		
Sodium (Na) (135–145 mEq/L)	118	135			
Chloride (Cl) (98–108 mEq/L)	103	102			
Calcium (Ca) (8.5–10.5 mg/dl)	8.8	8.8	14.4	13.4	16.6
Phosphorus (P) (2.5–4.5 mg/dl)	3.1		2.2		
Magnesium (Mg) (1.5–2.5 mEq/L)			1.3		

337 On admission, Mr. Peterson's lab tests suggested that he had decreased vascular volume, also called ECFV deficit or dehydration. (Refer to Chapter 5, Dehydration if needed.)

Which of his following lab results are indicative of decreased vascular volume?

( ) a. Hemoglobin 18.5 g
( ) b. Hematocrit 54%
( ) c. BUN 34 mg/dl
( ) d. Serum albumin 2.8 g/dl
( ) e. Serum potassium 3.1 mEq/L (hypokalemia)
( ) f. Serum sodium 118 mEq/L (hyponatremia)
( ) g. Serum chloride 103 mEq/L
( ) h. Serum calcium 8.8 mg/dl
( ) i. Serum phosphorus 3.1
( ) j. Serum lactate 17 mg/dl
( ) k. Serum uric acid 7.8 mg/dl

— — — — — — — — — — — — — — — — — —

a. X; b. X (very high normal); c. X; d. X; e. X (possible); f. X (possible); g. –; h. –; i. –; j. –; k. –.

---

338 Hemoglobin and BUN could be elevated because of (hemodilution/hemoconcentration) ____________________.

Protein, sodium, and potassium are shifted to the tumor site, thus (increasing/decreasing) ________________ oncotic pressure/colloid osmotic pressure. Will this have an effect on vascular fluid balance?

____________. Explain *______________________________.

— — — — — — — — — — — — — — — — — —

hemoconcentration; decreasing; Yes. Decreased oncotic pressure in the vascular space causes fluid loss or dehydration

---

339 On admission, Mr. Peterson's vital signs were T 100^{6} F, P 124, R 28, BP 96/60.

Which of his vital signs could be indicative of fluid volume deficit (vascular fluid):

( ) a. T 100^{6} F
( ) b. P 124
( ) c. R 28
( ) d. BP 96/60

— — — — — — — — — — — — — — — — — —

a. X; b. X; c. X (possible); d. X

---

340 IV albumin and packed RBCs were given to raise the serum oncotic pressure and to maintain vascular fluid and adequate blood pressure. He was started on hyperalimentation or TPN to improve his overall nutritional status prior to chemotherapy. His anorexia, nausea, and fatigue gradually lessened and his BP stabilized at 116/74–110/70.

The purpose of IV albumin and packed RBC administration is to *__________ __________, *____________________,

and *____________________.

- - - - - - - - - - - - - - - - -

raise the serum oncotic pressure; maintain vascular fluid; maintain adequate blood pressure.

---

341 Were Mr. Peterson's lab results three days after admission of normal values?

__________. If you answered no explain *____________________

____________________.

- - - - - - - - - - - - - - - - -

Yes. Creatinine is slightly elevated and serum sodium is low normal.

---

342 Ten days after he was admitted Mr. Peterson's condition had greatly improved and chemotherapy was scheduled to begin the next day. However, he became restless and irritable and by evening was disoriented and combative. He also became increasingly weak and vomited several times.

On his tenth day after admission were his serum electrolytes of normal

values? __________.

Name the electrolyte imbalances present:

1. ____________________.
2. ____________________.
3. ____________________.
4. ____________________.

- - - - - - - - - - - - - - - - -

No. 1. Hypokalemia. 2. Hypercalcemia. 3. Hypomagnesemia. 4. Hypophosphatemia

---

343 His immediate treatment consisted of the following:

1. Calcitonin 100 MRC units subcutaneously q12h.
2. Furosemide 20 mg IV push q6h.
3. Sodium phosphate ($Na_2 HPO_4$) 15 ml po, Tid.
4. Potassium and magnesium increased in TPN
5. IV rate increased to 150 ml/hr.

On the eleventh day the calcitonin dose was changed:

1. Calcitonin 300 MRC units subcutaneously q12h.
2. Mithramycin 1 mg IV push.

Why was calcitonin administered? * ____________________.

Why were mithramycin and calcitonin given on the eleventh day? * ____________________

____________________.

- - - - - - - - - - - - - - - - - -

To decrease the high serum calcium level.
The serum calcium level was still high and both hypocalcemic agents decreased the serum level.

---

344 To determine the cause of Mr. Peterson's hypercalcemia a bone scan and PTH levels were done. The bone scan was negative. This would indicate

* ____________________.

His PTH level of 455 pg Eq/ml (norm: 163–375 pg Eq/ml) indicated that the metastatic tumors in his lung were secreting PTH ectopically. This increased PTH secretion causes which two (2) electrolyte imbalances?

____________________ and ____________________.

- - - - - - - - - - - - - - - - - -

that hypercalcemia was *not* due to bone destruction;
Hypercalcemia and hypophosphatemia. Also hypomagnesemia

---

345 Mr. Peterson was started on a combination of chemotherapeutics (Cisplatin, cyclophosphamide/Cytoxan, and Doxorubicin HC1) to shrink the tumors and decrease the PTH secretion.

After five (5) days of chemotherapy his serum calcium was 9 mg/dl. Is this in normal range? ________________. He was alert, oriented, and eating and drinking well; soon he was discharged.

– – – – – – – – – – – – – – – – – –

Yes, on the lower normal side

---

346 A month later he was readmitted with a serum calcium level of 16.6 mg/dl. The type of imbalance present is ____________________.

A similar drug regime was followed and after his serum calcium level returned to normal he was discharged on daily calcitonin injections.

– – – – – – – – – – – – – – – – – –

hypercalcemia

CLINICAL EXAMPLE II

Steven Blackman, 15 years old, was rushed to the local emergency room by his parents when he awakened feeling weak and short of breath. His parents told the physician that he had complained of feeling tired for two or three weeks, had a sore throat and swollen glands, and two nosebleeds the week before. Steven's lungs were not congested, but the physician noted moderate lymphadenopathy and an enlarged liver and spleen. He was admitted to the adolescent unit with a suspected diagnosis of acute leukemia.

Vital signs were as follows:

T. 101^4 F, P. 124, R. 30, BP 100/66

Steven Blackman's laboratory results on admission, day 1, and day 6 are given in Table 79. Refer to the table as needed as you proceed with this clinical example.

TABLE 79. LABORATORY STUDIES: STEVEN BLACKMAN

Laboratory Tests	On Admission	Day 1	Day 6
Hematology:			
Hemoglobin (Hgb) (12.5–18 g/dl)	7		
Hematocrit (Hct) (36–54%)	18		
Platelets (150,000–400,000 mm^3)	56,000		
White blood cells (WBC) (5000–10000 mm^3)	31,000		
Differential:			
Blasts (0–5%)	46%		
Biochemistry:			
BUN (10–25 mg/dl)	30	28	41
Creatinine (Cr) (0.6–1.2 mg/dl)	1.8	1.7	2.6
Uric acid (3.5–7 mg/dl)	8.6		19
Lactate (serum) (6–16 mg/dl)	17		
Potassium (serum) (3.5–5.3 mEq/L)	5.7	4.1	5.9
Sodium (serum) (135–145 mEq/L)	140		
Calcium (serum) (4.5–5.5 mEq/L)	5.0		7.6
Magnesium (serum) (1.5–2.5 mEq/L)	2.0		
Chloride (Cl) (98–108 mEq/L)	103		
Phosphorus (serum) (2.5–4.5 mEq/L)	3.8		8.0
Arterial Blood Gases/ABGs:			
pH (7.35–7.45)	7.28	7.38	
pO_2 (70–90%)	90	92	
pCO_2 (35–45 mmHg)	34	38	
HCO_3 (24–28 mEq/L)	22	26	

---

347 On admission, Steven's hemoglobin and hematocrit were decreased because his bone marrow had been infiltrated by leukemic cells which led to decreased RBC production. This could be the reason for his ______________.

His platelet count was (high/low) ______________, a possible reason for his ______________.

His elevated WBCs are indicative of a serious problem.

- - - - - - - - - - - - - - - - - -

weakness or fatigue; low; nosebleeds. GOOD.

---

348 His ABGs indicate what acid-base disorder? * ______________ (refer to Chapter 3 if needed).

At his elevated serum lactate level the specific acid-base imbalance would be * ______________.

- - - - - - - - - - - - - - - - - -

Metabolic acidosis; his pH and HCO_3 are low.
lactic acidosis. This is the result of an abnormal carbohydrate metabolism in the blast cells (immature WBCs).

---

349 Steven's elevated BUN and creatinine could be caused by * ______________, because his uric acid, in particular, is elevated.

- - - - - - - - - - - - - - - - - -

renal insufficiency. Remember, if the BUN were slightly elevated and the creatinine, in normal range, this insufficiency could be due to ECFV deficit or dehydration. Renal insufficiency reduces the body's ability to excrete the excess lactic acid.

---

350 His potassium level indicated ______________. This could be cellular breakdown and acidotic state.

- - - - - - - - - - - - - - - - - -

hyperkalemia

---

351 Steven received Na HCO_3 in 5% D/W to correct acidosis. His serum potassium had to be carefully monitored while the acidotic state was being corrected.

Why? * ______________________________.

- - - - - - - - - - - - - - - - -

As the pH rises potassium will shift back into the cells and too much $NaHCO_3$ could cause alkalosis and hypokalemia

---

352 A bone marrow biopsy showed that the marrow had been almost entirely replaced by immature myeloblasts. This confirmed Steven's diagnosis of acute myelogenous leukemia (AML).

Steven was started on a chemotherapeutic regimen in an attempt to induce remission (absence of all leukemic cells).

High doses of chemotherapy result in the lysis of large numbers of malignant cells; therefore Steven was monitored for signs and symptoms of

* ______________________________.

- - - - - - - - - - - - - - - - -

tumor lysis syndrome

---

353 On day 6 he complained of flank pain and his urinary output decreased sharply. Lab values were indicative of tumor lysis syndrome.

Name the imbalances that are significant of this disorder.

______________, ______________,

______________ and ______________.

- - - - - - - - - - - - - - - - -

hyperuricemia; hyperkalemia; hypocalcemia; hyperphosphatemia

---

354 Two other lab values that were indicative of his decreased urine output were

* ______________ and * ______________.

- - - - - - - - - - - - - - - - -

elevated BUN and elevated creatinine

355 His uric acid was corrected with increased IV fluids, allopurinol, and $NaHCO_3$ (to alkalinize the urine promote uric acid excretion).

His nurses monitored I & O carefully and checked for edema, chest rales and level of consciousness (LOC). Explain why? * ______________________________

______________________________.

— — — — — — — — — — — — — — — —

decreased urinary output and increased IV fluids could result in fluid overload or ECFV excess.

NURSING ASSESSMENT

Assessment for and recognition of fluid and electrolyte imbalance in patients with cancer is especially difficult because the symptoms (e.g., anorexia, vomiting, fatigue, diarrhea, and muscle weakness) mimic those of chemotherapy, radiation, or general deterioration in advanced cancer.

Many fluid and electrolyte conditions in cancer patients can be reversed or controlled. It is essential that nurses assess them. Table 80 gives the nursing assessment for fluid and electrolyte imbalance in cancer patients. This table can be used in hospitals and clinics or at home. To use it as an assessment tool check the blanks in the nursing assessment column. For additional information or clarification use the comment column. Refer back to the table as needed.

TABLE 80. NURSING ASSESSMENT OF FLUID AND ELECTROLYTE IMBALANCE

Type of primary cancer: ______________________________.

Stage: ____________. Liver metastasis: ____________. Bone metastasis: ____________.

Observation	Nursing Assessment		Comments
Vital signs	Temperature	________	
	Pulse	________	
	Respiration	________	
	Blood Pressure	________	
		________	
	Heart Sounds	________	
	Peripheral pulses	________	
Intake	PO	________	
	IV infusions	________	
	Amounts	________	
Output	Amounts	________	
	Specific gravity	________	
	Urine osmolality	________	
	Polyuria	________	
	Oliguria	________	
	Anuria	________	

TABLE 80. (Continued)

Type of primary cancer: ______________________.

Stage: __________. Liver metastasis: __________. Bone metastasis: __________.

Observation	Nursing Assessment		Comments
Weight and skin changes	Daily weight	________	
	Skin turgor	________	
	Skin temperature	________	
	Edema	________	
	Ascites	________	
GI changes	Anorexia	________	
	Nausea	________	
	Vomiting	________	
	Diarrhea	________	
	Constipation	________	
	Bowel sounds	________	
	Abdominal distention	________	
	Abdominal cramps	________	
	Fistula	________	
	GI suction	________	
	Draining tube	________	
Respiratory changes	Dyspnea	________	
	Hyperpnea	________	
	Chest rales	________	
	Others	________	
Neurological changes	Headache	________	
	LOC changes	________	
	Irritability	________	
	Disorientation	________	
	Confusion	________	
	Paresthesia	________	
	Altered perception	________	
	Seizures	________	
	Coma	________	
State of being	Alert	________	
	Fatigue	________	
	Lethargic	________	
Muscular changes	Muscle weakness	________	
	Hyporeflexia	________	
	Hyper-reflexia	________	
	Muscle cramps	________	
	Twitching	________	
	Tetany signs	________	
Body chemistry and hematology changes	Hemoglobin	________	
	Hematocrit	________	
	Platelets	________	
	WBC	________	
	Differential	________	
	Electrolytes:		
	Potassium	________	
	Sodium	________	
	Calcium	________	
	Magnesium	________	
	Chloride	________	
	Phosphorus	________	

TABLE 80. (Continued)

Type of primary cancer: ______________________________.
Stage: ____________. Liver metastasis: ______________. Bone metastasis: ____________.

Observation	Nursing Assessment		Comments
	Serum osmolality	______	
	Protein	______	
	Albumin	______	
	BUN	______	
	Creatinine	______	
	Uric Acid	______	
	Lactate	______	
	ABGs:		
	pH	______	
	pO_2	______	
	pCO_2	______	
	HCO_3	______	
Chemotherapy	Drug	______	
	Dose, route	______	
	Side effects	______	
Radiation therapy	Dose	______	
	Times	______	
	Target area	______	

356 The following laboratory tests should be monitored:

( ) a. hemoglobin
( ) b. hematocrit
( ) c. electrolytes
( ) d. hormones, e.g., PTH
( ) e. phenylketonuria
( ) f. protein and albumin
( ) g. lactate
( ) h. uric acid
( ) i. lipoproteins
( ) j. BUN
( ) k. creatinine
( ) l. ABGs

a. X; b. X; c. X; d. X; e. –; f. X; g. X; h. X; i. –; j. X; k. X; l. X

CASE REVIEW

1. The three problems associated with primitive biochemical function of malignant cells are the following:

   a. ______________________.

   b. ______________________.

   c. ______________________.

2. Give the names of electrolyte imbalances associated with cancer and cancer therapy:

   a. Calcium ______________________.

   b. Calcium ______________________

      and phosphorus ______________________.

   c. Sodium ______________________.

   d. Potassium ______________________.

      ______________________.

   e. Phosphorus ______________________.

3. Hyperuricemia is a serious condition that results from massive cell destruction by radiation or chemotherapy. What is its effect on the renal tubules?

   *______________________. What could happen to the body? *______________________

   ______________________.

4. Symptoms of hyperuricemia include ______________________,

   ______________________, and ______________________.

5. Hypercalcemia is a serious condition. It is generally caused by *______________________

   ______________________ or *______________________

   ______________________. A very high calcium level could cause ______________________ and ______________________.

6. SIADH is another serious condition. Why? *______________________

   ______________________.

7. Give three (3) reasons why hypokalemia may occur:

   a. ______________________.

   b. ______________________.

   c. ______________________.

8. What type of acid-base imbalance is caused by anaerobic metabolism of glucose?

   *______________________.

   Explain how *______________________

   ______________________.

9. Name the four most common causes of fluid and electrolyte imbalance in cancer patients.

   a. ____________________.

   b. ____________________.

   c. ____________________.

   d. ____________________.

Mr. Peterson had been diagnosed 14 months before as having lung cancer. He was readmitted to the hospital because of severe weight loss, fatigue, and dyspnea.

10. Mr. Peterson's hemoglobin and hematocrit were elevated. This could be the result of ____________________.

11. His BUN and creatinine were slightly elevated. This could be caused by ____________________ or ____________________.

12. What effects do low serum protein and albumin levels have on the vascular fluid? ____________________. Why? * ____________________ ____________________.

On day 10 after admission Mr. Peterson's calcium level was high. He was restless, irritable, disoriented, combative, weak, and vomiting.

13. Two drugs Mr. Peterson received to decrease his serum calcium level were ____________________ and ____________________.

Steven Blackman, 15 years old, was diagnosed as having acute leukemia. The results of his hematology on admission indicated a low hemoglobin, hematocrit, and platelet count and an elevated WBC.

14. Steven's pH and HCO_3 were low and his serum lactate, slightly elevated. What acid-base disorder do you suspect? * ____________________.

15. His serum potassium level was 5.7 mEq/L. What potassium imbalance is present? ____________________. Why? * ____________________ ____________________.

16. When correcting acidosis with $NaHCO_3$ name the imbalance that could occur. * ____________________.

17. Steven was given chemotherapy in massive doses to destroy the large numbers of malignant cells. His uric acid was 19 mg/dl. What disorder frequently results from elevated uric acid? * ____________________.

    What imbalance did he have? ____________________.

18. Steven was given increased amounts of IV fluids, allopurinol, and $NaHCO_3$ (which alkalinizes the urine) to correct what imbalance? ________________________.

- - - - - - - - - - - - - - - - - -

1. a. cachexia
   b. abnormal (redundant) hormone secretion
   c. increased uric acid or breakdown of purine nucleic acids
2. a. hypercalcemia
   b. hypocalcemia; hyperphosphatemia
   c. hyponatremia
   d. hypokalemia; hyperkalemia
   e. hypophosphatemia
3. uric acid precipitates in the renal tubules and collecting ducts
   a small increased level could cause renal disorder and possible renal failure
4. hematuria, flank pain; nausea and vomiting.
   Also symptoms of renal failure.
5. bone destruction which results from metastasis or elevated PTH; confusion, disorientation, and cardiac arrest. Also brittle bones.
6. ADH may be secreted ectopically from neoplastic tumors, e.g., lung tumor.
   SIADH can cause severe water intoxication, hyponatremia, headaches, and behavioral changes
7. a. decreased dietary intake
   b. vomiting diarrhea
   c. excessive urinary output due to diuretics
   also ileostomy, colostom, and fistulas.
8. lactic acidosis (metabolic acidosis). anaerobic metabolism of glucose produces pyruvic acid which is converted during hypoxia to lactic acid.
9. a. cachexia
   b. tumor lysis syndrome
   c. uncontrolled cell growth
   d. ectopic hormone production
10. dehydration, fluid loss, or hemoconcentration
11. dehydration; renal insufficiency.
    If the BUN returns to normal after hydration the problem is dehydration.
12. decreased vascular volume.
    protein shifts to tumor site, thus decreasing oncotic pressure.
13. calcintonin; mithramycin
14. lactic acidosis
    If the results were based on ABGs, metabolic acidosis.
15. hyperkalemia
    cellular breakdown and acidotic state
16. hypokalemic alkalosis (potassium shifts back into cells)
17. tumor lysis syndrome
    hyperuricemia

18. high uric acid or hyperuricemia

NURSING DIAGNOSES

Fluid volume deficit related to decreased concentrating ability of renal tubules as evidenced by increased serum calcium levels, dehydration, and polyuria.

Potential for injury related to bone demineralization as evidenced by increased serum calcium levels.

Decreased activity tolerance related to excess serum calcium as evidenced by fatigue and muscular weakness.

Fluid volume deficit related to excessive excretion of sodium as evidenced by decreased serum sodium and a drop in blood pressure.

Fluid volume excess related to dilutional hyponatremia as evidenced by edema and ascites.

Alteration in sensory perception related to a serum sodium decrease as evidenced by restlessness, confusion, muscle twitching, and coma.

Alteration in fluid volume excess related to elevated levels of ADH as evidenced by hypoosmolality of plasma and hyponatremia.

Potential alteration in urinary elimination related to cell lysis and buildup of uric acid in the nephron as evidenced by elevated serum uric acid, uric acid crystals in urine, and increased BUN and creatinine.

Alteration in comfort related to renal calculi occluding ureter as evidenced by acute flank pain.

Alteration in nutrition of less than body requirements related to excessive fluid excretion in diarrhea or diuresis as evidenced by hypokalemia, abdominal distention, nausea, and vomiting.

Potential for injury related to hypokalemia as evidenced by weakness, fatigue and paresthesia.

Impaired physical mobility related to decreased serum magnesium as evidenced by hyperexcitability of neuromuscular function (e.g., muscle twitching, tremors, nystagmus, and cramps).

Impaired thought processes related to elevated serum hydrogen ion concentration as evidenced by disorientation, delirium, stupor, and coma.

Potential for physical injury related to hyperkalemia revealed by muscle weakness, paresthesias, and muscle cramps.

Alteration in cardiac output (decreased) related to elevated serum potassium as evidenced by ECG changes and cardiac arrhythmias.

Alteration in cerebral tissue perfusion related to red-blood-cell lysis which results from hypophosphatemia; as evidenced by irritability, seizures, coma.

Fluid volume deficit related to decreased serum protein as evidenced by a drop in blood pressure, elevated hemoglobin and hematocrit, and elevated BUN.

Potential alteration in urinary output related to hypocalcemia and hyperphosphatemia as evidenced by oliguria, azotemia, and tetany.

Potential for anxiety and fear related to the unknown (outcome) and death.

NURSING ACTIONS

General

1. Assess the patient with cancer frequently for signs and symptoms of fluid and electrolyte imbalance (see section on nursing assessment). Frequent assessment is needed during periods of uncontrolled cell growth, tumor lysis, and cachexia, and when ectopic hormone secretion is suspected and the patient is on chemotherapeutics.
2. Maintain optimal nutritional status by oral, tube, or parenteral feedings. When TPN is ordered flow rate and adverse reactions should be monitored closely.
3. Maintain adequate hydration, with PO fluids if appropriate or with IV fluid as ordered by the physician.
4. Report and assist in treating infections.
5. Maintain bedrest and decrease activity when indicated to lessen risk of falls or fractures. This is especially true when patient's serum calcium level is elevated and demineralization has occurred.
6. Elevate side rails and pad if patient is confused or at risk of seizures attendant on hypercalcemia, hypophosphatemia, hypomagnesemia, hyponatremia.
7. Medicate appropriately (following the physician's orders) for pain, nausea, or cardiac arrhythmia.
8. Notify the physician of significant changes in laboratory values ECG results, and signs and symptoms of imbalance.
9. Administer and monitor IV fluid, blood and blood products, and medications as ordered.
10. Reorient patient to time, place, and person if confusion or disorientation is apparent.
11. Institute a gradual increase in activity as conditions improve.
12. Explain in understandable language the conditions, progress, purpose, and procedural techniques to patient and family.
13. Allow the patient as much control as possible over his or her own care and environment.
14. Encourage the patient and family to ask questions and express their concerns.

Refer to preceding chapters for nursing actions related to patients with hypercalcemia, hypocalcemia, hyperkalemia, hypokalemia, hyponatremia, and hypomagnesemia.

SIADH

1. Make frequent assessments of I & O, urine specific gravity, LOC, breath sounds, weight, heart sounds, peripheral pulses, and edema.
2. Assess for nausea, vomiting, anorexia, weakness, and fatigue.
3. Monitor serum electrolytes and notify the physician of Na $<$120 mEq/L, K $<$3.5 mEq/L, Ca $<$8.5 mg/dl, or serum osmolality $<$280 mOsm/kg.
4. Monitor and report changes in LOC.

5. Report weight gain of greater than 2 kg/day.
6. Restrict fluid intake as ordered by physician.
7. Administer 3% saline IV and drugs (furosemide/Lasix, demeclocycline, or lithium carbonate) as ordered by the physician. The physician will most likely discontinue or decrease the dosage of vincristine and cyclophosphamide chemotherapy which can induce SIADH.

Hyperuricemia

1. Monitor urinary output and urine color, clarity, hematuria, and specific gravity.
2. Assess flank pain and medicate appropriately.
3. Assess and report signs of renal failure.
4. Monitor serum uric acid >7.0 mg/dl, urinary output <30 ml/hr, BUN >25 mg/dl, creatinine >1.2 mg/dl, or sudden weight gain with elevated BP, lung congestion, or edema. Notify the physician.
5. Administer allopurinol and Na HCO_3 and increase IV fluid rate as ordered before chemotherapy or radiation.
6. Offer fat-free milk and noncitrus juice (low in purines) if renal function is adequate.

Decreased Vascular Volume

1. Assess BP in supine and standing positions. Report to the physician a fall of >10 mm Hg, systolic, or >15 mm Hg, diastolic.
2. Monitor I & O, weight, serum electrolytes, serum protein, edema, and infection. Report significant changes to the physician.
3. Administer whole blood, packed RBCs, or albumin, as ordered.
4. Maintain optimal nutritional status, especially protein intake.

Hypophosphatemia

1. Assess weakness, fatigue, changes in LOC, anorexia, bone pain, edema, and neck-vein distention.
2. Monitor I & O, heart sounds, breath sounds, bowel sounds, ECG, and weight.
3. Notify the physician of serum phosphorus <2.5 mg/dl or <1.7 mEq/L or significant changes in serum calcium, potassium, magnesium, and pH.
4. Administer Na_2HPO_4 as ordered.
5. Offer fat-free milk and citrus juices (high in phosphorus and potassium).
6. Assess the signs of hyperphosphatemia after phosphorus replacement therapy is started.

Lactic Acidosis

1. Assess LOC, respiratory status, I & O, and cardiac arrhythmia.
2. Assess the signs and symptoms of renal insufficiency or hyperkalemia.
3. Monitor blood gases and serum electrolytes and report pH <7.35, serum potassium >5.3 mEq/L, and serum lactate >16 mg/dl to the physician.

4. Medicate for nausea, vomiting diarrhea, or cardiac arrhythmias as ordered by physician.
5. Administer IV $NaHCO_3$ as ordered to correct acidosis. It may be given PO with mild chronic lactic acidosis.
6. Administer calcium chloride or hyperosmolar glucose with insulin IV or Kayexalate in sorbitol PO or enema to correct hyperkalemia.
7. Monitor serum potassium levels carefully as acidosis is corrected. As potassium is excreted by the kidneys and potassium shifts back into cells hypokalemia may occur. IV potassium diluted in solution may be ordered to prevent or correct this problem.

CHRONIC OBSTRUCTIVE PULMONARY DISEASE

357 Chronic obstructive pulmonary disease (COPD) is a classification assigned to conditions associated with airway obstruction.

Examples of COPD are emphysema, chronic bronchitis, and asthma. Smoking is the leading cause of emphysema and bronchitis.

COPD is characterized by * ______________________.

- - - - - - - - - - - - - - - - - -

airway obstruction

---

358 Morphologic changes in emphysema are (1) thickening of bronchial walls caused by submucosal edema and excess mucous secretion; (2) loss of elastic recoil of lung tissue; and (3) destruction of the alveolar septa that promote overdistention and dead air space.

The leading cause of emphysema and bronchitis is ______________. Do you know another cause for COPD? * ______________________

______________________.

- - - - - - - - - - - - - - - - - -

smoking; hereditary (familial) trait for $alpha_1$ antitrypsin, chronic bacterial infection, air pollution, or inhalant of chemical irritants. VERY GOOD.

359 Name three (3) morphologic changes in COPD:

1. *________________________.
2. *________________________.
3. *________________________.

- - - - - - - - - - - - - - - -

1. thickening of bronchial walls
2. loss of elastic recoil of lung tissue
3. destruction to alveolar septa or overdistended alveoli

PHYSIOLOGIC FACTORS

Bronchitis and emphysema generally coexist. Table 81 lists the physiologic changes associated with COPD conditions. The rationale for physiologic changes is included. Refer to the table as needed.

TABLE 81. PHYSIOLOGICAL CHANGES IN CHRONIC OBSTRUCTIVE PULMONARY DISEASE (COPD)

Physiologic Changes	Rationale
Decreased elasticity of bronchiolar walls (loss of elastic recoil)	Loss of elastic recoil causes a premature collapse of airways with expiration. Alveoli become overdistended when air is trapped in the affected lung tissue and dead air space is increased. Overdistention leads to rupture and coalescence of several alveoli.
Alveolar damage	Chronic air trapping and airway inflammation lead to weakened bronchiolar walls and alveolar disruption. Coalescence of adjacent alveoli results in bullae (parenchymal air-filled spaces > 1 cm in diameter). The total area for gas exchange is greatly reduced and pulmonary hypertension may develop.
Mucous gland hyperplasia and increased mucous production	Oversecretion of mucous is commonly found in bronchitis and advanced emphysema. Increased mucous production can cause mucous plugs which lead to airway obstruction.
Inflammation of bronchial mucosa	Inflammatory infiltration and edema of the bronchial mucosa commonly occur in bronchitis but are also found in advanced emphysema. Edema and infiltration cause thickening of bronchiolar walls.
Airway obstruction	This condition is caused primarily by bronchioles narrowed by edema and mucous plugs. Obstruction is greatest on expiration. During inspiration bronchial lumena widen to admit air; the lumena collapse during expiration, however.
CO_2 retention Increased pCO_2 >45 mm Hg (hypercapnia)	Accumulation of carbon dioxide (CO_2) concentration in the arterial blood from inadequate gas exchange is the result of hypoventilation. CO_2 excess, >60 mm Hg, can lead to ventricular fibrillation.

TABLE 81. (Continued)

Physiologic Changes	Rationale
Norms: 35–45 mm Hg	
Respiratory acidosis	CO_2 retention results from hypoventilation. Water combines with CO_2 to produce carbonic acid, and with increased CO_2 retention respiratory acidosis occurs ($H_2O + CO_2$; $HCO_3 + H$). The arterial blood gases would be pH <7.35; pCO_2 >45 mm Hg.
Hypoxemia	Hypoxemia, or reduced oxygen (O_2), in the blood is frequently caused by airway obstruction and alveolar hypoventilation. Also the defects of thickened alveolar capillary membrane reduces O_2 diffusion.
Cor pulmonale (right-sided heart failure due to pulmonary hypertension)	Destruction of alveolar tissue leads to a reduction of the pulmonary capillary bed. Pulmonary hypertension occurs when 2/3 to 3/4 of the vascular bed is destroyed. The workload of the right ventricle is then increased, thus causing right ventricular hypertrophy and eventual CHF.
Increased red blood cell (RBC) count	Secondary polycythemia occurs as a compensatory mechanism with prolonged hypoxemia. Hemoglobin and hematocrit are increased to enhance O_2 transport.
Alpha$_1$ antitrypsin deficiency	A genetic predisposition to alpha$_1$ antitrypsin deficiency is present. Antitrypsin or trypsin inhibitor is produced in the liver. A deficit of antitrypsin allows proteolytic enzymes (released in the lungs from bacteria or phagocytic cells) to damage lung tissue. The result is emphysema.

---

360 Indicate which of the following physiological changes are contributing factors to beonchitis and emphysema:

( ) a. Increased elasticity of lung tissue.
( ) b. Overdistention of alveolar walls.
( ) c. Increased dead space.
( ) d. Increased mucous production.
( ) e. Effective gas exchange.
( ) f. Airway obstruction.

- - - - - - - - - - - - - - - - - -

a. –; b. X; c. X; d. X; e. –; f. X

---

361 Airway obstruction in COPD is caused primarily by thickening of the bronchiolar walls due to ________________ and * ________________

________________.

- - - - - - - - - - - - - - - - - -

edema and infiltration

---

362 Airway obstruction is greatest on (inspiration/expiration) ________________.

expiration

363 The normal value of pCO_2 is ________________ mm Hg.

The term for CO_2 retention is ________________. The pCO_2 would be

________________ mm Hg

35–45; hypercapnia; >45

364 The serious acid-base (A-B) imbalance that occurs in advanced COPD is

* ____.

Explain how this A-B imbalance occurs. * ____

____.

The pH would be ________________ and the pCO_2 would be

________________.

respiratory acidosis. CO_2 combines with water to produce carbonic acid and acidosis results. <7.35; >45 mmHg

365 The name for reduced O_2 in the blood is ________________.

Decreased arterial O_2 causes red blood cells to (increase/decrease) ____________.

The hemoglobin and hematocrit would be (increased/decreased) ____________.

Why? * ____.

hypoxemia; increase; increased. More RBC and hemoglobin are needed to carry oxygen.

366 Could a nonsmoker with an $alpha_1$ antitrypsin deficiency develop emphysema?

________________.

Give the reason for your answer. ______________________________

____.

Yes. Antitrypsin deficiency permits the proteolytic enzymes in the lungs to damage lung tissue

---

367 Cor pulmonale is right-sided heart failure caused by * ____________________

__________________. It can result in COPD when alveolar tissue destruction leads to * ______________________________________.

pulmonary hypertension; a reduction in the pulmonary capillary bed

---

Early signs and symptoms of bronchitis and/or emphysema are fatigue and dyspnea after exertion. Table 82 lists signs and symptoms of early-to-advanced COPD (bronchitis and emphysema). Refer to the table as needed.

TABLE 82. SIGNS AND SYMPTOMS OF COPD

Signs and Symptoms	Rationale
Chronic fatigue	Fatigue, an early sign of COPD is caused by hypoxia and the increased effort required to move air into and out of lungs.
Dyspnea	Difficulty in breathing and shortness of breath following exertion is an early sign of COPD. In advanced COPD dyspnea occurs with little or no exertion.
Vital Signs:	
BP increased	Increased blood pressure is due to increased sympathetic stimulation from stress.
P increased	Increased pulse rate results from poor oxygenation. The body's attempt to compensate for hypoxemia (decreased oxygen in the blood) by increasing the heart rate to carry more oxygen.
R labored and increased	Loss of elasticity of lung tissue causes the bronchioles to collapse during normal expiration, thus prolonging expiratory phase of respiration. Accessory respiratory muscles are used to improve alveolar ventilation and gas exchange.
Barrel-shaped chest (AP diameter > lateral diameter)	This is the result of loss of lung elasticity, chronic air trapping, and chest-wall expansion with chest rigidity. It may also be compounded by dorsal kyphosis which results from a bent-forward position to breathe. Shoulders are elevated and the neck appears shorten. Accessory muscles of respiration are used for breathing.

TABLE 82. (Continued)

Signs and Symptoms	Rationale
Cough (productive)	A cough is usually associated with bronchitis because of the excessive secretion of mucous glands. In emphysema a cough is associated with respiratory infection or cardiac failure. Bacterial growth in retained mucous secretions leads to repeated infections and a chronic cough.
Cyanosis	In advanced COPD marked cyanosis is due to poor tissue perfusion which results from hypoxemia. Signs of cyanosis may also appear when the hemoglobin is <5 g.
Clubbing of nails	Clubbing of nails is commonly seen in association with hypoxemia and polycythemia. It may be due to capillary dilation in an attempt to draw more oxygen to the fingertips.
Laboratory Results:	
Arterial Blood Gas (ABG) pH <7.35 pCO_2 >45 mm Hg HCO_3 >28 mEq/L pO_2 <70 mm Hg B.E. >+2 (respiratory acidosis with metabolic compensation).	Increased CO_2 retention and water cause an excessive amount of carbonic acid. As a result of too much carbonic acid in the blood acidosis develops and the pH is decreased. The pCO_2 is the respiratory component of ABG. A decreased pH and an increased pCO_2 indicate respiratory acidosis. The pO_2 may be normal or greatly reduced, depending on the degree of distortion of ventilation/perfusion ratio. An increased bicarbonate (HCO_3) indicates metabolic compensation to neutralize or decrease the acidotic state. A normal HCO_3 (24–28 mEq/L) indicates no compensation.
Hemoglobin (Hgb) and hematocrit (Hct) increased (hemoglobin may increase to 20g)	Increased Hgb and Hct are due to hypoxemia. More hemoglobin can carry more oxygen. Higher hemoglobin is a sign that cyanosis is more likely.
Electrolytes:	
Potassium, low to low normal Sodium, normal to slightly elevated	The serum potassium level may be between 3.0–3.7 mEq/L and could be the result of poor dietary intake from breathlessness, potassium-wasting diuretics, or chronic use of steroid (e.g., cortisone). Usually the sodium level is normal but could be elevated due to cardiac failure or excess IV saline infusions.

368 Two (2) early signs and symptoms of COPD are ______________ and
* ____.

— — — — — — — — — — — — — — — —

fatigue; dyspnea on exertion

369 Vital signs (VS) may be the following:

1. Blood pressure. ______________________.
2. Pulse rate. ______________________.
3. Respiration. ______________________.

The expiratory phase of respiration would be ______________________.

\- - - - - - - - - - - - - - - - - -

Increased; Increased; Labored
prolonged

---

370 A common characteristic of COPD is a barrel-shaped chest. Explain.

* ____.

\- - - - - - - - - - - - - - - - - -

Loss of lung elasticity and chest-wall expansion with chest rigidity or dorsal kyphosis from a bent position and using accessory respiratory muscles.

---

371 A cough is more common with what COPD problem? ______________________.

Respiratory infection is a complication of COPD. Explain why? * ______________________

____.

\- - - - - - - - - - - - - - - - - -

Bronchitis.
Bacteria grows in retained mucous secretions and respiratory infection results.

---

372 What major type of acid-base imbalance occurs in COPD? * ______________________

______________________.

Indicate the arterial blood gases that occur with this acid-base imbalance.
( ) a. pH 7.46
( ) b. pH 7.32
( ) c. pCO_2 55 mm Hg
( ) d. pCO_2 32 mm Hg

\- - - - - - - - - - - - - - - - - -

Respiratory acidosis
a. –; b. X; c. X; d. –

373 Explain the significance of an elevated bicarbonate/HCO_3 level in respiratory acidosis. *____.

— — — — — — — — — — — — — — — — —

Elevated HCO_3 level indicates metabolic compensation. Conservation of bicarbonate helps to decrease the acidotic state.

---

374 Which of the following laboratory results frequently occur in chronic COPD? Correct the incorrect responses.

( ) a. Hemoglobin decreased
( ) b. Hematocrit decreased
( ) c. Potassium low or low normal
( ) d. Sodium loss

— — — — — — — — — — — — — — — — —

a. —, increased to carry more oxygen; b. —, increased; c. X; d. —; sodium normal or slightly elevated

---

CLINICAL APPLICATIONS

Joseph Hall, age 54, has smoked two packs of cigarettes for the last 35 years. He has repeatedly been admitted to the hospital in the last 7 years for COPD or emphysema. The nurse assessed Mr. Hall's physiological status and noted dyspnea following exertion (breathlessness), barrel-shaped chest, and mild cyanosis. Joseph complained of constant fatigue. When checking his breath sounds the nurse noted a prolonged expiration rate. Vital signs were BP 150/86, P 94, R 26 labored.

---

375 Which of Mr. Hall's clinical signs and symptoms taken on admission would indicate COPD?

( ) a. Breathlessness
( ) b. Barrel-shaped chest
( ) c. Mild cyanosis
( ) d. Fatigue
( ) e. Prolonged expiration

— — — — — — — — — — — — — — — — —

a. X; b. X; c. X; d. X; e. X

---

376 A risk factor of COPD which can be linked to Mr. Hall's problem is

____________________.

\- - - - - - - - - - - - - - - - - -

smoking

The results of the laboratory studies ordered for Mr. Hall are given in Table 83.

TABLE 83. LABORATORY STUDIES: MR. HALL

Laboratory Tests	On Admission	Day 1	Day 2
Hematology:			
Red blood cells (4.5–6 million)	6.6	6.5	6.2
Hemoglobin (M: 13.5–18 g)	16.8	16.6	16.2
Hematocrit (M: 40–54%)	57.8	57.2	55.6
White Blood Cells (5–10 mm^3)	12.8	13.0	10.5
Biochemistry:			
Potassium (K) (3.5–5.0 mEq/L)	3.5	3.6	3.7
Sodium (Na) (135–145 mEq/L)	140	138	139
Chloride (Cl) (98–108 mEq/L)	107	106	106
Carbon dioxide (CO_2)	30	36	38
Arterial Blood Gases/ABG:			
pH (7.35–7.45)	7.24	7.32	7.34
pCO_2 (35–45 mm Hg)	73	68	60
pO_2 (70–100 mm Hg)	45	70 (with O_2)	76 (with O_2)
HCO_3 (24–28 mEq/L)	28	34	37
BE (–2 to +2)	+2	+6	+9

---

377 Mr. Hall's RBC, hemoglobin, and hematocrit were elevated because of

* ____________________________________.

\- - - - - - - - - - - - - - - - - -

poor oxygenation or hypoxemia

---

378 His potassium is low average. This could be the result of * ______________________________ ______________________________.

- - - - - - - - - - - - - - - - -

poor nutritional intake. He could be given a potassium-wasting diuretic for heart failure as a result of prolonged respiratory distress.

---

379 Sodium level may be normal or slightly elevated. His serum sodium value was in the (high/normal/low) range. ______________.

- - - - - - - - - - - - - - - - -

normal

---

380 The serum CO_2 is a bicarbonate determinant. An increased value (alkalosis) could be due to base excess from bicarbonate intake or to metabolic compensation.

In Mr. Hall's situation the cause was most likely * ______________________________ ______________.

- - - - - - - - - - - - - - - - -

metabolic compensation

---

381 On admission, his arterial blood gases/ABG indicated (respiratory alkalosis/ respiratory acidosis). * ______________________________ (with/without) ______________ metabolic compensation.

Explain his ABGs in response to your previous answer. * ______________________________ ______________________________.

- - - - - - - - - - - - - - - - -

respiratory acidosis; without
The pH is low, pCO_2 is high, and HCO_3 and BE are normal values; no compensation

---

382 On day 1 and day 2 his acid-base imbalance was *____________________________.

Is there metabolic compensation? ____________________. Explain. *______________

____.

\- - - - - - - - - - - - - - - - -

respiratory acidosis. Yes. The HCO_3 and BE were elevated. This is a compensatory mechanism that brings the pH close to normal value.

---

383 Oxygen was administered to Mr. Hall at 2 liters/minute with a nasal O_2 cannula (nasal prongs). A ventimask delivered 24, 28, 35, and 40% of oxygen. A non-breathing oxygen mask should *not* be used. Do you know why?

*____.

The rate of O_2 flow with a nasal cannula should be no greater than

_______________________.

\- - - - - - - - - - - - - - - - - -

It delivers high concentration of O_2, >90%. This could decrease the hypoxic respiratory drive and cause CO_2 narcosis
2 liters/minute.

---

384 Mr. Hall's pO_2 was (high/low) ____________________. His pO_2 indicates

____.

\- - - - - - - - - - - - - - - - - -

low; hypoxemia or low oxygen content in the blood

---

385 The nurse rechecked his breath sounds and noted rhonchi in the lower base of both lungs.

Mr. Hall's WBC was elevated. Rhonchi and elevated WBC could be indicative

of *____.

\- - - - - - - - - - - - - - - - - -

respiratory infection. This is the result of trapped mucous secretions and the presence of bacteria.

CLINICAL MANAGEMENT

Mr. Hall received bronchodilators and 2 liters/minute of oxygen. Breathing exercises were explained to him.

Table 84 lists methods of managing COPD.

TABLE 84. CLINICAL MANAGEMENT OF COPD

Management Methods	Rationale
Oxygen/O_2	Low-flow oxygen: 1–2 liters per minute with a nasal O_2 cannula or ventimask with 24 or 28% is suggested. Mechanical ventilators may be needed to decrease CO_2 retention and to aid in ventilation. Care should be taken to avoid CO_2 narcosis; O_2 that is too high decreases hypoxic respiratory drive.
Hydration	Fluid intake should be increased to 3–4 liters per day to liquify secretions and ease in expectoration *unless* cor pulmonale and/or CHF are present.
Bronchodilators: Isoproterenol/Isuprel Metaproterenol/Alupent Terbutaline/Brethine Aminophylline Theophylline products	The purpose of these agents is to dilate bronchial tubes (bronchioles), to expectorate mucous, and to improve ventilation. Following use of a bronchodilator the patient should deep-breath and cough. Bronchodilators can be administered through nebulizers (pressurized aerosols or IPPB with low-flow O_2 or compressed air), intravenously in IV fluids (Aminophylline), or orally (Theophylline products). Side effects of these drugs are tachycardia, cardiac arrhythmia, and nausea/vomiting.
Antibiotics	When a respiratory infection is present antibiotics are given intravenously (diluted in 100 ml of solution) or orally.
Chest physiotherapy	Chest clapping loosens the thick, tenacious mucous secretions that must be "coughed up". Deep breathing and coughing should follow.
	Diaphragmatic breathing improves tidal volume and increases alveolar ventilation.
	Pursed-lip breathing prevents airway collapse so that trapped air in the alveoli can be expelled.
Exercise	Walking and stationary bicycling improve respiratory status and state of well-being.
Relaxation techniques	Practicing relaxation techniques decreases anxiety, fear, and panic. Decreased dyspnea can result from relaxation.

---

386 Low-flow oxygen is frequently needed to decrease hypoxemia. When a nasal O_2 cannula is used, the flow should be ______________________.

If a high concentration of oxygen is delivered what could happen to the respiratory drive? *______________________________________.

- - - - - - - - - - - - - - - - -

1–2 liters per minute; High concentration of O_2 would decrease hypoxic respiratory drive.

387 Why is hydration important in the management of COPD?

*______________________________.

Increased fluid intake should be contraindicated when ______________

and/or ______________ are present.

— — — — — — — — — — — — — — — — — —

to liquify secretions and ease in expectoration
cor pulmonale or CHF

---

388 Bronchodilators are used for the following purposes:

a. *______________________________.

b. *______________________________.

c. *______________________________.

— — — — — — — — — — — — — — — — — —

a. to dilate the bronchial tubes/bronchioles; b. to expectorate mucous;
c. to improve ventilation

---

389 The following are side effects that may result from constant use or overuse of bronchodilators:

( ) a. Bradycardia
( ) b. Tachycardia
( ) c. Nausea, vomiting
( ) d. Cardiac arrhythmias
( ) e. Hypotension
( ) f. Skin rash

— — — — — — — — — — — — — — — — — —

a. –; b. X; c. X; d. X; e. –; f. –

390 Give examples of chest physiotherapy and their purposes:

a. *______________________________.

Purpose *______________________________.

b. *______________________________.

Purpose *______________________________.

c. *______________________________.

Purpose *______________________________.

- - - - - - - - - - - - - - - - - -

a. chest clapping. To loosen thick, tenacious mucous secretions.
b. diaphragmatic breathing. To increase alveolar ventilation.
c. pursed-lip breathing. To prevent airway collapse

---

391 Use of a relaxation technique performed daily can improve ventilation. Explain how. *______________________________

______________________________.

- - - - - - - - - - - - - - - - - -

It decreases anxiety, fear, and panic related to breathlessness

---

CASE REVIEW

Mr. Joseph Hall, age 54, has had numerous admissions for severe dyspnea related to emphysema. His clinical signs, symptoms, and findings are stated under clinical applications.

1. Emphysema is classified as a *______________________________

   ______________________________. Other lung disorders under this classification are ______________________ and ______________________.
2. Name three (3) physiologic changes that occur in COPD:

   a. *______________________________.

   b. *______________________________.

   c. *______________________________.
3. What is the major risk factor in COPD? ______________________.

4. Name the protein produced in the liver that inhibits proteolytic enzymes in the lung. ________________. A deficit of this protein causes ________________ and the disease ________________.
5. Signs and symptoms of COPD are the following:
   a. ________________.
   b. ________________.
   c. ________________.
   d. ________________.
6. Mr. Hall's RBC, hemoglobin, and hematocrit values were elevated. Give the reason. * ________________.

Mr. Hall's ABG on admission were pH 7.35, pCO_2 73 mm Hg, and HCO_3 28 mEq/L. On day 2 his ABG were pH 7.34, pCO_2 60 mm Hg, HCO_3 37 mEq/L, BE +9.

7. Mr. Hall's acid-base imbalance on admission was * ________________ ________________. Was there metabolic compensation? ________________.
8. On day 2 his ABG indicated * ________________. Was there metabolic compensation? ________________. Explain * ________________ ________________.

Mr. Hall received 2 liters per minute of oxygen and ampicillin in IV fluids.

9. Patients with emphysema should not be given a high concentration of oxygen. Why? * ________________.
10. Why was Mr. Hall given ampicillin? * ________________ ________________.
11. Name three (3) nursing actions that assisted Mr. Hall with his breathing.
    a. ________________.
    b. ________________.
    c. ________________.

- - - - - - - - - - - - - - - -

1. chronic obstructive lung disease (COPD); bronchitis; asthma
2. a. decreased elasticity of lung tissue or loss of elastic recoil.
   b. alveoli overdistention and damage.
   c. excess mucous production.

others are edema of the bronchial mucosa and hypoxemia
3. smoking
4. antitrypsin; damage to lung tissue; emphysema
5. a. fatigue
   b. dyspnea
   c. barrel-shaped chest
   d. coughing
   others are cyanosis and abnormal ABG.
6. poor oxygenation or hypoxemia
7. respiratory acidosis. No
8. respiratory acidosis
   Yes
   HCO_3 and BE are elevated to decrease acidotic state
9. delivery of a high concentration of oxygen decreases hypoxic respiratory drive
10. His WBC was elevated indicating a possible infection (respiratory).
11. a. chest clapping
    b. teaching diaphragmatic breathing
    c. teaching pursed-lip breathing
    others are explaining relaxation technique, mild exercise, and hyration.

NURSING DIAGNOSES

Ineffective breathing patterns related to CO_2 retention and poor gas exchange secondary to COPD.

Ineffective airway clearance related to excess mucous secretions and the collapse of the bronchial tubes secondary to COPD.

Impaired gas exchange related to alveoli damage and the collapse of bronchial tubes (bronchioles).

Potential alteration in tissue perfusion related to hypoxemia.

Potential alteration in comfort related to breathlessness and muscle pain (diaphragm, intercostal).

Potential for ineffective coping related to breathlessness and life style.

Activity intolerance related to breathlessness and fatigue.

Knowledge deficit related to a lack of understanding of breathing exercises and of following a drug regime.

Potential for injury (lungs) related to smoking and respiratory infection.

Potential for self-care deficit related to the inability to take part in the activities of daily living because of dyspnea or breathlessness.

Potential alteration in the family process related to the inability to deal with the family member's chronic disease and/or to participate in the care-giving role.

Anxiety related to breathlessness, dependence on others, and the treatment regime.

Potential alteration in cardiac output related to breathlessness and hypoxemia secondary to COPD.

Potential alteration in nutrition related to breathlessness.

Potential for impaired physical mobility related to breathlessness.

NURSING ACTIONS

1. Explain the major physiologic changes associated with COPD; i.e., decreased elasticity of lung disease, alveoli overdistention and damage, and airway obstruction.
2. Observe signs and symptoms of COPD; i.e., changes in vital signs, dyspnea, fatigue, barrel-shaped chest, and cyanosis.
3. Check laboratory results (hematology, electrolytes, and arterial blood gases) and report abnormal findings.
4. Check breath sounds for rhonchi and rales. Provide chest physiotherapy (chest clapping) for rhonchi and have patient deep-breath and cough.
5. Teach the patient how to do breathing exercises; i.e., pursed-lip breathing (to prevent airway collapse) and diaphragmatic breathing (to increase alveolar ventilation).
6. Teach the patient not to get overfatigued and to avoid smoking, use of chemical irritants (bleaches, paints, and aerosol hair spray), people with respiratory infections, air pollution, excess dust, pollen, and extreme heat or cold weather, all of which increase breathlessness.
7. Instruct the patient to use bronchodilators as directed. Overuse of pressurized bronchodilator aerosol can cause a rebound effect.
8. Monitor fluid and food intake. Hydration is important to liquify tenacious mucous secretions. Frequent small feedings may be necessary.
9. Assist the patient with the activities of daily living (ADL) as needed. Encourage the patient to try mild exercises in the afternoon or when breathlessness is not severe. Avoid early mornings when mucous secretions are increased and after meals when energy is needed for digestion.
10. Encourage the patient to select and engage in a relaxation technique; help in the selection.
11. Teach the patient to recognize early signs of respiratory infection; i.e., change in sputum, elevated temperature, and coughing.
12. Explain the treatment and nursing care to the patient and family members and answer questions or refer them to the physician.
13. Be supportive of patient and family members.

CONGESTIVE HEART FAILURE

392 Heart failure is the inability of the heart to pump adequate supply of blood (pump failure) to meet the body's metabolic needs. It is the result of increased stress on the heart and is secondary to major disease entities, e.g., arteriosclerotic heart disease, hypertension, pulmonary diseases, kidney diseases, and hyperthyroidism.

Congestive heart failure (CHF) is circulatory congestion related to pump failure.

With congestive heart failure, there is circulatory congestion related to the heart's inability to pump *____________________.

- - - - - - - - - - - - - - - - - -

adequate supply of blood

---

393 Congestive heart failure is frequently (primary/secondary) ________________ ________________ to major disease entities.

- - - - - - - - - - - - - - - - - -

secondary

PHYSIOLOGIC FACTORS

Table 85 gives the physiologic factors associated with congestive heart failure and the compensatory mechanisms to prevent heart failure. Left-sided and right-sided heart failure are included as part of the physiologic factors. Study the table and refer to it as necessary.

TABLE 85. PHYSIOLOGIC CHANGES ASSOCIATED WITH CONGESTIVE HEART FAILURE

Physiologic Factors	Rationale
Cardiac reserve (decreased) ↓	Decreased cardiac reserve is inability of heart to respond to increased burden, e.g., fever, exercise, or excitement.
Cardiac compensation	The heart, in early heart failure, will compensate for loss of cardiac reserve. The heart increases its cardiac output through ventricular dilatation, ventricular hypertrophy, and tachycardia.
Ventricular dilatation	Muscle fibers of myocardium increase in length and so ventricle enlarges to augment its output. Heart muscle will stretch to certain point and then cease to increase heart contractility. Dilated heart will need more oxygen, however, with "normal" coronary blood flow, heart muscle will not receive sufficient O_2.
Ventricular hypertrophy	There is increased thickening of ventricular wall, which increases weight of heart. Ventricular hypertrophy mostly follows dilatation, and hypertrophy will aid in heart contractility. Hypertrophied heart works harder than normal heart and has greater O_2 need.
Tachycardia	Increased heart rate is least effective of three compensatory mechanisms. Heart rate will increase to point that ventricles are unable to fill adequately.
Cardiac decompensation	It occurs when three compensatory mechanisms fail to maintain heart function and adequate circulation. Symptoms develop with activity.
Left-sided heart failure	It generally results from left ventricular damage to myocardium.
	Heart at first is unable to eject full blood flow from ventricle. Three compensatory mechanisms come into play.
	With compensatory mechanism failure, residual blood remains in dilated ventricle.
	Left atrium dilates and atrial hypertrophy results. When atrium is unable to receive blood from pulmonary veins, pulmonary congestion or pulmonary edema occurs.
	It frequently is due to hypertension, myocardial infarction (heart attack), rheumatic fever affecting aortic valve, or syphilis.
	Symptoms of left-sided heart failure are similar to symptoms of overhydration (Chapter 5).
Right-side heart failure	It generally results from increased pressure in pulmonary vascular system. Right ventricle tries to pump blood to congested lungs, thus meeting resistance. Blood and fluids are "backed up" in venous circulation, thus causing congestion in gastrointestinal tract, liver, and kidneys. Peripheral edema also occurs.
	Right-sided heart failure generally follows left-sided heart failure; however, occasionally it is independent of left-sided failure.
	Symptoms of right-sided heart failure include liver congestion and enlargement, fullness in abdomen, and peripheral edema of lower extremities, mostly refractory and pitting.

394 With congestive heart failure, there is a(n) (increase/decrease) ________ in cardiac reserve.

What is cardiac reserve? * ____

____.

- - - - - - - - - - - - - - - - - -

decrease; ability of the heart to respond to increased burden

---

395 In early heart failure, the heart will compensate in order to meet oxygen and circulatory needs. The three methods by which the heart compensates are * ________________________________, * ________________________________, and ________________________________.

- - - - - - - - - - - - - - - - - -

ventricular dilatation; ventricular hypertrophy; tachycardia

---

396 Physiologically, how does ventricular dilatation occur? * ________________________________

____

With ventricular dilatation, does the heart need (more/less) oxygen? ________ ________________.

- - - - - - - - - - - - - - - - - -

The muscle fibers of the myocardium increase in length and so the ventricle enlarges to increase output and circulation
more

---

397 Explain the rationale for ventricular hypertrophy in CHF. * ________________________________

____

____.

- - - - - - - - - - - - - - - - - -

There is an increased thickening of the ventricle wall, which increases heart contractility

---

398 Tachycardia is * ______________________.

Tachycardia will increase in rate until the ventricles are * ______________________ ______________________.

- - - - - - - - - - - - - - - - - -

a fast heart beat or fast pulse rate (frequently over 100);
unable to fill adequately

---

399 What is cardiac decompensation? * ______________________ ______________________.

- - - - - - - - - - - - - - - - - -

It is the failure of the three compensatory mechanisms to maintain heart function

---

400 Left-sided heart failure generally results from * ______________________ ______________________.

When the compensatory mechanisms fail with left-sided failure, what happens to the ventricle? * ______________________ ______________________ and atrium * ______________________ ______________________.

- - - - - - - - - - - - - - - - - -

left ventricle damage
Ventricle remains dilated with residual blood; dilates and atrial hypertrophy results

---

401 What type of edema occurs from left-sided heart failure? ______________________ ______________________.

Name four symptoms of left-sided heart failure (overhydration):

a. * ______________________. c. * ______________________.

b. ______________________. d. * ______________________.

- - - - - - - - - - - - - - - - - -

pulmonary

a. constant, irritating cough b. dyspnea
c. engorged veins d. moist rales

---

402 Right-sided heart failure generally results from *____________________

____________________________________.

What occurs when the right ventricle fails adequately to pump blood to the lungs? *____________________________________

____

____.

— — — — — — — — — — — — — — — — — —

increased pressure in the pulmonary vascular system. Blood and fluids are "backed up" in the venous circulation, increasing venous pressure and causing GI, liver, and kidney congestion. Peripheral edema in low extremities also results

---

403 Explain the occurrence of right-sided heart failure in relation to left-sided heart failure. *____________________________________

____.

What type of peripheral edema occurs? *____________________

____________________________. Should it be assessed in the morning or in the evening? *____________________________

— — — — — — — — — — — — — — — — — —

Right-sided heart failure generally follows left-sided heart failure.
Refractory or nondependent edema (Chapter 5, Frame 50); also pitting edema in the morning

---

CLINICAL APPLICATIONS

Mrs. Allen, age 68, was admitted to the hospital with congestive heart failure. She had shortness of breath when walking up a flight of stairs. The nurse assessed Mrs. Allen's physiologic status and noted irritating cough, dyspnea on exertion, moist rales in the lungs, hand vein engorgement in upward position after 30 seconds, and swelling in the ankles and feet. Her BP was 154/96 and pulse was 110. Her ECG showed ventricular hypertophy.

---

404 Mrs. Allen had two compensatory physiologic factors present for maintaining cardiac function, which were ______________________________ and *______________________________.

Were the compensatory mechanisms effective? ____________ Explain why.

*______________________________.

______________________________.

- - - - - - - - - - - - - - - - - -

tachycardia; ventricular hypertrophy
No. Most likely, ventricular dilatation was present; symptoms of congestive heart failure were still present

---

405 According to Mrs. Allen's symptoms, which type(s) of heart failure was (were) present?

( ) a. Left-sided heart failure
( ) b. Right-sided heart failure

- - - - - - - - - - - - - - - - - -

a. X; b. X

---

406 The nursing assessment of Mrs. Allen to determine pulmonary congestion includes:

a. *______________________________.

b. ______________________________.

c. *______________________________.

d. *______________________________.

- - - - - - - - - - - - - - - - - -

a. irritating cough
b. dyspnea
c. moist rales
d. hand vein engorgement

407 Swelling in the feet and ankles was indicative of ________________-sided heart failure.

— — — — — — — — — — — — — — — —

right

Table 86 gives the laboratory results for Mrs. Allen on day of admission, second, and fourth days. Be able to state which laboratory results are normal and which are not. Explain the abnormal laboratory findings.

TABLE 86. LABORATORY STUDIES OF MRS. ALLEN

Laboratory Tests	On Admission	Day 1	Day 4
Hematology			
Hemoglobin (12.9–17.0 g)	12.5		
Hematocrit (40–46%)	40		
WBC (white blood count)	8200		
Biochemistry			
BUN (blood urea nitrogen) (10–25 mg/dl)*	28	24	18
Plasma/serum CO_2 †	22	24	24
Plasma/serum chloride (98–108 mEq/L)	107	106	107
Plasma/serum sodium (135–146 mEq/L)	151	148	143
Plasma/serum potassium (3.5–5.3 mEq/L)	3.6	3.8	4.0

*mg/100 ml = mg/dl

†*Plasma* and *serum* are used interchangeably.

---

408 Mrs. Allen's serum sodium was ________________________.
Sodium retention can cause Mrs. Allen's extracellular fluid volume to (rise/decrease) ________________________.

What type of fluid imbalance was present? ________________________.

— — — — — — — — — — — — — — — —

elevated or increased; rise; edema

409 Mrs. Allen's low-average serum potassium may be due to (increase/decrease

________________ in ECF. Explain why.

* ________________________________

________________________________.

Give another reason why Mrs. Allen's serum potassium may be low-average.

* ________________________________

________________________________.

— — — — — — — — — — — — — — — — —

increase. Potassium may be diluted due to an increase of ECF
Lack of food intake containing potassium or cellular breakdown from insufficient circulation. If she were receiving diuretics, this might cause low serum K

CLINICAL MANAGEMENT

The "three D's" are frequently employed in the management of congestive heart failure. They are:

1. Diet
2. Digitalization
3. Diuretics

The clinical management for CHF in Mrs. Allen's case incorporated the "three D's."

410 *Diet*

Mrs. Allen was placed on a low-sodium diet and her fluid intake was limited to 1200 ml (300 ml below daily requirement).

By limiting salt and water intake this would:

( ) 1. increase edema
( ) b. prohibit further increase of edema
( ) c. decrease water intoxication

— — — — — — — — — — — — — — — — —

a. –; b. X; c. –

411 *Digitalization*

Digitalization is the process of increasing the serum level of digitalis to achieve the desired physiologic effect.

Digitalis preparations, which are under the classification of cardiotonics, will slow down the ventricular contractions and make them more forceful. Examples of digitalis preparations are Digoxin, Digitoxin, Gitaligin, Cedilanid, and digitalis leaf. Digoxin is frequently the cardiotonic of choice for digitalizing patients with poor cardiac output.

Digitalis is classified as a ____________. This drug will slow down the *____________ and make the heart beat *____________.

Mrs. Allen was digitalized with Digoxin and then placed on a daily maintenance dose of Digitoxin 0.2 mg. This would (increase/decrease) ____________ cardiac output. Blood circulation would then be ____________.

The urinary output would be ____________.

It is important that you remember the toxic effects of digitalis preparations, which are pulse below 60, nausea, vomiting, and anorexia.

— — — — — — — — — — — — — — — — —

cardiotonic; ventricular contractions; more forcefully; increase; improved or increased; increased

---

412 *Diuretics* (Review Chapter 2, Table 7, and the section on diuretics.)

Diuretics are used for the excretion of sodium and water. Many diuretics will excrete not only sodium, water, and chloride, but also the valuable electrolyte ____________.

— — — — — — — — — — — — — — — — —

potassium

---

413 Frequently, physicians will prescribe a potassium-sparing diuretic with a potassium-wasting diuretic to prevent excessive loss of what ion?

____________.

— — — — — — — — — — — — — — — — —

potassium

---

414 Identify diuretics that are potassium-wasting and potassium-sparing by placing K-W for potassium-wasting diuretics and K-S for potassium-sparing diuretics.

____ a. hydrochlorothiazide (HydroDIURIL)

____ b. triamterene (Dyrenium)

____ c. furosemide (Lasix)

____ d. mannitol

____ e. spironolactone (Aldactone)

— — — — — — — — — — — — — — — — —

a. K-W; b. K-S; c. K-W; d. K-W; e. K-S

---

415 Give at least five symptoms of hypokalemia-potassium deficit. (Refer to Chapter 2 if necessary.)

a. *________________________________.

b. ________________________.

c. ________________________.

d. *____.

e. *________________________________.

— — — — — — — — — — — — — — — — —

a. muscular weakness
b. dizziness
c. arrhythmia
d. silent ileus—decrease peristalsis
e. abdominal distention

CASE REVIEW

Mrs. Allen, age 68, was in congestive heart failure on admission. The clinical assessment of her symptoms and findings were stated under clinical applications.

1. In early heart failure, name the three compensatory mechanisms that would assist Mrs. Allen's cardiac output to maintain circulation.

   a. *____.

   b. *____.

   c. ____.

2. Was Mrs. Allen in cardiac compensation or decompensation?

   ______________________. Explain your answer. *______________________

   ____.

3. Does the heart need (more/less) ______________ blood when there is ventricular dilatation and hypertrophy?
4. With right-sided heart failure, will the venous (hydrostatic) pressure be

   (increased/decreased)? ______________________. Explain why.

   *____.

   ____.

The nurse assessed Mrs. Allen's physiologic status and identified symptoms of left-sided heart failure and right-sided heart failure.

5. Mrs. Allen's symptoms of left-sided heart failure included:

   a. *____.

   b. *____.

   c. *____.

   d. *____.

   e. *____.

   f. *____.
6. The symptoms of left-sided heart failure are similar to symptoms of

   ____.

7. Mrs. Allen's symptom of right-sided heart failure was *______________________

   ____.

8. Clinical management for Mrs. Allen consisted of the "three D's," which included

   ______________________________, ______________________________

   ______________________________ and ______________________________.
9. The diuretic Mrs. Allen was receiving was HydroDIURIL. Is this a (potassium-

   wasting/potassium-sparing) ______________________________________

   diuretic? Which potassium imbalance can occur? (hypokalemia/hyperkalemia)?

   ____.

10. If Mrs. Allen's serum potassium was below average, what effect would this have on digitoxin? (Review Chapter 2 if necessary.)
11. Give three symptoms of digitalis intoxication.

    a. *______________________________________.

b. ______________________________.

c. *______________________________.

1. a. ventricular dilatation
   b. ventricular hypertrophy
   c. tachycardia
2. decompensation. The compensatory mechanisms failed to maintain cardiac output, since symptoms were present
3. more
4. increased. The blood is backed up in the venous system, causing increased pressure
5. a. irritating cough
   b. dyspnea on exertion
   c. moist rales
   d. hand vein engorgement
   e. pulse 110—tachycardia
   f. ventricular hypertrophy
6. overhydration
7. swelling in the ankles and feet
8. diet; digitalization; diuretics
9. potassium-wasting; hypokalemia
10. Hypokalemia will enhance the action of any digitalis preparation, making the Digitoxin stronger (cumulative action can occur)
11. a. bradycardia—pulse ↓ 60 or arrhythmia, or both
    b. anorexia
    c. nausea and vomiting

NURSING DIAGNOSES

Alteration in fluid volume excess related to cardiac decompensation secondary to left-sided and right-sided heart failure.
Alteration in tissue perfusion related to cardiopulmonary insufficiency.
Potential for electrolyte imbalance related to drug therapy (diuretics).
Potential for ineffective breathing patterns related to fluid retention in lung tissues.
Potential for impaired skin integrity related to fluid accumulation in the extremities and buttocks.
Potential for self-care deficit due to the inability to perform activities of daily living.

NURSING ACTIONS

1. Recognize the major physiologic factors associated with congestive heart failure, which are decreased cardiac reserve (inability to respond to stress, fever, exercise), ventricular dilatation and hypertrophy, and increased heart rate.

2. Assess for signs and symptoms of left-sided heart failure (overhydration or pulmonary edema), i.e., constant, irritated cough, dyspnea, neck and/or hand vein engorgement, chest rales.
3. Assess for signs and symptoms of right-sided heart failure, i.e., pitting peripheral edema, liver enlargement, fullness of abdomen.
4. Teach patient to assess for peripheral edema in the morning and not at night. Evening peripheral edema could be due to venous stasis, e.g., varicose veins.
5. Teach patient not to use table salt to season foods. Salt contains sodium, which can cause water retention.
6. Instruct the patient to eat foods rich in potassium (fruits, vegetables) if he or she is taking a potassium-wasting diuretic and Digoxin. Hypokalemia can enhance the action of digoxin and could cause digitalis toxicity (slow, irregular pulse, nausea/vomiting).
7. Assess for signs and symptoms of hypokalemia (serum potassium deficit), i.e., dizziness, muscular weakness, abdominal distention, diminished peristalsis, and arrhythmia, if patient has been receiving potassium-wasting diuretics for several months.

CIRRHOSIS OF THE LIVER

PHYSIOLOGIC FACTORS

416 Liver (hepatic) disease is frequently associated with sodium and water retention caused by increased portal pressure and increased aldosterone secretion.

Water is retained in excess to sodium. With hepatic disease, there is a sodium and water retention caused by *_______________ _______________ and *_______________.

— — — — — — — — — — — — — — — — —

increased portal pressure; increased aldosterone

---

417 Water is retained in _______________ to sodium.

— — — — — — — — — — — — — — — — —

excess

418 Aldosterone (an adrenal cortical hormone) has a sodium- and water-retaining effect and a potassium-excreting effect.

With an increase of aldosterone secretion, the serum potassium would be

___________________________.

\- - - - - - - - - - - - - - - - - -

decreased

---

419 In spite of the presence of an excess of total body sodium, hyponatremia results

because water is retained in *____________________________________

____.

\- - - - - - - - - - - - - - - - - -

excess to sodium

---

420 Other contributing causes of hyponatremia are:

a. Sodium moves into the intracellular space, replacing potassium, which leaves the cells because of dehydration, malnutrition, or diuresis
b. Prolonged use of potent diuretics
c. Low (or restricted) sodium diet

Hyponatremia frequently occurs with liver dysfunction because water is retained in excess to sodium, sodium shifts into the intracellular spaces, prolonged use of *__________________________, and/or

*__________________________________.

\- - - - - - - - - - - - - - - - - -

potent diuretics; low sodium diet

421 The four contributing causes of hyponatremia are:

1. *____________________.
2. *____________________.
3. *____________________.
4. *____________________.

– – – – – – – – – – – – – – – – –

1. water is retained in excess to sodium
2. sodium moves into the intracellular spaces
3. prolonged use of potent diuretics
4. low sodium diet

---

422 In addition to potassium loss resulting from increased aldosterone secretion, dehydration, and ____________________, diuresis leads to further potassium loss.

– – – – – – – – – – – – – – – – –

malnutrition

---

423 In Chapter 5, the section on edema, ascites was defined. Can you recall its meaning? *____________________

____________________.

– – – – – – – – – – – – – – – – –

An accumulation of fluid in the peritoneal cavity (abdomen)

---

424 With cirrhosis, ascites may develop suddenly or insidiously.

Ascites is *____________________

____________________.

– – – – – – – – – – – – – – – – –

an accumulation of fluid in the peritoneal cavity

One of the major complications of cirrhosis of the liver is the development of ascites. It occurs frequently with cellular liver damage and portal hypertension. The portal circulation, which is the liver's circulatory system, becomes affected when there is liver cellular damage. The blood cannot circulate through the liver sufficiently, thus the portal pressure is greatly increased.

Since ascites is a major problem in fluid and electrolyte imbalance and since it frequently accompanies severe cirrhosis of the liver, emphasis will be placed on ascites.

Study Table 87 carefully, noting where there is an increase or decrease in the physiologic factors resulting in ascites, and the rationale for these physiologic factors. Refer to this table as needed.

TABLE 87. PATHOGENESIS OF ASCITES

Physiologic Factors		Rationale
Portal obstruction hypertension	Increased ↑	Portal obstruction resulting in portal vein hypertension will not of itself cause ascites. When ascites does accompany portal hypertension, its presence can be explained by associated liver damage. Surgical relief of portal hypertension will relieve ascites without its fluid accumulating elsewhere in body. Thus, portal hypertension influences fluid accumulation in abdomen, but fundamental cause as ascites lies in damage to cellular structure of liver.
Capillary permeability	Increased ↑	Capillary permeability is increased due to permeability defect of capillary endothelium, which contributes to transudation of fluid from portal system into abdomen.
Plasma osmotic pressure	Decreased ↓	With increased liver congestion and failure of liver to synthesize albumin, protein-rich fluid will leave capillaries and pass into abdominal cavity, thus lowering plasma osmotic pressure (Hypoproteinemia and hypoalbuminemia result.)
Retention of sodium and water	Increased ↑	When plasma volume is reduced, an increased hormonal response (aldosterone) occurs, which will decrease urinary output and cause retention of sodium and water.

---

425 Place I for increased and D for decreased beside the physiologic factors associated with ascites.

____ 1. Portal obstruction and hypertension

____ 2. Capillary permeability

____ 3. Plasma osmotic pressure

____ 4. Retention of sodium and water

— — — — — — — — — — — — — — — — —

1. I; 2. I; 3. D; 4. I

---

426 Portal hypertension influences fluid accumulation in the abdomen, but the fundamental cause of ascites lies in *____

____.

— — — — — — — — — — — — — — — — —

damage to the cellular structure of the liver

---

427 With ascites, there is a permeability defect of capillary endothelium that contributes to *____

____.

Will permeability increase or decrease? ______________________________.

— — — — — — — — — — — — — — — — —

the transudation of fluid; increase

---

428 Because of increased liver congestion and failure of the liver to synthesize albumin, the protein fluid will leave the ______________________________ and pass into the *____.
Will the plasma osmotic pressure be increased or decreased? ________________
______________________________. The result will be (hypoproteinemia/hyperproteinemia) ____.

— — — — — — — — — — — — — — — — —

capillaries; abdominal cavity; decreased; hypoproteinemia

---

429 The hormonal response (aldosterone) will cause:

a. *__________________________________.

b. *__________________________________.

\- - - - - - - - - - - - - - - - - -

a. decrease in urinary output; b. retention of sodium and water

---

Place I for increased and D for decreased beside the phsyiologic factors as they occur with ascites. Give the rationale for each physiologic factor. After completing this review, refer to the table for additional information.

________ Portal obstructions and hypertension

*__________________________________

____

__________________________________.

\- - - - - - - - - - - - - - - - - -

I. Portal hypertension influences ascites, but its presence is associated with liver damage

---

________ Capillary permeability

*__________________________________

____

__________________________________.

\- - - - - - - - - - - - - - - - - -

I. A permeability defect of capillary endothelium contributes to ascites

---

________ Plasma osmotic pressure

*__________________________________

____

__________________________________.

\- - - - - - - - - - - - - - - - - -

D. Due to liver congestion, protein fluid leaves the capillaries and passes into the abdomen

---

________ Retention of sodium and water

* ____

____

____.

- - - - - - - - - - - - - - - - -

I. Increased hormonal response (aldosterone). This decreases urinary output and causes retention of Na and H_2O

CLINICAL APPLICATIONS

---

430 Persons with ascites eventually become refractory in regard to diuretic agents. Frequently, paracentesis (surgical puncture of the abdominal cavity for relieving fluid) is required to relieve symptoms of pressure or respiratory distress, or both.

With ascites, there is a tendency for the sufferer to become refractory to

________________________.

- - - - - - - - - - - - - - - - -

diuretics

---

431 *Paracentesis* means a * ____________________________________

____.

- - - - - - - - - - - - - - - - -

surgical puncture of the abdominal cavity

---

432 Paracentesis is generally required to relieve symptoms of ________________

__________________ or * __________________________________.

- - - - - - - - - - - - - - - - -

pressure; respiratory distress

433 Repeated paracenteses result in a great loss of protein, electrolytes, and water.
As a result of repeated paracenteses, what three substances may be lost?

______________, ______________, and ______________.

- - - - - - - - - - - - - - - - - -

protein; electrolytes; water

---

434 The aftereffect of abdominal paracentesis is an antidiuresis of water, with hemodilution of sodium in the blood. Following this, there is a rapid outpouring of fluid into the abdominal cavity with hemoconcentration and a drop in blood volume. This leads to a greater retention of sodium.

Abdominal paracentesis (is/is not) __________ the permanent cure for ascites.

After an abdominal paracentesis, antidiuresis of water results, causing hemodilution of *______________________________.

- - - - - - - - - - - - - - - - - -

is not; sodium in the blood

---

435 After hemodilution of serum sodium following a paracentesis, fluid then pours into the *______________. Would the blood volume be increased or decreased? ______________ As the result of the latter, would hemoconcentration be present? ______________.

The nurse should observe for symptoms of serum sodium (excess/deficit) ______________.

- - - - - - - - - - - - - - - - - -

abdominal cavity; decreased; yes; excess

436 Repeated paracenteses will cause a great loss in ______________________

______________________, and ______________________.

The removal of a large volume of fluid by paracentesis causes a rapid shift of fluid from the plasma into the * ______________________.
Symptoms of circulatory collapse (shock symptoms) should be observed following the removal of a large volume of abdominal fluid.

Name at least five symptoms of shock.

1. * ______________________.
2. * ______________________.
3. * ______________________.
4. * ______________________.
5. * ______________________.

- - - - - - - - - - - - - - - - -

water; protein; electrolytes; abdominal cavity (third space fluid)

1. pallid, cold, clammy skin
2. fast pulse rate
3. apprehension and restlessness
4. fall in blood pressure
5. respirations are shallow and rapid
6. others, i.e., weakness, oliguria

CLINICAL EXAMPLE

437 Mr. Moore, age 58, was admitted to the medical floor of the hospital complaining of shortness of breath with no chest pain. He had massive ascites with distended veins over the abdomen and 4+ pitting leg edema. Mr. Moore's shortness of breath would most likely be the result of ______________.
His records show that his admissions in the past 8 years have been due to cirrhosis of the liver with ascites and peripheral edema. Congestive heart failure frequently accompanies severe liver damage with ascites; therefore, a cardiotonic is given. Mr. Moore was placed on diuretics and a cardiotonic drug (Digoxin).

Diuretics were ordered to increase *______________
______________.

Digoxin was given to *______________
______________.

(Refer to section on CHF if necessary.)

- - - - - - - - - - - - - - - - -

ascites
fluid loss via urinary output
strengthen the heart beat and improve circulation of fluid

The laboratory studies of Mr. Moore in Table 88 show how his laboratory results deviate from the norms at the time of his illness.

TABLE 88. LABORATORY STUDIES OF MR. MOORE

Laboratory Tests	On Admission	2 Weeks Later	3 Weeks Later	4 Weeks Later
Hematology				
Hemoglobin (12.9–17.0 g)	10.4		12.6	
Hematocrit (40–46%)	34		41	
WBC (white blood count) (5000–10,000/cu mm)	8000			
Biochemistry				
BUN (blood urea nitrogen) (10–25 mg/dl)	47	55	132	190
Plasma/serum CO_2 (50–70 vol %)	44	48		29
(22–32 mEq/L)	20	22		13
Plasma/serum chloride (98–108 mEq/L)	105	95	99	91

TABLE 88. (Continued)

Laboratory Tests	On Admission	2 Weeks Later	3 Weeks Later	4 Weeks Later
Plasma/serum sodium (135–146 mEq/L)	136	124	133	119
Plasma/serum potassium (3.5–5.3 mEq/L)	4.9	4.6	3.9	4.5
Plasma albumin (3.2–5.6 Gm/dl)	1.4			

mg/100 ml = mg/dl

Plasma and *serum* are used interchangeably.

---

438 Mr. Moore's hemoglobin level on admission was 10.4 g, which could be indicative of secondary anemia or (hemodilution/hemoconcentration)

______________________.

The BUN on his admission was 47 mg/dl, which would mean that

*____

____

— — — — — — — — — — — — — — — — —

hemodilution;
he was dehydrated due to a lack of fluid.

---

439 Two weeks after admission, Mr. Moore's BUN became alarmingly increased.

What would this mean in regard to his kidneys? *______________________

____

____.

— — — — — — — — — — — — — — — — —

The inability of the kidneys to excrete the urea, or kidney failure, or both.

---

440 Mr. Moore's serum CO_2 on admission was 20 mEq/L, which could mean he was

in a mild ______________ state.

— — — — — — — — — — — — — — — — —

acidotic

---

441 What does Mr. Moore's plasma albumin indicate (hyperalbuminemia or hypoalbuminemia)? ________________.

- - - - - - - - - - - - - - - - -

hypoalbuminemia

---

442 Explain where Mr. Moore's albumin could be found? * ________________________________.

- - - - - - - - - - - - - - - - -

In the peritoneal cavity or abdomen along with sodium, potassium, and other electrolytes

---

CLINICAL MANAGEMENT

Mr. Moore received for clinical management:

1. Diuretics
   a. Lasix (furosemide) 160–200 mg (Stat doses—immediate)
   b. Daily doses of Aldactone (spironolactone), an aldosterone antagonist and potassium-sparing diuretic

   Lasix (furosemide) is a potent diuretic that acts on the proximal and distal tubules and ascending limb of Henle's loop. If given in excessive amounts, furosemide can lead to a profound diuresis with water and electrolyte depletion.

   Aldactone (spironolactone) inhibits the production of aldosterone (the hormone that causes sodium and water retention and potassium excretion). Therefore, Aldactone promotes sodium and water excretion and inhibits potassium excretion.

   Hypokalemia can cause hepatic coma or liver failure. Low serum potassium has a tendency to increase ammonium accumulation, which precipitates hepatic toxicity.

2. Digoxin—0.25 mg daily
3. Low sodium diet—1.5 g
4. Limited fluid intake
5. Daily weights

443 Mr. Moore received Lasix by ____________ doses. It is a potent diuretic that acts on what areas of the kidneys? * ____

____.

Large and continuous doses of Lasix can lead to a profound diuresis, causing depletion of ____________ and ____________________.

Aldactone promotes excretion of ____________ and ____________ and inhibits excretion of ____________________.

Digoxin was given daily to * ____________________ and * ____________________.

Low sodium diet and limited fluid intake would decrease the body's ____________ and ____________.

— — — — — — — — — — — — — — — — —

stat
proximal and distal tubules and the loop of Henle
water; electrolytes
sodium; water; potassium
strengthen heart beat; improve circulation
sodium; water

---

444 Mr. Moore did not respond well to Lasix and Aldactone. Paracentesis was then indicated when his ascites became refractory to ____________________.

Paracentesis was done on the tenth day after admission and during the third week.

Following the removal of a large volume of fluid by paracentesis, the nurse should observe for symptoms of * ____

____.

— — — — — — — — — — — — — — — — — —

diuretics; circulatory collapse (shock)

445 On several occasions Mr. Moore received 300 ml of 3% saline (hypertonic) intravenously over 3 hours. Fifteen minutes after the fluid was started IV, Lasix 200 mg was given. The purpose of a hyperosmolar solution and a potent diuretic would be to draw the fluid from the * ________________

into the * ________________ for urinary excretion.

Once the major symptoms have been relieved through diuretics, paracentesis, and saline IV administration, the physician will be able to treat the basic problem, which is cirrhosis of the liver.

- - - - - - - - - - - - - - - - - -

abdominal cavity; intravascular spaces

CASE REVIEW

Mr. Moore, age 58, was diagnosed as having cirrhosis of the liver with ascites and peripheral edema. He had been in congestive heart failure, so he received Digoxin and diuretics (Lasix and Aldactone). Diuretics were mostly for the removal of fluids from the abdominal area, i.e., ascites. He had been on a low sodium diet (1 g). His serum sodium was low, 119-136 mEq/L.

1. Hyponatremia occurs in liver diseases in spite of the presence of an excess of total body sodium because water is * ________________ .
2. Name three other causes that can contribute to Mr. Moore's hyponatremia.
   a. * ________________ .
   b. * ________________ .
   c. * ________________ .
3. Mr. Moore had ascites, which is * ________________ .
4. Portal hypertension influences fluid accumulation in the abdomen, but the fundamental cause of ascites lies in * ________________ .
5. With ascites, there is a capillary permeability (increase/decrease) ________ and a plasma osmotic pressure (increase/decrease) ________.
6. Lasix (furosemide) is a (potassium-sparing/potassium-wasting) ________ diuretic. Aldactone (spironolactone) is a potassium- ________

_____________ diuretic. Why are these two types of diuretics frequently given

together? * ____

____.

Mr. Moore was not responding to the diuretics. He was having difficulty in breathing due to his ascites. A paracentesis was performed.

7. Mr. Moore had a paracentesis done to alleviate the symptoms of:

   a. ____.

   b. * ____.

8. A paracentesis was indicated, for he became refractory to which group of drugs?

   ____.

9. Repeated paracenteses result in a great loss of ____________________,

   ____________________, and ____________________.

10. Following an abdominal paracentesis, antidiuresis of water results; then there is a

    outpouring of fluid into * ____________________, resulting in hemoconcentra-

    tion and a drop in * ____________________.

11. The nurse checked Mr. Moore's laboratory results. She noticed a low serum

    sodium. The nurse should observe for symptoms of ____________________.
    Name four of the symptoms.

    a. * ____.

    b. * ____.

    c. ____.

    d. * ____.

- - - - - - - - - - - - - - - - -

1. retained in excess to sodium
2. a. sodium moves into the intracellular spaces
   b. prolonged use of potent diuretics
   c. low sodium diet
3. an accumulation of fluid in the peritoneal cavity
4. damage to the cellular structure of the liver
5. increase; decrease
6. potassium-wasting; sparing
   Aldactone will retain potassium, thus preventing excessive potassium loss (or similar answer)
7. a. pressure
   b. respiratory distress
8. diuretics

9. protein; electrolytes; water
10. the abdominal cavity; blood volume
11. hyponatremia
    a. abdominal cramps
    b. muscular weakness
    c. headaches
    d. nausea and vomiting

NURSING DIAGNOSES

Fluid volume deficit related to fluid shift to the third space (ascites).

Potential for electrolyte imbalance related to drug therapy (diuretics) and fluid shift with electrolytes to the third space (ascites).

Ineffective breathing patterns related to a large volume of fluid in the peritoneal cavity (ascites) which causes pressure on the diaphragm.

Potential for impaired skin integrity related to fluid accumulation in the interstitial space (tissues) in extremities and buttocks.

Alteration in nutrition related to ascites and edema in the extremities.

Potential activity intolerance related to ascites and edema in the extremities.

Potential alteration in the family process related to the family member's social problem.

NURSING ACTIONS

1. Recognize the physiologic factors associated with cirrhosis of the liver, such as portal hypertension, increased capillary permeability, loss of protein (which lowers the serum colloid osmotic pressure), and sodium and water retention from excess aldosterone secretion. From the above physiologic factors, ascites occurs.
2. Check the laboratory results and report abnormal findings. Serum potassium, sodium, and chloride may be low. Sodium and water are retained; however; hyponatremia (sodium deficit) may occur due to sodium shift to cells and to the abdominal cavity, low sodium diet, and diuretics.
3. Monitor Mr. Moore's urine output. Urine output of less than 600 ml per 24 hours could indicate hypovolemia from fluid shift into the abdominal cavity, kidney (renal) impairment, limited fluid intake, or excess aldosterone secretions.
4. Teach Mr. Moore to eat foods that are nutritional and high in protein. Increasing Mr. Moore's serum protein level will increase his serum colloid osmotic pressure, which would promote abdominal fluid to shift to the vascular area.
5. Assess for signs and symptoms of hypokalemia when Mr. Moore is receiving a potent diuretic (Lasix) and digoxin. Symptoms include dizziness, arrhythmia, muscular weakness, abdominal distention, and diminished peristalsis (bowel sounds). Hypokalemia can enhance the action of Digoxin and cause digitalis toxicity.
6. Monitor intravenous fluids and observe for overhydration (hypervolemia), especially when administering 3% saline intravenously. The hyperosmolar solution

pulls fluid from the abdominal cavity and the drug, Lasix, causes diuresis. Mr. Moore was refractory to Lasix, thus overhydration could easily occur.

7. Observe Mr. Moore for signs and symptoms of shock after paracentesis. Removal of large volumes of abdominal fluid can cause vascular fluid to shift into the abdomen. Symptoms of shock are cold, clammy skin; rapid pulse rate; apprehension and restlessness; rapid, shallow respirations; and a drop in blood pressure.

DIABETIC ACIDOSIS

PHYSIOLOGIC FACTORS

446 The inability of the body to utilize glucose results in an increasing concentration of sugar in the blood.

The elevation of blood sugar increases glucose concentration in the glomerular filtrate of the kidneys. This increased solute load in the kidneys requires extra fluid from the rest of the body; therefore, polyuria (increased amount of urine) occurs.

In diabetic acidosis, explain in your words why polyuria occurs.

* ____

____

____

____.

- - - - - - - - - - - - - - - - - -

An increased glucose concentration will cause an increase in intravascular fluid, and so glucose and water are excreted, or a similar response

---

447 When the concentration of glucose in the glomerular filtrate exceeds the renal threshold for tubular reabsorption, glycosuria results. Increased glucose concentration acts as an osmotic diuretic that causes diuresis.

Do you know what *glycosuria* means? * ______________________.

Do you know what *diuresis* means? * ______________________.

- - - - - - - - - - - - - - - - - -

sugar in the urine—Good!
excess urine excretion

---

448 Osmotic diuresis inhibits reabsorption of sodium causing excessive sodium loss with glycosuria.

An increased blood sugar will cause:

( ) a. osmotic diuresis
( ) b. polyuria
( ) c. glycosuria
( ) d. sodium excretion
( ) e. sodium reabsorption
( ) f. fluid reabsorption

- - - - - - - - - - - - - - - -

a. X; b. X; c. X; d. X; e. –; f. –

---

449 Elevated blood sugar will also increase the hyperosmolality of the extracellular fluid.

The hyperosmolality of the extracellular fluid leads to a withdrawal of fluid from the cells, thus equalizing the extracellular and intracellular osmolality.

The fluid from the cells will dilute the extracellular sodium concentration, producing (hypernatremia/hyponatremia) ________________.

The migration of intracellular fluid into the extracellular fluid will result in:

( ) a. cellular dehydration
( ) b. cellular hydration

- - - - - - - - - - - - - - - -

hyponatremia
a. X; b. –

---

450 Failure to metabolize glucose leads to an increased fat utilization for energy, producing ketonic acids (products of fat catabolism), which accumulate in the blood. Ketosis occurs, which is an excess number of ketone bodies in the blood.

Failure of glucose metabolism will cause:

( ) a. fat utilization for energy
( ) b. excessive ketone bodies due to fat catabolism

Excessive number of ketone bodies in the body is known as ________________.

Ketosis will lead to metabolic ________________.

- - - - - - - - - - - - - - - -

a. X; b. X; ketosis; acidosis

---

451 In renal excretion, the ketones (strong acids) will combine with the cation sodium, causing a sodium depletion. Ketone bodies (ketones) are excreted as ketonuria. The additional solute load of ketones in the glomerular filtrate will result in:

( ) a. a decreased loss of water in the formation of ketonuria
( ) b. an increased loss of water in the formation of ketonuria

\- - - - - - - - - - - - - - - - - -

a. –; b. X

---

452 Polyuria can result from:

( ) a. glycosuria
( ) b. anuria
( ) c. ketonuria

With the loss of water the solute concentration of the blood ______________,

and the blood volume ______________.

\- - - - - - - - - - - - - - - - - -

a. X; b. –; c. X; increases; decreases

---

453 As a result of metabolic acidosis, nausea and vomiting occur, thus causing a severe fluid and electrolyte imbalance.

There is an increase of water loss by way of the lungs due to Kussmaul breathing (hyperventilating–deep, rapid breathing).

Dehydration occurs from:

( ) a. nausea and vomiting
( ) b. Kussmaul breathing
( ) c. oliguria
( ) d. polyuria

\- - - - - - - - - - - - - - - - - -

a. X; b. X; c. –; d. X

---

454 The failure of cellular utilization of glucose causes potassium to leave the cells. The serum potassium may therefore show normal or high serum values.

There is a serum potassium loss due to vomiting and renal excretion, but the hemoconcentration can cause the serum potassium to show

* ____.

\- - - - - - - - - - - - - - - - - -

normal or high serum values

---

455 With severe dehydration, the renal blood flow will become inadequate, urinary output will be decreased, and the excretion of potassium will be lessened.

Therefore, the blood levels of potassium will be ______________.

\- - - - - - - - - - - - - - - - -

elevated or increased

Table 89 shows the distribution of electrolytes in plasma/serum in normal state and in early and late stages of diabetic acidosis. Note in particular the increase in ketone bodies (products of fat catabolism) in late acidosis. Know what happens to the bicarbonate ions when there is an increase in ketones. You are not expected to know the effects of the cations. Refer to this table as needed.

TABLE 89. SERUM ELECTROLYTES IN DIABETIC ACIDOSIS

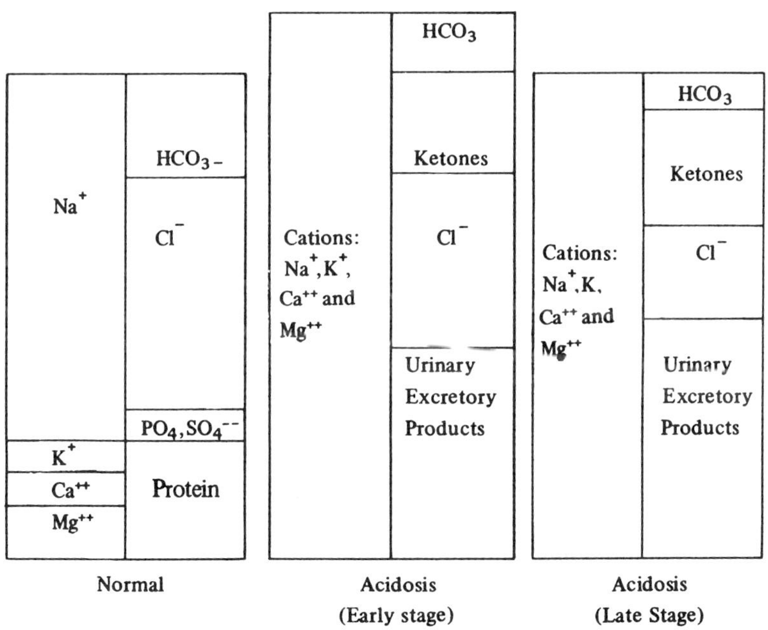

Adapted with permission from Stratland: *Fluid and Electrolytes in Practice*, ed 3. Philadelphia, JB Lippincott Co, p 268.

---

456 The center column in Table 89 shows that the ketones increase at the expense of the * ______________________.

Because of dehydration, the serum osmolality is ______________________,

and the serum levels of electrolytes will be ______________________.

— — — — — — — — — — — — — — — — — —

bicarbonate ions; increased; increased

---

457 With an increased number of ketones (Table 89, last column) the alkali reserve (bicarbonate) will be ______________________.

— — — — — — — — — — — — — — — — — —

decreased

---

458 With a decrease in the number of bicarbonates and an increase in ketones (fatty acid) would (metabolic alkalosis/metabolic acidosis) occur? ______________.

* ____

— — — — — — — — — — — — — — — — — —

Metabolic acidosis

CLINICAL APPLICATIONS

---

459 Dehydration is one of the major symptoms and concerns for persons in diabetic acidosis.

When there is a marked intracellular and extracellular fluid depletion, the end result is:

( ) a. decreased hemoconcentration
( ) b. increased hemoconcentration
( ) c. decreased blood volume
( ) d. increased blood volume

— — — — — — — — — — — — — — — — — —

a. –; b. X; c. X; d. –

Table 90 gives the laboratory results and the symptoms related to dehydration in diabetic acidosis. Actually, you may not find a patient in diabetic acidosis displaying all the listed laboratory results or all the listed observed symptoms, but this table should serve as a guide for assessment.

There is no horizontal relationship between the two columns in the table. Take time to study this table. Quiz yourself on the signs and symptoms and then proceed to the frames. Refer to the table as needed.

TABLE 90. LABORATORY RESULTS AND OBSERVED SIGNS AND SYMPTOMS RELATED TO DEHYDRATION IN DIABETIC ACIDOSIS

Laboratory Test Results	Observed Signs and Symptoms
Blood Chemistry	Thirst (avid)
Hyperglycemia	Polyuria (due to excess blood sugar and ketone bodies)
Decreased serum CO_2 (bicarbonate determinant)	Specific gravity > 1.030
Potassium normal or hyperkalemia	Nausea and vomiting
Hyponatremia	Decreased fluid intake
Hypochloremia—excreted with Na and water	Vital signs
Hypomagnesemia (cellular dehydration)	Pulse: rapid, thready
Hypophosphoremia	Respiration: deep, rapid, vigorous (Kussmaul breathing)
Hematology	Blood pressure: decreased
Elevated hematocrit	Dry mucous membrane
Elevated hemoglobin	Skin dry (poor skin turgor)
Elevated leukocyte count	Lips: dry and parched
Arterial Blood Gases (ABG):	Sunken eyes and soft eyeballs
pH decreased	Disorientation, confusion
HCO_3 decreased (bicarbonate ions are replaced with ketones, which are strong acids)	
Urine	
Glycosuria	
Ketonuria	

460 Check the laboratory findings of dehydration in diabetic acidosis:

( ) a. hypoglycemia
( ) b. hyperglycemia
( ) c. hyponatremia
( ) d. hyperchloremia
( ) e. decreased CO_2 combining power
( ) f. elevated hematocrit count
( ) g. low hematocrit count
( ) h. elevated leukocyte count
( ) i. glycosuria
( ) j. ketonuria
( ) k. serum pH 6.9
( ) l. hypermagnesemia

— — — — — — — — — — — — — — — — — —

a. –; b. X; c. X; d. –; e. X; f. X; g. –; h. X; i. X; j. X; k. X; l. –

---

461 Check the physical symptoms of dehydration in diabetic acidosis:

( ) a. disorientation, confusion
( ) b. thirst
( ) c. slow pulse rate
( ) d. polyuria
( ) e. nausea and vomiting
( ) f. Kussmaul breathing
( ) g. dyspnea
( ) h. increased blood pressure
( ) i. pulse fast and thready
( ) j. dry mucous membrane
( ) k. moist skin
( ) l. sunken eyes and soft eyeballs

— — — — — — — — — — — — — — — — — —

a. X; b. X; c. –; d. X; e. X; f. X; g. –; h. –; i. X; j. X; k. –; l. X

---

462 Would (hypervolemia/hypovolemia) ________________ be present with diabetic acidosis?

— — — — — — — — — — — — — — — — — —

hypovolemia

463 Check the following laboratory findings and symptoms that relate to dehydration in diabetic acidosis, and correct the incorrect responses:
( ) a. hypernatremia
( ) b. hyperglycemia
( ) c. thirst
( ) d. nausea and vomiting
( ) e. disorientation, confusion
( ) f. bradycardia—slow pulse
( ) g. polyuria
( ) h. decreased blood pressure
( ) i. hypochloremia
( ) j. hypomagnesemia
( ) k. hyperphosphoremia
( ) l. decreased CO_2 combining power
( ) m. hypovolemia
( ) n. elevated hematocrit
( ) o. elevated leukocyte count
( ) p. sunken eyes and soft eyeballs
( ) q. decreased leukocyte count
( ) r. Kussmaul breathing

— — — — — — — — — — — — — — — — —

a. –, hyponatremia; b. X; c. X; d. X; e. X; f. –, fast, thready pulse rate; g. X; h. X; i. X; j. X; k. –, hypophosphoremia; l. X; m. X; n. X; o. X; p. X; q. –, increased leukocyte count; r. X

CLINICAL MANAGEMENT

464 In the first 24 hours, 80% of the total water and salt deficit should be replaced. For the other electrolytes, there is less urgency since the rate of assimilation of the intracellular electrolytes is limited. Administration of potassium must be included, but *not in early treatment* (unless indicated) since serum potassium could be increased to toxic level.

With diabetic acidosis, 80% of (salt and water/potassium and magnesium) ________________ should be replaced in the first 24 hours.

Cellular assimilation of electrolytes is (faster/slower) ________________ than extracellular.

— — — — — — — — — — — — — — — — —

salt and water
slower

Table 91 outlines the restoration for extracellular fluid and intracellular fluid deficits. ECF is restored directly from IV therapy. Potassium replacement frequently begins by adding KCl to the second liter of fluids unless the serum potassium level remains elevated. ICF replacement occurs in two days, but cellular electrolyte replacement takes 2 to 5 days or longer. Study the table carefully, noting when the listed elements are restored. Refer to the table as needed.

TABLE 91. RESTORATION OF ECF AND ICF DEFICITS

Restoration of Extracellular Fluid Deficits	Restoration of Intracellular Fluid Deficits
Large deficiencies should be made up in 24–48 hours.	As serum osmolality drops with glucose utilization, fluid will enter cells. This frequently occurs in first two days.
Na Cl HCO_3 Water	From around 2nd to 5th days, K, Mg, PO_4, and water are replenished in the cells. Around the 4th to 7th day, nitrogen replacement begins.

---

465 Potassium replacement should start approximately 6–8 hours after the first dose of insulin has been administered (subcutaneous or IV) and the acidotic state is being corrected. Serum potassium levels should be taken frequently. Potassium moves back into cells as fluid and the acidotic state are corrected.

If potassium is not given as acidosis is corrected, would the serum potassium level be (high/low)? ______________.

- - - - - - - - - - - - - - - - - -

low. Potassium moves back into the cells leaving a serum K deficit.

---

466 Restoration of intracellular fluid deficit is somewhat (slower/faster) __________ than extracellular fluid deficit.

Around the second to fifth day, the electrolytes __________, __________, and __________ are replenished in the cells. Complete potassium replacement takes several days.

Nitrogen replacement begins around the *______________ day.

- - - - - - - - - - - - - - - - - -

slower; potassium, magnesium, phosphate; fourth to seventh

467 The longer the acidosis persists, the more resistant the person is likely to be to insulin. Insulin is a hormone secreted by the pancreas and is essential for oxidation and utilization of blood sugar (glucose).

Only regular, unmodified, or crystalline insulin may be administered in intravenous fluids.

If acidosis persists, the person may require (more/less) ____________ insulin.

The following types of insulin can be administered intravenously:

( ) a. NPH
( ) b. regular
( ) c. unmodified
( ) d. PZI
( ) e. crystalline

— — — — — — — — — — — — — — — —

more
a. –; b. X; c. X; d. –; e. X

---

468 As acidosis is corrected, insulin is utilized more rapidly by the body for metabolizing sugar (glucose). Can (hypoglycemia/hyperglycemia) ____________ result? Can you explain why? * ____________

____

____.

— — — — — — — — — — — — — — — —

hypoglycemia. When acidosis is corrected, there is less resistance to insulin. The insulin that had been previously administered (nonfunctional insulin due to acidosis) will then metabolize the sugars for cellular use, causing low blood sugar (hypoglycemia)

---

469 The nurse should observe the patient for symptoms of hypoglycemia, also known as insulin reaction or hypoglycemia reaction. These symptoms include cold, clammy skin; nervousness; weakness; dizziness; tachycardia; low blood pressure; and slurred speech. The blood sugar is frequently 50 mg/dl or lower.

Do you know how insulin reaction can be corrected quickly?

*____

____.

— — — — — — — — — — — — — — — — —

You may need to consult a nursing text; however, a glass of orange juice with a packet or two of sugar can raise the blood sugar rapidly

---

470 When people are treated for diabetic acidosis with large doses of insulin, the nurse should observe for what type of reaction that might occur? ______________

______________________.

Too much insulin will cause symptoms similar to shock. Name three symptoms.

______________________, ______________________, and ______________

______________________.

— — — — — — — — — — — — — — — — —

insulin reaction; tachycardia, nervousness, and weakness. Also low blood pressure, dizziness, and slurred speech.

---

471 Once insulin and fluid therapy are started, the blood sugar should be carefully monitored. If the blood sugar drops to or below 180 mg/dl it is suggested that 5% or 10% dextrose in water be given to prevent *______________________

____.

— — — — — — — — — — — — — — — — —

hypoglycemia or insulin reaction

---

472 People with diabetes who are ill are advised to go to bed because rest reduces metabolism. This decreases fat and protein catabolism.

These people should also be protected from overheating and chilling. If they are in a state of vascular collapse, extra heat should not be applied since it will increase vasodilatation and intensify the failure of the circulation.

Rest reduces metabolism in an ill diabetic person; therefore, rest will decrease the chance of ____________ and ________________ catabolism.

In the state of vascular collapse, extra heat may cause further ____________ ______________________.

— — — — — — — — — — — — — — — —

fat; protein
vasodilatation

---

473 Orange juice and broth are the fluids that are frequently given after nausea and vomiting have ceased. Broth is a good source of sodium, potassium, and water and is usually tolerated when served hot. Orange juice is also a good source of potassium as well as of sugar and water. Ginger ale is also helpful after cessation of nausea and vomiting.

The oral fluids that are frequently tolerated soon after nausea and vomiting are:
( ) a. broth
( ) b. milkshake
( ) c. orange juice
( ) d. ginger ale

Broth is rich in the following fluid and electrolytes:
( ) a. calcium
( ) b. potassium
( ) c. water
( ) d. sodium
( ) e. sulfate

Orange juice is rich in:
( ) a. sugar
( ) b. potassium
( ) c. calcium
( ) d. water

— — — — — — — — — — — — — — — —

a. X; b. –; c. X; d. X
a. –; b. X; c. X; d. X; e. –
a. X; b. X; c. –; d. X

CLINICAL EXAMPLE

---

474 Mrs. Thompson arrived in the emergency room in a semicomatose state. Prior to admission, she had been vomiting and had complained of being weak. The family stated she had had a severe cold with a fever for weeks. They felt the vomiting was due to a viral infection.

The mucosa in her mouth was dry. Vomiting and dry mucosa would indicate

______________________________.

Her respirations were rapid and deep, which would be an indication of:

( ) a. Kussmaul breathing

( ) b. dyspnea

Her heart sinus rhythm was sinus tachycardia (pulse rate 120). Her breath had a very sweet smell. The family stated she did not have diabetes mellitus, but there was a familial history of it.

In the emergency room, a stat blood chemistry was done and a retention catheter was inserted. The blood sugar was 476 mg/dl, the normal range being 70–110 mg/dl. This would indicate a (hypoglycemic/hyperglycemic) ________________ state. The serum CO_2 combining power was very low, which would indicate an

______________________________ state.

\- - - - - - - - - - - - - - - -

dehydration; a. X; b. –; hyperglycemic; acidotic

Table 92 gives the laboratory studies of Mrs. Thompson, which show how her results deviated from the normal values at the time of her illness.

TABLE 92. LABORATORY STUDIES FOR MRS. THOMPSON

Laboratory Tests	On Admission	1st Day			2nd Day	3rd Day
Hematology						
Hemoglobin (12.9–17.0 g)	17.8					
Hematocrit (40–46%)	52					
Biochemistry						
BUN (blood urea nitrogen) (10–25 mg/dl)*	15					
Sugar feasting—postprandial (under 150 mg/dl)	476	825	458	382	60	144
Acetone	+1 / 1:10	+1 / 1:10	Trace / 1:8			
Plasma/serum CO_2† (50–70 vol %) / 22–32 mEq/L	7 / 3	10 / 4	14 / 6	18 / 8	34 / 15	44 / 20
Plasma/serum chloride (98–108 mEq/L)	104		130	132	133	110
Plasma/serum sodium (135–146 mEq/L)	137		151	159	164	145
Plasma/serum potassium (3.5–5.3 mEq/L)	4.8		2.7	3.2	4.2	4.5

*mg/100 ml = mg/dl.

†"Plasma" and "serum" are used interchangeably.

475 Her urinalysis was as follows:

color, dark yellow
specific gravity, 1.024
reaction, acid
albumin, +3
sugar, +4
WBC, many

Her specific gravity shows

( ) a. a very high range
( ) b. a high average range
( ) c. a low range
( ) d. an indication of an increased amount of products in the urine

The +4 sugar in the urine would indicate (hypoglycemia/hyperglycemia) ______________________.

The +3 albumin in the urine would indicate:

( ) a. normal range
( ) b. pathologic involvement, i.e., kidney cell injury

- - - - - - - - - - - - - - - - -

a. –; b. X; c. –; d. X
hyperglycemia
a. –; b. X

---

476 Mrs. Thompson's hemoglobin and hematocrit counts were:

( ) a. normal
( ) b. below normal
( ) c. above normal
( ) d. an indication of mild edema
( ) e. an indication of mild dehydration

- - - - - - - - - - - - - - - - -

a. –; b. –; c. X; d. –; e. X

477 The feasting blood sugars (blood drawn after eating) on admission and the first day were:
( ) a. normal
( ) b. below normal
( ) c. above normal
( ) d. an indication of hyperglycemia
( ) e. an indication of hypoglycemia
The second day, her blood sugar was 60 mg/dl, which would indicate a

______________________________ reaction.

- - - - - - - - - - - - - - - - -

a. –; b. –; c. X; d. X; e. –
hypoglycemic or insulin

---

478 Mrs. Thompson's serum CO_2 combining power was (normal/very low/very high)

____________________.

Her CO_2 combining power would indicate:
( ) a. metabolic acidosis
( ) b. metabolic alkalosis
( ) c. a bicarbonate loss
( ) d. a bicarbonate increase

- - - - - - - - - - - - - - - - -

very low; a. X; b. –; c. X; d. –

---

479 On admission, Mrs. Thompson's serum chloride, sodium, and potassium were in

(high/low/normal) ____________ range.
On the first day, the laboratory studies indicated:
( ) a. hyperchloremia
( ) b. hypochloremia
( ) c. hypernatremia
( ) d. hyponatremia
( ) e. hyperkalemia
( ) f. hypokalemia

- - - - - - - - - - - - - - - - -

normal range
a. X; b. –; c. X; d. –; e. –; f. X

480 On admission, Mrs. Thompson's laboratory results were hemoglobin 17.8 g, hematocrit 52%, BUN 15 mg/dl, feasting blood sugar 476, acetone +1 in 1:10 dilution, serum CO_2 7% and 3mEq/L, serum chloride 104 mEq/L, serum sodium 137 mEq/L, and serum potassium 4.8 mEq/L.

Indicate whether these results are high, low, or normal. If high or low, give one condition that each abnormality might indicate.

hematocrit. * ______________________.

hemoglobin. * ______________________.

BUN. * ______________________.

feasting blood sugar. * ______________________.

acetone. * ______________________.

serum CO_2. * ______________________.

serum chloride. * ______________________.

serum sodium. * ______________________.

serum potassium. * ______________________.

- - - - - - - - - - - - - - - - -

high, dehydration;	high, hyperglycemia;	normal range;
high, dehydration;	high, fat catabolism;	normal range;
normal range;	low, metabolic acidosis;	normal range

Table 92 explains Mrs. Thompson's medical treatment for the first three days, which included IV therapy, electrolytes, and insulin. The amount of electrolytes she received varied according to her serum electrolytes. Also, her insulin varied according to her blood sugar. On the second day she did not receive insulin because her blood sugar was low.

Study this table carefully and then proceed to the frames. Refer to the table as needed.

TABLE 93. FLUID, ELECTROLYTES, AND INSULIN REPLACEMENTS FOR MRS. THOMPSON

Date	Intravenous Therapy	Electrolytes	Insulin
Emergency room	2000 ml normal saline	$NaHCO_3$ ampules (sodium bicarbonate)	Regular insulin U.50 subcutaneously
1st Day	1000 ml 5% dextrose/saline	KCl 40 mEq/L (potassium chloride) $NaHCO_3$ amp 1	1. regular insulin U.50 subcutaneously 2. regular insulin U.50 IV 3. *1 hour later:* regular insulin U.50 subcutaneously qh X 4 (total 200 U. in 4 hours) 4. *4 hours later:* regular insulin U.50 subcutaneously regular insulin U.25 IV
	1000 ml normal saline	$NaHCO_3$ amp 1 KCl 60 mEq/L	
2nd Day	1000 ml 5% dextrose in water	$NaHCO_3$ amp 1 KCl 80 mEq/L	
	In 4 hours 500 ml 2.5% dextrose in water	$NaHCO_3$ amp 1	
	In 4 hours 500 ml 2.5% dextrose in water		
	Keep vein open with 500 ml 2.5% dextrose in water		
3rd Day	Oral fluids		NPH insulin U.20 to start

481 In diabetic acidosis there is frequently a serum sodium decrease before treatment due to:

( ) a. fluid intake
( ) b. vomiting
( ) c. urine excretion

In the emergency room, Mrs. Thompson received saline, $NaHCO_3$, and insulin. The sodium bicarbonate was to:

( ) a. increase the serum CO_2 combining power
( ) b. decrease the serum CO_2 combining power
( ) c. increase the bicarbonate in the plasma
( ) d. decrease the bicarbonate in the plasma

– – – – – – – – – – – – – – – –

a. –; b. X; c. X;
a. X; b. –; c. X; d. –

---

482 The first day she was given potassium chloride because * ______________________ ______________________.

She was given regular insulin U.50 on an hourly basis. This was to:

( ) a. lower her elevated blood sugar
( ) b. increase her blood sugar
( ) c. utilize the free glucose in her blood circulation

– – – – – – – – – – – – – – – –

her serum potassium was low
a. X; b. –; c. X

---

483 The second day she did not receive normal saline IV. This was because her

* ______________________________ and

* ______________________________ were elevated.

She did not receive insulin the second day since her blood sugar was __________.

– – – – – – – – – – – – – – – –

serum sodium; serum chloride
low

484 Mrs. Thompson received normal saline to replace the ________________ and ________________ losses due to ________________ and * ________________.

She received $NaHCO_3$ since her serum CO_2 was ________________.

She received KCl since her serum potassium was ________________.

She received regular insulin to * ________________

and * ________________.

She also received additional sodium and chloride from the electrolyte drugs ________________ and ________________.

- - - - - - - - - - - - - - - - - -

sodium; chloride;
vomiting; urine excretion;
(very) low;
(very) low;
lower her blood sugar; to metabolize the glucose;
$NaHCO_3$; KCl

CASE REVIEW

Mrs. Thompson, age 52, unknown diabetic, was admitted to the emergency room in a semicomatose state. Her respirations were deep and rapid, and her breath had a sweet smell. She had been urinating frequently. She had been vomiting for several days. Her laboratory results were Hgb 17.8 g, Hct 52, sugar feasting or postprandial 476 mg/dl, serum CO_2 3 mEq/L, serum sodium 137 mEq/L, serum chloride 104 mEq/L, and serum potassium 4.8 mEq/L.

1. According to Mrs. Thompson's history, give her three clinical symptoms of hyperglycemia:

   a. * ________________.

   b. * ________________.

   c. * ________________.

2. Her two clinical symptoms that indicated dehydration were * ________________ and * ________________.

   What two laboratory results also indicated dehydration? ________________ and ________________.

3. Her feasting sugar or postprandial blood sugar was * ________________.

Normal range for fasting blood sugar is *____________________

and for feasting sugar is *____________________.

4. Mrs. Thompson's increased blood sugar causes the body fluids to be (hypo-molar/hyperosmolar) ____________________; thus, osmotic diuresis results.
5. Polyuria can occur from ketonuria and ____________________.
6. Mrs. Thompson's serum CO_2 was markedly ____________________.

   What type of acid-base imbalance would be present? *____________________

   ____________________.
7. Her acidosis is the result of fat catabolism, producing *____________________.

   This type of acidosis is referred to as *____________________.

In the emergency room, Mrs. Thompson received 2 liters of normal saline (0.9% NaCl), $NaHCO_3$, and insulin. The nurse checked her laboratory results and noted that her blood sugar remained high and her serum CO_2 remained low. Her electrolytes were Na 151 mEq/L, Cl 130 mEq/L, and K 2.7 mEq/L.

8. Her elevated serum sodium and chloride may be due to *____________________

   ____________________.
9. The low serum potassium level may be due to *____________________

   ____________________.
10. Potassium should be administered *__________ hours after correction of acidosis and hyperglycemia have been started.

In the emergency room, Mrs. Thompson received 50 units of regular insulin. The first day of admission she received a total of 375 units of regular insulin and 2 liters of normal saline solution (1 liter contained 5% dextrose). Also, the first day she received KCl–100 mEq/L in 2 liters of IV fluids.

Her laboratory results the second day were blood sugar 60 mg/dl, serum CO_2 15 mEq/L, and 20 mEq/L, serum Na 164 mEq/L, Cl 133 mEq/L, and K 4.2 mEq/L.

11. Blood sugars need to be monitored frequently. When the blood sugar level drops to *____________________ 5% or 10% dextrose in water should be given.
12. The nurse should observe for symptoms of hypoglycemia, such as __________

    ____________________, *____________________ and

    ____________________, ____________________.
13. Mrs. Thompson's serum sodium and serum chloride levels were still elevated the second day. This might have been due to *____________________

    ____________________.

1. a. respirations were deep and rapid (Kussmaul breathing)
   b. sweet smelling breath
   c. frequent urination
2. frequent urination; prolonged vomiting
   hemoglobin; hematocrit
3. 476 mg/dl; 70–110 mg/dl (fasting); 150 mg/dl or lower (feasting)
4. hyperosmolar
5. glycosuria
6. decreased; metabolic acidosis
7. ketone bodies or ketosis (strong acid); diabetic acidosis
8. the 2 liters of normal saline solutions. One liter of normal saline (0.9% NaCl) supplies 154 mEq/L of Na^+ and 154 mEq/L of Cl^-
9. rehydration causing dilution of potassium. Also, some of the potassium may be returning to the cells with the correction of diabetic acidosis
10. 6 to 8 hours
11. 180 mg/dl or lower
12. nervousness; cold and clammy skin; tachycardia; slurred speech.
    Others: hunger; dizziness; low blood pressure
13. 2 liters of normal saline solutions administered first day

NURSING DIAGNOSES

Fluid volume deficit related to hyperglycemia and osmotic diuresis (polyuria).
Alterations of electrolyte, chemistry, and hematology related to polyuria, hyperglycemia, and ketosis secondary to diabetic acidosis.
Potential alteration in fluid volume excess related to excessive IV infusions and renal insufficiency.
Potential for ineffective breathing patterns related to Kussmall breathing secondary to diabetic acidosis.
Potential for injury to cells and tissues related to ineffective use of glucose and infection secondary to diabetic acidosis.
Anxiety related to outcome of ketoacidotic state and ineffective breathing patterns.
Alteration in nutrition related to insufficient nutrient intake and absorption.
Potential for ineffective coping related to chronic disorder and complications.
Knowledge deficit related to a lack of understanding of chronic disorder and its management.

NURSING ACTIONS: DIABETES

1. Observe for signs and symptoms related to dehydration in diabetic acidosis: disorientation; confusion; extreme thirst; polyuria; nausea and vomiting; rapid, thready pulse; low blood pressure; deep, rapid, vigorous breathing (Kussmaul breathing); dry mucous membranes; dry, warm skin; and sunken eyes.
2. Identify abnormal laboratory results associated with diabetic acidosis. Blood results: hyperglycemia; elevated hematocrit and hemoglobin; decreased serum

CO_2, decreased pH; decreased arterial HCO_3; normal serum potassium level or hyperkalemia; hyponatremia; hypochloremia; and hypomagnesemia. Urine results: glycosuria and ketonuria.

3. Check urine frequently or as order for sugar and ketone bodies using Clinitest and Acetest tablets or check blood with chemistrip.
4. Administer regular (crystalline) insulin intravenously. All other insulins can not be administered in IV solution.
5. Observe for signs and symptoms of hypoglycemia reaction (insulin reaction) from possible overcorrection of hyperglycemia. The symptoms include cold, clammy skin, nervousness, weakness, dizziness, tachycardia, low blood pressure, and slurred speech.
6. Report abnormal blood glucose levels. Blood sugar < 70 mg/dl indicates hypoglycemia and > 110 mg/dl indicates hyperglycemia.
7. Check for signs and symptoms of acidosis (metabolic/diabetic), i.e., Kussmaul breathing, flushed skin, decreased serum CO_2, decreased pH, and decreased HCO_3.
8. Teach patient to test urine for sugar and acetone (ketone bodies), or test blood with chemistrip, eat foods on the prescribed diet, use the exchange list, and take the prescribed insulin dose daily.
9. Observe for signs and symptoms of hypokalemia and hyperkalemia. Symptoms of hypokalemia are malaise, dizziness, arrhythmias, hypotension, muscular weakness, abdominal distention, and diminished peristalsis. Hypokalemia can occur as the acidotic state is corrected. Symptoms of hyperkalemia are tachycardia, and then bradycardia, abdominal cramps, oliguria, numbness, and tingling in extremities.
10. Monitor intravenous fluids and adjust flow rate according to orders. If IV fluids are to run fast, observe for symptoms of overhydration.

REFERENCES

Abbott Laboratories: *Fluid and Electrolytes.* North Chicago, 1970, pp 31-36, 39-42.

Bouchard-Kurtz R, Speese-Owens N: *Nursing Care of the Cancer Patient,* ed. 4. St. Louis, CV Mosby Co, 1981.

Brundage D: *Nursing Management of Renal Problems,* ed 2. St. Louis: CV Mosby Co, 1980, pp 51-68, 78-95.

Brunner LS, Suddarth DS: *Textbook of Medical-Surgical Nursing,* ed. 5. Philadelphia. JB Lippincott Co, 1984, pp. 520–527.

Burgess A: *Fluid & Electrolyte Balance.* New York, McGraw-Hill Book Co, 1979, pp 333-398 (Chap 12 and 14).

Burns N: *Nursing and Cancer,* Philadelphia, WB Saunders Co, 1982.

Carnes HE (ed): Acute renal failure. *Therapeutic Notes* 71: 8-12, January 1964 (Parke, Davis and Co).

Carnes HE (ed): Treatment of burns. *Therapeutic Notes* 234-239, October 1963 (Parke, Davis and Co).

Carozza V: Ketoacidotic crisis: Mechanism and management. *Nursing '73* 13-14, May 1973.

Chernecky CC, Ramsey PW: *Critical Nursing Care of the Client with Cancer,* Norwalk, CN, Appleton-Century-Crofts, 1984, pp 1–51.

Christopher KL: The use of a model for hemodynamic balance to describe burn shock. *Nurs Clin North Am* 15 (3): 617-627, September 1980.

Cunningham SG: Fluid and electrolyte disturbances associated with cancer and its treatment. *Nurs Clin North Am*, 17(4): 579–591, December 1982.
Detecting parathyroid disorders. *Endocrine Disorders*, Springhouse PA, Springhouse Corp, 1984, pp 91–99.
Earley L, Gottschalk C (ed): *Strauss and Welt's Diseases of the Kidney*, ed. 3, Vol II. Boston, Little, Brown and Co, 1979.
Fajans S: What is diabetes? Definition, diagnosis and course. *Med Clin North Am* 55: 793–804, 1971.
Fitzgibbons JP: Fluid, electrolyte, and acid-base management in the acutely traumatized patient. *Orthop Clin North Am* 57 (1): February 1977.
Flear CTG: Electrolyte and body water changes after trauma. *J Clin Pathol* (*Suppl*) 23 (4): 16-31, 1970.
Fochtman D, Foley G (ed). *Nursing Care of the Child with Cancer*. Boston, Little, Brown, and Co, 1982.
Garb S: *Laboratory Tests in Common Use*, ed 6. New York, Springer Publishing Co, 1976.
Groer MW, Shekleton ME: *Basic Pathophysiology*, ed. 2. St Louis, CV Mosby Co, 1983, pp 219–232.
Gutch C, Stoner M: *Review of Hemodialysis for Nurses and Dialysis Personnel*, St Louis: CV Mosby Co, 1978.
Harrington JD, Brener ER: *Patient Care in Renal Failure*. Philadelphia, WB Saunders Co, 1973.
Hayter J: Emergency nursing care of the burned patient. *Nurs Clin North Am* 13 (2): 223-234, June 1978.
Hochman HI, et al.: Dehydration, diabetic ketoacidosis and shock in the pediatric patient. *Pediatr Clin North Am* 26 (4): 803-826, November 1979.
Jones CA, Richards KE: Burns: Fluids resuscitate, in *Monitoring Fluid and Electrolytes Precisely: Nursing Skillbook*. Horsham, PA: Intermed Communications, Inc, 1978, pp 155-162.
Juliani L: Assessing Renal Function. *Nursing '78* 8 (1): 34-35, January 1978.
Kee JL: *Laboratory and Diagnostic Tests with Nursing Implications*. Norwalk, CN, Appleton-Century-Crofts Co, 1983.
Kee JL, Gregory AP: The ABC's (and mEq's) of fluid balance in children. *Nursing '74*, 4 (6), 28-36, June 1974.
Lancaster L (ed): *The Patient with End Stage Renal Disease*, New York, John Wiley & Sons, 1984.
Leaf A, Cotran R: *Renal Pathophysiology*, ed 3. New York, Oxford University Press, 1985.
Lindsey AM, Piper BF, Carrieri V: Malignant cells and ectopic hormone production. *Oncology Nursing Forum*, 8(3): 13–15, Summer 1981.
Lorentz WB: On pediatric fluid loss. *Emergency Medicine*, February 1977, pp 23-29, 35, 37-39, 45-46.
Lubash GD: Acute renal failure. *Hosp Med* 1: 14-19, April 1965.
Luckmann J, Sorensen KC: *Medical-Surgical Nursing*, ed 2. Philadelphia, WB Saunders Co, 1980, pp 799-812, 990-998, 1501-1505, 1544-1577.
Managing the adult diabetic patient—roundtable. *Patient Care* 9: 16-49, June 1, 1975
Metheney NM, Snively WD: *Nurses' Handbook of Fluid Balance*, ed. 4. Philadelphia, JB Lippincott Co, 1983, pp 315–333.
Oakley WG: *Diabetes and Its Management*. London, Blackwell Scientific Publications, 1973.
O'Neill M: Peritoneal dialysis. *Nurs Clin North Am* 1: 319-323, June 1966.
Papper S: *Clinical Nephrology*, Boston, Little, Brown Co, 1978.
Papper S: The effects of age in reducing renal function. *Geriatrics*, May 1973, pp 83-87.
Perkin R, Levin DL: Common fluid & electrolyte problems in the pediatric intensive care unit. *Pediatr Clin North Am* 27 (3): 567-586, August 1980.
Pickerning L, Robbins D: "Fluid, electrolyte, and acid-base balance in the renal patient. *Nurs Clin North Am* 15 (3): 577-592, September 1980.
Phipps WJ, Long BC. Woods NF: *Medical-Surgical Nursing*, ed. 2. St. Louis, CV Mosby Co, 1983, pp 321–1325.
Poe CM, Radford AI: The challenge of hypercalcemia in cancer. *Oncology Nursing Forum*, 12(6): 29–34, November/December 1985.

Price SA, Wilson LM: *Pathophysiology*, ed. 2. New York, McGraw-Hill Book Co, 1982, pp 433–439.

Pruitt BA: Fluid and electrolyte replacement in the burned patient. *Surg Clin North Am* 58 (6): 1291-1310, December 1978.

Roberts SL: Renal assessment: A nursing point of view. *Heart Lung* 8 (1): 105-113, January/February 1979.

Rudolph A: *Pediatrics*, ed 16. New York, Appleton-Century-Crofts Co, 1977.

Schoengrund L, Balzer P (ed): *Renal Problems in Critical Care*, New York, John Wiley & Sons, 1985.

Shiris GT: Postoperative, post-traumatic management of fluids. *Bull NY Acad Med* 55 (2): 248-256, February 1979.

Silver H, et al.: *Handbook of Pediatrics*, ed 10. California, Lange, 1973, pp 627-34.

Stark JL: Renal failure: Imbalances inevitable, in *Monitoring Fluid and Electrolytes Precisely: Nursing Skillbook.* Horsham, PA, Intermed Communications, Inc, 1978, pp 117-124.

Statland H: *Fluid and Electrolytes in Practice*, ed 3, Philadelphia, JB Lippincott Co, 1963, pp 63-75, 254-263, 276-280, 281-284.

Taber CW: *Taber's Cyclopedic Medical Dictionary*, ed. 15. Philadelphia, FA Davis Co, 1985.

"Teaming up to send the end-stage COPD patient home. *Nursing 84* 14(1): 65–76, January 1984.

Travenol Laboratories, Inc: *Guide to Fluid Therapy.* Deerfield, IL, 1970, pp 111-112.

Trusk CW: Hemodialysis for acute renal failure. *Am J Nurs* 65: 80-85, February 1965.

Vaughan V, McKay RJ (eds): *Nelson's Textbook of Pediatrics*, ed 11. Philadelphia, WB Saunders Co, 1979, pp 262-299.

Whaley LF, Wong DL: *Nursing Care of Infants & Children*, ed 2. St. Louis, CV Mosby Co, 1983.

Widmann FK: *Clinical Interpretation of Laboratory Tests*, ed 9. Philadelphia, FA Davis Co, 1983.

Wittaker A: Acute renal dysfunction. *Focus on Critical Care* 12(3): 12–17, June 1985.

APPENDIX A

Osmolality of Solutions

The table below gives three categories of fluids with examples and the purposes for using these solutions.

Osmolality is osmotic pull exerted by all particles per unit of water, expressed as osmols or milliosmols per kilogram of water.

Osmolarity is osmotic pull exerted by all particles per unit of solution, expressed as osmols or milliosmols per liter of solution.

Hypo-osmolar (hypotonic)	0.45% NACL, 5% D/W (when used continuously) Will cause H_2O to flow into the cells by osmosis. Useful for daily maintenance and helpful for excretory needs.
Danger:	Can cause H_2O intoxication—headaches, nausea and vomiting, excessive perspiration, CNS behavioral changes.
Iso-osmolar (isotonic)	5% dextrose/W, normal saline—0.9% NaCl lactated ringers's, 2½% dextrose/.45% NaCl, 5% dextrose/.2% NaCl (saline) Osmolality is same as plasma. Useful for daily maintenance and for dehydration. Helps to establish renal function—urinary output.
Hyper-osmolar (hypertonic)	10% dextrose/W, 5% dextrose/0.45% NaCl or in normal saline, 5% or 10% dextrose in lactated ringer's Osmolality is higher than plasma. Will draw water out of the cells, shrinking them. Helpful in removing water out of cells, i.e., cerebral edema. Helpful for maintenance and replacement of fluid. Helpful in increasing caloric intake.
Danger:	Dehydration could occur.

APPENDIX B

Fluid, Electrolyte, and Acid-Base Assessment Tool

This assessment tool can be used for assessing fluid, electrolyte, and acid-base imbalances in all clinical settings.

A. Nursing History of the Clinical Problem
   1. Chief complaint
   2. History of the present illness
      a. Date of Onset
      b. Duration
      c. Effects on bodily function
      d. Treatment and medications—effective or noneffective
   3. Past history of clinical condition

B. Fluid Imbalance Assessment
   1. Skin turgor
   2. Mucous membrane
   3. Vital signs—P ↑, BP ↓, T sl ↑
   4. Insensible fluid loss
      a. Diaphoresis
      b. Hyperventilation
   5. Body weight loss—2½ lb = 1 liter of water
   6. Intake and output
      a. Intake: Oral; IV fluids
      b. Output: Urine; stool; GI fluids; blood; wound
   7. Edema
   8. Behavioral changes—confusion and irritability (ECFV ↓ AND ICFV ↑)
   9. Neck and/or hand vein engorgement
   10. Chest sounds—rales

11. Laboratory results
    a. Serum osmolality ↓ 280 mOsm/L—overhydration
       ↑ 295 mOsm/L—dehydration
    b. Hgb and Hct
    c. BUN: 10-25 mg/dl (range)
    d. Creatinine: 0.7-1.4 mg/dl

C. Electrolyte Imbalance Assessment
1. Serum electrolytes
    a. Potassium: 3.5-5.3 mEq/L
    b. Sodium: 135-146 mEq/L
    c. Calcium: 4.5-5.5 mEq/L or 8.5 to 10.5 mg/dl
    d. Magnesium: 1.8-2.4 mEq/L
    e. Chloride: 98-108 mEq/L
2. Signs and symptoms of hypo-hyperkalemia, hypo-hypernatremia, hypo-hypercalcemia
3. Urine osmolality (range): 50-1400 mOsm/L
   (500-800 mOsm/L)
   Urine specific gravity: 1.010-1.030
4. Urine electrolytes
    a. Potassium: 25-120 mEq/ 24 hr
    b. Sodium: 40-220 mEq/ 24 hr
    c. Calcium: 50-150 mg/ 24 hr
    d. Chloride: 110-250 mEq/ 24 hr
       Example: urine Na 20 mOsm/L/24 hr—body retaining Na even if serum Na is low. Commonly seen with CHF and cirrhosis.
5. Vital signs—pulse
6. ECG
7. Behavioral changes—potassium imbalance
8. Drug therapy—diuretics, steroids, digitalis, antibiotics
9. Continuous GI suctioning

D. Acid-Base Imbalance Assessment
1. Arterial blood gases
    a. pH
    b. PCO_2 and HCO_3 and/or BE
2. Serum CO_2
3. Overbreathing or underbreathing
4. Chest sounds
5. Assisted ventilator
6. Behavioral changes
7. Intubations (Levine tube)
8. Electrolyte imbalance—↓ K—hypokalemic alkalosis
   (loss of K and HCl)

Glossary

abdomen the portion of the body lying between the chest and the pelvis.
acid any substance that is sour in taste and that will neutralize a basic substance.
acid metabolites see metabolites.
ACTH abbreviation for adrenocorticotropic hormone. A hormone secreted by the hypophysis or pituitary gland. It stimulates the adrenal cortex to secrete cortisone.
afterload resistance in the vessels against which the ventricle ejects blood during systole.
aged age of 65 years or older.
albumin simple protein. It is the main protein from the blood.
serum Main function is to maintain the colloid osmotic pressure of the blood.
alimentary pertaining to nutrition; the alimentary tract is a digestive tube from the mouth to the anus.
alkaline any substance that can neutralize an acid and that, when combined with an acid, will form a salt.
alveolus air sac or cell of the lung.
amphoteric ability to bind or release excess H^+.
anesthetic an agent causing an insensibility to pain or touch.
anion a negatively charged ion.
anorexia a loss of appetite.
anoxia oxygen deficiency.
anuria a complete urinary suppression.
aortic arch the arch of the aorta soon after it leaves the heart.
aphasia a loss of the power of speech.
arrhythmia irregular heart rhythm.
arterioles minute arteries leading into a capillary.
arteriosclerosis pertaining to thickening, hardening, and loss of elasticity of the walls of the blood vessels.
artery a vessel carrying blood from the heart to the tissue.
ascites an accumulation of serous fluid in the peritoneal cavity.
atrium a chamber. In the heart it is the upper chamber of each half of the heart.
atrophy decrease in size of structure.
azotemia an excessive quantity of nitrogenous waste products in the blood.

biliary pertaining to or conveying bile.
blood intravascular fluid composed of red and white blood cells and platelets.
bradycardia a slow heart beat.
bronchiectasis dilatation of a bronchus.
BUN abbreviation for blood urea nitrogen. Urea is a by-product of protein metabolism.

capillary a minute blood vessel connecting the smallest arteries (arterioles) with the smallest veins (venules).
carbonic anhydrase inhibitor an agent used as a diuretic that inhibits the enzyme carbonic anhydrase.
cardiac output amount of blood ejected by the heart each minute.
cardiac reserve capacity of the heart to respond to increased burden.
carotid sinus a dilated area at the bifurcation of the carotid artery that is richly supplied with sensory nerve endings of the sinus branch of the vagus nerve.
cation a positively charged ion.
cerebrospinal fluid fluid found and circulating through the brain and spinal cord.
cirrhosis a chronic disease of the liver characterized by degenerative changes in the liver cells.
colloid gelatin-like substance, e.g., protein.
colloid osmotic pressure pressure exerted by nondiffusible substances.
congestive heart failure (CHF) circulatory congestion related to pump failure.
contraindication nonindicated form of therapy.
conversion table
1 kilogram (kg) = 2.2 pounds (lb)
1 gram = 1000 milligrams (mg) or 15 grains (gr)
1 liter (L) = 1 quart or 1000 milliliters (ml)
1 cubic centimeter (cc) = 1 milliliter (ml); 100 ml = dl
1 drop (gtt) = 1 minim (m)
qh = every hour
X = times
cortisone a hormone secreted by the adrenal cortex.
creatinine end product of creatine (amino acid).
crystalloids diffusible substances dissolved in solution will pass through a selectively permeable membrane.
CVP abbreviation for central venous pressure. It is the venous pressure in the vena cava or the heart's right atrium.
cyanosis a bluish or grayish discoloration of the skin due to a lack of oxygen in the hemoglobin of the blood.

decompensation failure to compensate.
deficit lack of.
dependent edema see edema.
dermis the true skin layer.
dextran colloid hyperosmolar solution.
dextrose a simple sugar, also known as glucose.
diabetes mellitus a disorder of carbohydrate metabolism due to an inadequate production or utilization of insulin.
diabetic acidosis an excessive production of ketone bodies (acid) due to a lack of insulin and inability to utilize carbohydrates.
dialysate an isotonic solution used in dialysis that has similar electrolyte contents to plasma or Ringer's solution, with the exception of potassium.
diaphoresis excessive perspiration.
diffusion the movement of each molecule along its own pathway irrespective of all other molecules; going in various directions.
disequilibrium syndrome rapid shift of fluids and electrolytes during hemodialysis which causes CNS disturbances.
dissociation a separation.
diuresis an abnormal increase in urine excretion.

diuretic a drug used to increase the secretion of urine
  potassium-sparing diuretic retains potassium and excretes other electrolytes.
  potassium-wasting diuretic excretes potassium and other electrolytes.
diverticulum a sac or pouch in the wall of an organ.
dry weight normal body weight without excess water.
duodenal pertaining to the duodenum, which is the first part of the small intestine.
dyspnea a labored or difficult breathing.
edema an abnormal retention of fluid in the interstitial spaces.
  dependent edema fluid present in the interstitial spaces due to gravity (frequently found in extremities after being in standing or sitting position).
  nondependent edema fluid present in the interstitial spaces, but not necessarily due to gravity along, e.g., cardiac, liver, or kidney dysfunction.
  pitting edema depression in the edematous tissue.
  pulmonary edema fluid throughout the lung tissue.
  refractory edema fluid in the interstitial spaces that does not respond to diuretics.
e.g. for example.
electrolyte a substance that, when in solution, will conduct an electric current.
endothelium flat cells that line the blood and lymphatic vessels.
enzyme a catalyst, capable of inducing chemical changes in other substances.
epidermis an outer layer of the skin.
erythrocytes red blood cells.
erythropoietin factor secreted by the kidneys that stimulates bone marrow to produce red blood cells.
excess too much.
excretion an elimination of waste products from the body.
excretory pertaining to excretion.
extra outside of.
extracellular fluid volume shift (ECF shift) shift of fluid within the ECF compartment from intravascular to interstitial spaces or from the interstitial to the intravascular spaces.

febrile pertaining to a fever.
flatus gas in the alimentary tract.

generic name reflects the chemical family to which a drug belongs. The name never changes.
globulin a group of simple proteins.
glomerulus a capillary loop enclosed within the Bowman's capsule of the kidney.
glucose formed from carbohydrates during digestion and frequently called dextrose.
glycogen a stored form of sugar in the liver or muscle that can be converted to glucose.
glycosuria sugar in the urine.
gtts abbreviation for drop.

hematocrit the volume of red blood cells or erythrocytes in a given volume of blood.
hemoconcentration increase in number of red blood cells and a decrease in plasma volume.
hemodilution an increase in the volume of blood plasma due to a lack of red blood cells or an excess of intravascular fluid.
hemoglobin conjugated protein consisting of iron-containing pigment in the erythrocyte.
hemolysis the destruction of red blood cells with liberation of hemoglobin.
heparin anticoagulant.
hepatic coma liver failure.
hernia a protrusion of an organ through the wall of a cavity.
  inguinal protrusion of the intestine at the inguinal opening.
homeostasis uniformity or stability. State of equilibrium of the internal environment.
hormone a chemical substance originating in an organ or gland, which travels through the blood and is capable of increasing body activity or secretion.

ADH abbreviation for antidiuretic hormone; a hormone to lessen urine secretion.
hydrostatic pressure pressure of fluids at equilibrium.
hyperalimentation intravenous administration of a hyperosmolar solution of glucose, protein, vitamins, and electrolytes to promote tissue synthesis.
hyperbaric oxygenation oxygen under pressure carried in the plasma.
hypercalcemia a high serum calcium.
hyperchloremia a high serum chloride.
hyperglycemia an increase in the blood sugar.
hyperkalemia a high serum potassium.
hypernatremia a high serum sodium.
hyperosmolar increased number of osmols per solution. Increased solute concentration as compared to plasma.
hypertension high blood pressure.
essential a high blood pressure that develops in the absence of kidney disease. It is also called primary hypertension.
hypertonic a higher solute concentration than plasma.
hypertrophy increased thickening of a structure.
hyperventilation increased breathing or respiration.
hypervolemia an increase in blood volume.
hypocalcemia a low serum calcium.
hypochloremia a low serum chloride.
hypoglycemia low blood sugar.
hypokalemia a low serum potassium.
hyponatremia a low serum sodium.
hypo-osmolar decreased number of osmols per solution. Decreased solute concentration as compared to plasma.
hypophysis the pituitary gland.
hypotension a low blood pressure.
hypotonic a lower solute concentration than plasma.
hypovolemia a decrease in blood volume.

i.e. that is.
incarcerated constricted, as an irreducible hernia.
infusion an injection of a solution directly into a vein.
insensible perspiration water loss by diffusion through the skin.
insulin a hormone secreted by the beta cells of the islets of Langerhans found in the pancreas. It is important in the oxidation and utilization of blood sugar (glucose)
inter between.
intervention action.
intra within.
ion a particle carrying either a positive or negative charge.
ionization separation into ions.
ionizing separating into ions.
iso-osmolar same number of osmols per solution as compared to plasma.
isotonic same solute concentration as plasma.

ketone bodies oxidation of fatty acids.
ketonuria excess ketones in urine.
Kussmaul breathing hyperactive, abnormally vigorous breathing.

lassitude weariness.
lidocaine also known as Xylocaine. A drug used as a surface anesthetic. It also can be used to treat ventricular arrhythmias.
lymph an alkaline fluid. It is similar to plasma except that its protein content is lower.

lymphatic system the conveyance of lymph from the tissues to the blood.

malaise uneasiness, ill feeling.
medulla the central portion of an organ, e.g., adrenal gland.
membrane a layer of tissue that covers a surface or organ or separates a space.
mercurial diuretic a drug that affects the proximal tubules of the kidneys by inhibiting reabsorption of sodium.
metabolism the physical and chemical changes involved in the utilization of particular substances.
metabolites the by-products of cellular metabolism or catabolism.
milliequivalent the chemical activity of elements.
milligram measures the weight of ions.
milliosmol 1/1000th of an osmol. It involves the osmotic activity of a solution.
molar 1 gram molecular weight of a substance.
myocardium the muscle of the heart.

narcotic a drug that depresses the central nervous system, relieves pain, and can induce sleep.
necrosis destroyed tissue.
nephritis inflammation of the kidney.
nephrosis degenerative changes in the kidney.
neuromuscular pertaining to the nerve and muscle.
nondependent edema see edema.
nonvolatile acid fixed acid resulting from metabolic processes, excreted by the kidneys.

oliguria a diminished amount of urine.
osmol a unit of osmotic pressure.
osmolality osmols or milliosmols per kilogram of water.
osmolarity osmols or milliosmols per liter of solution.
osmosis the passage of a solvent through a partition from a solution of lesser solute concentration to one of greater solute concentration.
osmotic pressure the pressure or force that develops when two solutions are of different concentrations and are separated by a selectively permeable membrane.
oxygenation the combination of oxygen in tissues and blood.
oxyhemoglobin hemoglobin carrying oxygen.

packed cells red blood cells (RBCs).
pallor or pallid pale.
PAP the abbreviation for pulmonary artery pressure. It measures the pressure in the pulmonary artery.
paracentesis the surgical puncture of a cavity, e.g., abdomen.
parathyroid an endocrine gland secreting the hormone parathormone, which regulates calcium and phosphorus metabolism.
parenteral therapy introduction of fluids into the body by means other than the alimentary tract.
patency the state of being opened.
PCWP the abbreviation for pulmonary capillary wedge pressure. Its reading indicates the pumping ability of the left ventricle. Left ventricular end diastolic pressure (LVEDP) is the best indicator of ventricular function. PCWP reflects LVEDP.
perfusion passing of fluid through body space.
pericardial sac fibroserous sac enclosing the heart.
pericarditis inflammation of the pericardium or the membrane that encloses the heart.
peristalsis wavelike movement occurring with hollow tubes such as the intestine for the movement of contents.
peritoneal cavity a lining covering the abdominal organs with the exclusion of the kidneys.
permeability capability of fluids and/or other substances, e.g., ions, to diffuse through a human membrane.

selectively permeable membrane refers to the human membrane.
semipermeable membrane refers to artificial membranes.

phlebitis inflammation of the vein.

physiologic pertaining to body function.

pitting edema see edema.

plasma intravascular fluid composed of water, ions, and colloid. Plasma is frequently referred to as serum.

plasmanate commercially prepared protein product used in place of plasma.

pleural cavity the space between the two pleuras.

pleurisy inflammation of the pleura or the membranes that enclose the lung.

polyionic many ions or ionic changes.

polyuria an excessive amount or discharge of urine.

porosity the state of being porous.

portal circulation circulation of the blood through the liver.

postoperative following an operation.

potassium-sparing diuretics see diuretics.

potassium-wasting diuretics see diuretics.

prednisone synthetic hormonal drug resembling cortisone.

preload pressure of the blood that fills the left ventricle during diastole.

pressoreceptors sensory nerve ending in the aorta and carotid sinus, which when stimulated will cause a change in the blood pressure.

pressure gradient the difference in pressure which makes the fluid flow.

protein nitrogenous compounds essential to all living organisms.
plasma relates to albumin, globulin, and fibrinogen.
serum relates to albumin and globulin.

pruritus itching.

psychogenic polydipsia psychologic effect of drinking excessive amounts of water.

pulmonary artery pressure see PAP

pulmonary capillary wedge pressure see PCWP

pulmonary edema see edema.

pulse pressure the difference between the systolic pressure and the diastolic pressure.

rales pertaining to rattle. It is the sound heard in the chest due to the passage of air through the bronchi which contain secretions or fluid.

rationale the reason.

reabsorption the act of absorbing again an excreted substance.

refractory edema see edema.

retention retaining or holding back in the body.

sclerosis hardening of an organ or tissue.

selectively permeable membrane see permeability.

semi-Fowler's position 45° elevation.

semipermeable membrane see permeability.

sensible perspiration the loss of water on the skin due to sweat gland activity.

serous cavity a cavity lined by a serous membrane.

serum consists of plasma minus the fibrinogen. It is the same as plasma except that after coagulation of blood, the fibrinogen is removed. Serum is frequently referred to as plasma.

sign an objective indication of disease.

solute a substance dissolved in a solution.

solvent a liquid with a substance in solution.

specific gravity a weight of a substance, e.g., urine. Water has a specific gravity of 1.000. The specific gravity of urine is higher.

steroid an organic compound. It is frequently referred to as an adrenal cortex hormone.

stress effect of a harmful condition or disease(s) affecting the body.
stroke volume amount of blood ejected by the left ventricle with each contraction.
sympathetic nervous system a part of the autonomic nervous system. It can act in an emergency.
symptom subjective indication.

tachycardia a fast heart beat.
tetany a nervous affection characterized by tonic spasms of muscles.
thrombophlebitis inflammation of a vein with a thrombus or a blood clot.
total protein nutrition (TNP) also known as hyperalimentation.
trade name the name given to a drug by its manufacturer.
transudation the passage of fluid through the pores of a membrane.
trauma an injury.

urea the final product of protein metabolism that is normally excreted by the kidneys.
uremia a toxic condition due to the retention of nitrogenous substances (protein by-products) that cannot be excreted by the kidneys.

Valsalva's maneuver procedure in which individual takes deep breath and bears down to increase intrathoracic pressure for prevention of air injection.
vasoconstriction constriction of blood vessels.
vasodilatation dilatation of blood vessels.
vasomotor pertaining to the nerves having a muscular contraction or relaxation control of the blood vessel walls.
vasopressors drugs given to contract muscles of the blood vessel walls to increase the blood pressure.
vein a vessel carrying unoxygenated blood to the heart.
ventilation the circulation of air.
  pulmonary the inspiration and expiration of air from the lungs.
ventricles the lower chambers of the heart.
venules minute veins moving from capillaries.
vertigo dizziness.
volatile acid acid excreted as a gas by the lungs.

Bibliography

BOOKS

Abbott, Laboratories: *Fluid and Electrolytes.* North Chicago, 1970.

Best CH, Taylor NB: *The Physiological Basis of Medical Practice,* ed 9. Baltimore, Williams and Wilkins Co, 1973.

Bland JW: *Clinical Metabolism of Body Water and Electrolytes.* Philadelphia, WB Saunders Co, 1963.

Borg N et al.: *Core Curriculum for Critical Care Nursing.* Philadelphia, WB Saunders Co, 1981.

Bouchard-Kurtz R, Speese-Owens N: *Nursing Care of the Cancer Patient,* ed. 4. St. Louis, CV Mosby Co, 1981.

Brundage D: *Nursing Management of Renal Problems,* ed 2. St. Louis, CV Mosby Co, 1980.

Brunner LS, Suddarth, DS: *Textbook of Medical-Surgical Nursing,* ed 5. Philadelphia, JB Lippincott Co, 1984.

Burgess A: *Fluid and Electrolyte Balance.* New York, McGraw-Hill Book Co, 1979, pp 333–398 (Chap 12 and 14).

Burns N: *Nursing and Cancer,* Philadelphia, WB Saunders Co, 1982.

Chernecky CC, Ramsey PW: *Critical Nursing Care of the Client with Cancer,* Norwalk CT, Appleton-Century-Crofts Co, 1984, pp 1–51.

Coco CD: *Intravenous Therapy,* St. Louis, CV Mosby Co, 1980.

Detecting parathyroid disorders. *Endocrine Disorders.* Springhouse PA, Springhouse Corp, 1984, pp 91–99.

Fochtman D, Foley G (ed): *Nursing Care of the Child with Cancer.* Boston, Little, Brown, and Co, 1982.

French R: *Guide to Diagnostic Procedures,* ed 5. New York, McGraw-Hill Book Co, 1980.

Gard S: *Laboratory Tests in Common Use,* ed 6. New York, Springer Publishing Co, 1976.

Groer MW, Shekleton ME: *Basic Pathophysiology,* ed. 2. St Louis, CV Mosby Co, 1983, pp 219–232.

Gutch C, STOner M: *Review of Hemodialysis for Nurses and Dialysis Personnel.* St Louis, CV Mosby Co, 1978.

Guyton AC: *Textbook of Medical Physiology,* ed 5. Philadelphia, WB Saunders Co, 1976.

Harrington JD, Brener ER: *Patient Care in Renal Failure.* (Saunders Monographs in Clinical Nursing). Philadelphia, WB Saunders Co, 1973.

Jacob SW, Francone CA: *Structure and Function in Man,* ed 3. Philadelphia, WB Saunders Co, 1974.

Kee JL: *Laboratory and Diagnostic Tests with Nursing Implications,* Norwalk, CT, Appleton-Century-Crofts Co, 1983, pp 45–46.

Krause MV, Mahan LK: *Food, Nutrition and Diet Therapy.* Philadelphia, WB Saunders Co, 1979.

Lancaster L (ed): *The Patient with End-Stage Renal Disease,* New York, John Wiley & Sons, 1984.

Leaf A, Cotran, R: *Renal Pathophysiology*, ed 3. New York, Oxford University Press, 1985.
Luckman J, Sorensen KC: *Medical-Surgical Nursing.* Philadelphia, WB Saunders Co, 1980.
Maxwell H, Kliman CR (ed): *Clinical Disorders of Fluid and Electrolyte Metabolism,* New York, McGraw-Hill Book Co, 1972.
McGaw Laboratories: *Guide to Parenteral Fluid Therapy*, Glendale, CA, 1963.
Metheney NM, Snively WD: *Nurses' Handbook of Fluid Balance*, ed. 4. Philadelphia, JB Lippincott Co, 1983, pp 315–333.
Miller BF, Keane CB: *Encyclopedia and Dictionary of Medicine, Nursing, and Allied Health*, ed 3. Philadelphia, WB Saunders Co, 1983, pp 325, 810.
Mitchell H: *Nutrition in Health and Disease*, ed 16. Philadelphia, JB Lippincott Co, 1976.
Montag M, Swenson RPS: *Foundamentals in Nursing Care*, ed. 3. Philadelphia, WB Saunders Co, 1959.
Oakley WG: *Diabetes and Its Management.* London, Blackwell Scientific Publications, 1973.
Papper S: *Clinical Nephrology*. Boston, Little, Brown Co, 1978.
Perry A, Potter P: *Shock*. St Louis, CV Mosby Co, 1983.
Pestana C: *Fluids and Electrolytes in the Surgical Patient.* Baltimore, Williams & Wilkins Co, 1977.
Phipps WJ, Long BC. Woods NF: *Medical-Surgical Nursing*, ed. 2. St. Louis, CV Mosby Co, 1983, pp 321–1325.
Price SA, Wilson LM: *Pathophysiology*, ed. 2. New York, McGraw-Hill Book Co, 1982, pp 433–439.
Rodman MJ, Smith DW: *Clinical Pharmacology in Nursing.* Philadelphia, JB Lippincott Co, 1980.
Rudolph A: *Pediatrics*, ed 16. New York, Appleton-Century-Crofts Co, 1977.
Sager, DP, Bomar SK: *Intravenous Medications.* Philadelphia, JB Lippincott Co, 1980.
Schoengrund L, Balzer P (ed): *Renal Problems in Critical Care*, New York, John Wiley & Sons, 1985.
Silver H et al.: *Handbook of Pediatrics*, ed 10. California, Lange, 1973, pp 627–634.
Spencer RT et al.: *Clinical Pharmacology*. Philadelphia, JB Lippincott Co, 1983, pp. 834–838.
Snively WD: *Sea Within.* Philadelphia, JB Lippincott Co, 1960.
Statland H: *Fluid and Electrolytes in Practice*, ed 3. Philadelphia, JB Lippincott Co, 1963.
Taber CW: *Taber's Cyclopedic Medical Dictionary*, ed 15. Philadelphia, FA Davis Co, 1985.
Thompson MA: *Shock syndrome*. Reading MA, Addison-Wesley Publishing Co, 1978.
Tilkian AM, Conover MH: *Clinical Implications of Laboratory Tests.* St. Louis, CV Mosby Co, 1975.
Travenol Laboratories, Inc: *Guide to Fluid Therapy.* Deerfield, ILL, 1970.
Vaughan V, McKay RJ (ed): *Nelson's Textbook of Pediatrics*, ed 11. Philadelphia, WB Saunders Co, 1979.
Whaley LF, Wong DL: *Nursing Care of Infants & Children*, St. Louis, CV Mosby Co, 1983.
Widmann FK: *Clinical Interpretation of Laboratory Tests.* ed 9. Philadelphia, FA Davis Co, 1983.
Wiener MB, Pepper GA: *Clinical Pharmacology and Therapeutics in Nursing*, ed 2. New York, McGraw-Hill Book Co, 1985, pp 325–350.
Wilson ED et al.: *Principles of Nutrition*, ed 3. New York, John Wiley & Sons, 1975.
Wyngaarden J, Smith, L: *Cecil Textbook of Medicine*, ed 16. Philadelphia, WB Saunders Co, 1982, pp 481–486, 1131–1134, 1318–1332.

PERIODICALS

Anderson MA, Aker, SN, Hickman, RO: The double-lumen Hickman catheter. *Am J Nurs* 82(2): 272–273, February 1982.
Baker WL: Hypophosphatemia. *Am J Nurs* 85(9): 998–1003, September 1985.
Barrows JJ: Shock demands drugs. *Nursing '82* 12(2), 34–41, February 1982.
Bielski MT, Molander DW: Laennec's cirrhosis. *Am J Nurs* 65(8): 82–86, 1965.
Birdsall C: When is TPN safe? *Am J Nurs* 83(1): 73, January 1983.

Blood Volume. *Pitoclinic,* Ames Co., 11: 5–10, March 1964.
Bobb J: What happens when your patient goes into shock. *RN* 47(3), 26–29, March 1984.
Bolte H: Treatment of acute and chronic hypokalemia. *Acta Cardiol [Suppl] (Brux)* 17: 213-215, 1973.
Borgen L: Total parenteral nutrition. *Am J Nurs* 78 (2): 224-228, February 1978.
Burgess RE: Fluids and electrolytes. *Am J Nurs* (10): 90-95, 1965.
Cardin S: Acid-base balance in the patient with respiratory disease. *Nurs Clin North Am* 15 (3): 593-601, September 1980.
Carnes HE (ed): Acute renal failure. *Therapeutic Notes* 71: 8-12, January 1964 (Parke, Davis and Co).
Carnes HE (ed): Treatment of burns. *Therapeutic Notes* 70: 234-239, October 1963 (Parke, Davis and Co).
Carozza V: Ketoacidotic crisis: Mechanism and management. *Nursing '73* 3: 13-14, May 1973.
Christopher KL: The use of a model for hemodynamic balance to describe burn shock. *Nurs Clin North Am* 15 (3): 617-627, September 1980.
Cohen S, Boyce B, King TKC: Blood-gas and acid base concepts in respiratory care—programmed instruction. *Am J Nurs* 76 (6) P.I.: 1-30, 1976.
Colley, R, Phillip K: Helping with hyperalimentation. *Nursing '73* 3 (7): 6-17, 1973.
Crowell CE (ed): Programmed Instruction—potassium imbalance. *Am J Nurs* 67: 343-366, 1967.
Cunningham SG: Fluid and electrolyte disturbances associated with cancer and its treatment. *Nurs Clin North Am,* 17(4): 579–591, December 1982.
Downing SR, Watkins FL: The patient having peritoneal dialysis. *Am J Nurs* 66: 1572-1577.
Elbaum N: Detecting and correcting magnesium imbalance. *Nursing '77* 7: 34-35, August 1977.
Fajans S: What is diabetes? Definition, diagnosis and course. *Med Clin North Am* 55: 793-804, 1971.
Felver L: Understanding the electrolyte maze. *Am J Nurs* 80 (9): 1591-1595, September 1980.
Fitzgibbons JP: Fluid, electrolyte, and acid-base management in the acutely traumatized patient. *Orthop Clin North Am* 57 (1): pp 627-647, February 1977.
Flear CTG: Electrolyte and body water changes after trauma. *J Clin Pathol [Suppl]* 23 (4): 16-31, 1970.
Fox B, Stegall B: Take precautions now. *Nursing '85* 15(5): 48–49, May 1985.
Geolot DH, McKinney NP: Administering parenteral drugs. *Am J Nurs* 75 (5): 788-793, 1975.
Grant M, Kubo W: Assessing a patient's hydration status. *Am J Nurs* 75: 1306-1311, 1975.
Grollman A: Diuretics. *Am J Nurs* 65: 84-89, 1965.
Hayter J: Emergency nursing care of the burned patient. *Nurs Clin North Am* 13 (2): 223-234, June 1978.
Heidland A, Hennemann HM, Rockel A: The role of magnesium and substances promoting the transport of electrolytes. *Acta Cardiol [Suppl] (Brux)* 17: 52-67, 1973.
Hochman HI, et al.: Dehydration, diabetic ketoacidosis and shock in the pediatric patient. *Pediat Clin North Am* 26 (4): 803-826, November 1979.
Jenkinson VM: Congestive heart failure, in *Basic Medical-Surgical Nursing.* Dubuque, IA, WC Brown Co, 1966, pp 166–171, 492.
Jones CA, Richards KE: Burns: Fluids resuscitate, in *Monitoring Fluid and Electrolytes Precisely: Nursing Skillbook.* Horsham, PA, Intermed Communications, Inc, 1978, pp 155-162.
Josephson A, Kliman A, Shively J: Transfusions. *Patient Care,* August 15, 1971, pp 118-139.
Kee JL, Gregory AP: The ABC's (and mEq's) of fluid balance in children. *Nursing '74* 4 (6): 28-36, June 1974.
Kerkovits G: Anatiarrhythmics and electrolytes. *Acta Cardiol [Suppl] (Brux)* 17: 155-157, 1973.
Keyes JL: Blood-gas analysis and the assessment of acid-base status. *Heart Lung* 5 (2): 247-255, 1976.
Keyes JL: Blood-gas and blood-gas transport. *Heart Lung* 3 (6): 945-954, 1974.
Kleinhenz TJ: Preload and Afterload. *Nursing '85* 15(5), 50–55, May 1985.
Kurdi WJ: Refining your IV therapy techniques. *Nursing '75* 5 (11): 40-47, 1975.

Lancour J: ADH and aldosterone: How to recognize their effects. *Nursing '78* 8 (9): 36–41, September 1978.
Levenstein BP: Intravenous therapy: A nursing specialty. *Nurs Clin North Am* 1: 259-265, 1966.
Lindsey AM, Piper BF, Carrieri V: Malignant cells and ectopic hormone production. *Oncology Nursing Forum*, 8(3): 13–15, Summer 1981.
Lorentz WB: On pediatric fluid loss. *Emergency Medicine,* February 1977, pp 23-29, 35, 37-39, 45-46.
MacLeod S: The rational use of potassium supplements. *Postgrad Med* 57 (2): 123-127, 1975.
Makoff D: Common fluid and electrolyte disorders in the cardiac patient. *Geriatrics,* November 1972, pp 67-76.
Managing the adult diabetic patient—roundtable. *Patient Care* 9: 16-49, June 1, 1975.
Meador B: Cardiogenic shock. *RN* 45(4), 38–42, April 1982.
Menzel LK: Clinical problems of fluid balance. *Nurs Clin North Am* 15 (3): 549-576.
Morris DG: The patient in cardiogenic shock. *Cardiovasc Nurs* 5: 15–17, 1969.
Moyer J, Mills L: Vasopressor agents in shock. *Am J Nurs* 75: 620-625, 1975.
Nanji A: Drug-induced electrolyte disorders. *Drug Intelligence and Clinical Pharmacy* 17: 175–185, March 1983.
Narins RG et al.: Diagnostic strategies in disorders of fluid, electrolyte and acid-base homeostasis. *Am J Med* 72: 496–518, March 1982.
Niemczura J: Eight Rules to remember when caring for the patient with swan-ganz catheter. *Nursing '85* 15(3), 38–45, March 1985.
Nursing care of patients in shock, Part 2. *Am J Nurs* 82(9), 1401–1403, September 1982.
O'Donnell TF, Belkin SC: The pathophysiology, monitoring, and treatment of shock. *Orthop Clin North Am* 9 (3): 589-610, July 1978.
O'Neill M: Peritoneal dialysis. *Nurs Clin North Am* 1: 309-323, June 1966.
Papper S: The effects of age in reducing renal function. *Geriatrics* May 1973, pp 83-87.
Perkin R, Levin DL: Common fluid & electrolyte problems in the pediatric intensive care unit. *Pediat Clin North Am* 27 (3): 567-586, August 1980.
Pickerning L, Robbins D: Fluid, electrolyte, and acid-base balance in the renal patient. *Nurs Clin North Am* 15 (3): 577-592, September 1980.
Poe CM, Radford AI: The challenge of hypercalcemia in cancer. *Oncology Nursing Forum*, 12(6): 29–34, November/December 1985.
Pruitt BA: Fluid and electrolyte replacement in the burned patient. *Surg Clin North Am* 58 (6): 1291-1310, December 1978.
Rice V: Shock management, Part II. Pharmacologic interventions. *Critical Care Nurse*, 5(1), 42–56, 1985.
Rice V: Shock, a clinical syndrome. *Critical Care Nurse* 1(5), 34–43, September/October 1981.
Roberts SL: Renal assessment: A nursing point of view. *Heart Lung* 8 (1): 105-113, January/February 1979.
Rose M: Shock: Fluids restore circulation, in *Monitoring Fluid and Electrolytes Precisely: Nursing Skillbook*. Horsham, PA: Internal Communications, Inc, 1978, pp 149-154.
Schakenback LH, Dennis M: And now, a quad-lumen IV catheter. *Nursing '85* 15(11): 50–51, November 1985.
Sharer JE: Reviewing acid-base balance. *Am J Nurs* 75 (6): 980-983, 1975.
Shiris GT: Postoperative, post-traumatic management of fluids. *Bull NY Acad Med* 55 (2): 248-256, February 1979.
Shock. Hospital Focus. (Knoll Pharmaceutical Co) 1-6, October 1, 1962.
Simeone FA: Shock: Its nature and treatment. *Am J Nurs* 66: 1386-1394, 1966.
Snider MA: Helpful hints on IV's. *Am J Nurs* 74: 1978-1981, 1974.
Stark JL: Renal failure: Imbalances inevitable, in *Monitoring Fluid and Electrolytes Precisely: Nursing Skillbook*. Horsham, PA, Intermed Communications, Inc, 1978, pp 117-124.
Strickland WM: Replacement therapy in traumatic shock. *Seminar Report* 6: 2-7, Spring 1961 (Merck, Sharp and Dohme).
Stude C: Cardiogenic shock. *Am J Nurs* 74: 1636-1640, 1974.

Suki WN: Disposition and regulation of body potassium: An overview. *Am J Med Sci* 272: 31–41, July/August 1976.
"Teaming up to send the end-stage COPD patient home. *Nursing '84* 14(1): 65–76, January 1984.
Transfusion: What blood component does your patient really need? *Nursing Update* 4 (3): 1-11, 1973.
Tripp A: Hyper- and hypocalcemia. *Am J Nurs* 76: 1142-1145, 1976.
Trusk CW: Hemodialysis for acute renal failure. *Am J Nurs* 65: 80–85, February 1965.
Ungvarski PJ: Parenteral therapy. *Am J Nurs* 76: 1974-1977, 1976.
Weiss H: Electrolyte disturbances in respiratory diseases. *Geriatrics,* October 1973, pp 151–154.
Wiley L: Shock—different kinds and different problems. *Nursing '74* 4 (5): 43–53, 1974.
Wilhelm L: Helping your patient settle in with TPN. *Nursing '85* 15(4): 60, April 1985.
Wilson JA: Infection control in intravenous therapy. *Heart Lung* 5 (3): 430-436, 1976.
Wittaker A: Acute renal dysfunction. *Focus on Critical Care* 12(3):12–17, June 1985.
Wool, NL et al: Hickman catheter placement simplified. *Am J Surg* 145:283–284, February 1983.

Index